T0396183

Nanocomposites for Environmental, Energy, and Agricultural Applications

Woodhead Publishing Series in Composites Science and Engineering

Nanocomposites for Environmental, Energy, and Agricultural Applications

Edited by

Chidambaram Thamaraiselvan

Woei Jye Lau

Antonio Juárez Maldonado

Nur Hidayati Othman

Woodhead Publishing is an imprint of Elsevier
50 Hampshire Street, 5th Floor, Cambridge, MA 02139, United States
125 London Wall, London EC2Y 5AS, United Kingdom

ISBN: 978-0-443-13935-2 (print)

ISBN: 978-0-443-13936-9 (online)

For information on all Woodhead Publishing publications
visit our website at https://www.elsevier.com/books-and-journals

Publisher: Matthew Deans
Acquisitions Editor: Gwen Jones
Editorial Project Manager: Rafael G. Trombaco
Production Project Manager: Prasanna Kalyanaraman
Cover Designer: Greg Harris

Typeset by MPS Limited, Chennai, India

Contents

List of contributors

Nur Hashimah Alias School of Chemical Engineering, College of Engineering, Universiti Teknologi MARA, Shah Alam, Selangor Darul Ehsan, Malaysia

Sacide Alsoy Altinkaya Department of Chemical Engineering, Izmir Institute of Technology, İzmir, Turkey

Angelin Arumugam Department of Biotechnology, Karunya Institute of Technology and Sciences, Coimbatore, Tamil Nadu, India

Amizon Azizan School of Chemical Engineering, College of Engineering, Universiti Teknologi MARA, Shah Alam, Selangor, Malaysia

Adalberto Benavides-Mendoza Department of Horticulture, Universidad Autónoma Agraria Antonio Narro, Saltillo, Mexico

Gregorio Cadenas-Pliego Applied Chemistry Research Center, Saltillo, Mexico

Ji Woong Chang Department of Chemical Engineering, Kumoh National Institute of Technology, Gumi-si, Gyeongsangbuk-do, Republic of Korea

Paola Durán Scientific and Technological Bioresource Nucleus, Universidad de La Frontera, Temuco, Chile

Paola Fincheira Centro de Excelencia en Investigación Biotecnológica Aplicada al Medio Ambiente, CIBAMA-BIOREN, Universidad de La Frontera, Temuco, Chile

Vasileios Fotopoulos Department of Agricultural Sciences, Biotechnology & Food Science, Cyprus University of Technology, Limassol, Cyprus

Carlos Alberto Garza-Alonso Facultad de Agronomía, Universidad Autónoma de Nuevo León, Gral. Escobedo, Mexico

Gholamreza Gohari Department of Horticultural Science, Faculty of Agriculture, University of Maragheh, Maragheh, Iran; Department of Agricultural Sciences, Biotechnology & Food Science, Cyprus University of Technology, Limassol, Cyprus

Yolanda González-García Faculty of Agronomy, Center for Protected Agriculture, Autonomous University of Nuevo Leon, General Escobedo, Mexico

Tortella Gonzalo Centro de Excelencia en Investigación Biotecnológica Aplicada al Medio Ambiente, CIBAMA-BIOREN, Universidad de La Frontera, Temuco, Chile; Departamento de Ingeniería Química, Facultad de Ingeniería y Ciencias, Universidad de La Frontera, Temuco, Chile

Rafidah Jalil Forest Products Division, Forest Research Institute Malaysia (FRIM), Kepong, Selangor, Malaysia

Antonio Juárez-Maldonado Department of Botany, Autonomous Agrarian University Antonio Narro, Saltillo, Coahuila, Mexico

Govindan Kadarkarai Department of Civil, Construction and Environmental Engineering, Marquette University, Milwaukee, WI, United States

Kamil Kayode Katibi Department of Process and Food Engineering, Faculty of Engineering, Universiti Putra Malaysia, Serdang, Selangor, Malaysia; Department of Physics, Faculty of Science, Universiti Putra Malaysia, Serdang, Selangor, Malaysia; Department of Agricultural and Biological Engineering, Faculty of Engineering and Technology, Kwara State University, Malete, Nigeria

Shareena Fairuz Abdul Manaf School of Chemical Engineering, College of Engineering, Universiti Teknologi MARA, Shah Alam, Selangor, Malaysia

George A. Manganaris Department of Agricultural Sciences, Biotechnology & Food Science, Cyprus University of Technology, Limassol, Cyprus

Fauziah Marpani School of Chemical Engineering, College of Engineering, Universiti Teknologi MARA, Shah Alam, Selangor Darul Ehsan, Malaysia

Muhammad Shafiq Mat Shayuti School of Chemical Engineering, College of Engineering, Universiti Teknologi MARA, Shah Alam, Selangor Darul Ehsan, Malaysia

Julia Medrano-Macías Department of Horticulture, Universidad Autónoma Agraria Antonio Narro, Saltillo, Mexico

Maryam Haghmadad Milani Department of Biology, Faculty of Basic Sciences, University of Maragheh, Maragheh, Iran

Raja Mohan Department of Food Technology, Excel Engineering College, Pallakkapalayam, Komarapalayam, Tamil Nadu, India

Álvaro Morelos-Moreno CONAHCYT, Universidad Autónoma Agraria Antonio Narro, Saltillo, Mexico

Anjani P. Nagvenkar School of Chemical Sciences, Goa University, Goa, India

Samidha S. Narvekar School of Chemical Sciences, Goa University, Goa, India

Emilio Olivares-Sáenz Faculty of Agronomy, Center for Protected Agriculture, Autonomous University of Nuevo Leon, General Escobedo, Mexico

Nur Hidayati Othman School of Chemical Engineering, College of Engineering, Universiti Teknologi MARA, Shah Alam, Selangor Darul Ehsan, Malaysia

Javiera Parada Centro de Excelencia en Investigación Biotecnológica Aplicada al Medio Ambiente, CIBAMA-BIOREN, Universidad de La Frontera, Temuco, Chile

Marissa Pérez-Alvarez Applied Chemistry Research Center, Saltillo, Mexico

Fabián Pérez-Labrada Department of Botany, Autonomous Agrarian University Antonio Narro, Saltillo, Coahuila, Mexico

Nivea Raghavan Electronics and Communication Department, Saveetha Institute of Medical Sciences and Technology, Chennai, Tamil Nadu, India

Norhasyimi Rahmat School of Chemical Engineering, College of Engineering, Universiti Teknologi MARA, Shah Alam, Selangor, Malaysia

Abiyazhini Rajendran Department of Physics, Bharathiar University, Coimbatore, Tamil Nadu, India

Luz Leticia Rivera-Solís Doctorate in Sciences in Protected Agriculture, Universidad Autónoma Agraria Antonio Narro, Saltillo, Mexico

Olga Rubilar Centro de Excelencia en Investigación Biotecnológica Aplicada al Medio Ambiente, CIBAMA-BIOREN, Universidad de La Frontera, Temuco, Chile; Departamento de Ingeniería Química, Facultad de Ingeniería y Ciencias, Universidad de La Frontera, Temuco, Chile

Zulfiqar Ali Sahito Ministry of Education (MOE), Key Laboratory of Environmental Remediation and Ecosystem Health, College of Environmental and Resources Science, Zhejiang University, Hangzhou, P. R. China

Suhaila Mohd Sauid School of Chemical Engineering, College of Engineering, Universiti Teknologi MARA, Shah Alam, Selangor, Malaysia

Rahida Wati Sharudin School of Chemical Engineering, College of Engineering, Universiti Teknologi MARA, Shah Alam, Selangor, Malaysia

Uma Maheshwari Subramanian Department of Physics, Mother Teresa Women's University, Kodaikanal, Tamil Nadu, India

Sakthivel Thangavel Department of Chemical Engineering, Kumoh National Institute of Technology, Gumi-si, Gyeongsangbuk-do, Republic of Korea

Khairul Faezah Md Yunos Department of Process and Food Engineering, Faculty of Engineering, Universiti Putra Malaysia, Serdang, Selangor, Malaysia

Section 1

Environmental applications

Fundamentals and applications of carbon nanotube-based carbocatalysts for water treatment

Govindan Kadarkarai[1], *Angelin Arumugam*[2], *Uma Maheshwari Subramanian*[3] *and Raja Mohan*[4]
[1]Department of Civil, Construction and Environmental Engineering, Marquette University, Milwaukee, WI, United States, [2]Department of Biotechnology, Karunya Institute of Technology and Sciences, Coimbatore, Tamil Nadu, India, [3]Department of Physics, Mother Teresa Women's University, Kodaikanal, Tamil Nadu, India, [4]Department of Food Technology, Excel Engineering College, Pallakkapalayam, Komarapalayam, Tamil Nadu, India

1.1 Introduction

Expeditious industrial growth and anthropogenic activities (industrial effluent discharge, vast mining activities, agricultural operations, and sanitary effluents) are accelerating a variety of contaminants into water bodies (Vandana et al., 2022). Spontaneous release of traditional inorganic contaminants (heavy metals, trace elements, mineral acids, and inorganic salts) and emerging organic pollutants (pharmaceuticals, personnel care products, polyfluorinated organic substances, pesticides, herbicides, insecticides, etc.) has become a cause of acute environmental concerns. The environmental deterioration through water contamination has severely interrupted the balance of harmonious co-existence between the ecosystem and human well-being. These unprecedented and undue toxic compounds in the water bodies are far beyond the natural decomposition ability of the ecosystem. Therefore, state-of-the-art remediation technologies for aqueous pollutants have been broadly explored.

Conventional water treatment methods such as adsorption, coagulation/flocculation, filtration, membrane separation, ion exchange, extraction, oxidation, and biological methods are widely employed to eradicate and reduce water pollutant levels. This standard treatment approach has also exhibited that the decontamination or decomposition of pollutants may be ineffective due to their slow reaction kinetics, poor selectivity, formation of secondary pollution, and complex water matrix. Further, the novel toxic chemicals also disrupt the functioning of conventional water treatment plants and negatively influence water quality parameters (turbidity, temperature, and pH), ultimately compromising the treatment plant's efficiency (Das et al., 2014). So, these present conventional approaches are unable to augment

Nanocomposites for Environmental, Energy, and Agricultural Applications. DOI: https://doi.org/10.1016/B978-0-443-13935-2.00001-2

their water purifying efficiency to offer hundred percent clean and secure water and have failed to meet the World Health Organization (WHO) standards of discharge criteria.

The beginning of nanotechnology has provided various nanomaterials such as fullerenes, nanofibers, nanowires, zeolites, and diverse nanoparticles (NPs) for water treatment. These nanoscaled carbonaceous materials have attracted the limelight as rising stars in material communities (Duan et al., 2019). It may be ascribed to the earth's abundance of carbon, tunable structure, higher robustness, rich functionalities, and featured electronic configuration. These carbon allotropes are revealed as promising metal-free catalysts to facilitate different chemical reactions with appreciable catalytic efficiency, chemical stability, durability, and high selectivity.

The metal-free carbocatalysts are divided into bulk carbonaceous materials (activated carbon (AC), carbon fiber, and biochar), 1D (OD, fullerene, nanodiamonds), 2D (DD, graphene), and 3D (hexagonally ordered mesoporous carbon (CMK-3)). Bulk carbonaceous materials have intrinsic drawbacks due to their complex structure as well as nonstoichiometry. Consequently, 2D nanoscaled carbon with a simplified configuration has received much attention for its characteristic outer surface chemistry (functionalities and defective level), pore size, and specific surface area (Su et al., 2013). Worldwide, more than sixty-five multinational companies/industries which include C12 Quantum Electronics, OCSiAl, Chasm Technologies, Cnano Technology, and Nanocop, have been preparing carbon nanotube (CNT) on a large scale for various catalytic applications in material science and composite substances, electronics, optics, energy storage device modeling of automotive parts, water filters, boat hulls, sporting goods, actuators, thin-film electronics, coatings, and electromagnetic shields. CNTs justify the properties of durability, strength, and conductivity (thermal and electrical) and seem to be more lightweight than conventional materials. It is mainly used as an additive to make textiles stain-resistant. The strength and flexibility create an application in aerospace for fuel efficiency and lightweightness. The gaming industry also enters the market for growth in CNT manufacturing. The global CNT manufacturers market report valued US$6.63 billion in 2021 and is projected to reach US$28.74 billion by 2030. Details of key CNT manufacturing companies/industries are presented as follows.

Nanocyl' is a Belgian CNT specialist company that manufactures high-quality multi-walled CNTs (MWCNTs). Founded in 2022, it leads in production and is also a privately owned SE company. Its global competitors are Nanolab, Raymor Industries, NanoTechlabs, Unidym, nanocomp Technologies, Nano-C, and OSCiAL. "CHASM Advanced Materials" is a US-based company launched in 2015 that is a leading developer and manufacturer of electronic and battery materials based on carbon. They specialize in hybrid carbon materials [single-walled CNTs (SWCNTs)] to enable highly conductive, transparent, flexible, and printed circuits. "ARKEMA" is a French chemicals company founded in 2004 and headquartered in Columbes, Paris. Specialist in providing coating solutions based on carbon. "Carbon Solutions was founded in 1998 in California, US. It manufactures single-walled CNTs." "Hyperion Catalysis International Corporation" aims to develop different morphologies of carbons. Hyperion's technology is a conductive and vapor-grown MWCNTs. "Nano-C Inc." is a core technology of energy and an environmentally efficient combustion-based process for manufacturing nanoscale carbon structures. It has raised $5.5 million in private funding, $3.5 million in grant funding, and $2.9 million through a 3-year National Institute of Standards and Technology project. "NoPo Nanotechnologies Private Limited" was founded in 2011 with the objective of developing technologies for space. A few core R&D-based start-ups are emerging in India. The company has developed a new HiPCO process to produce high-quality SWCNTs applied in electronics composites, thermal coatings, and space. "Cheap Tubes" is the leading supplier of high-quality, low-cost CNTs and graphene nanomaterials for research and industry.

Among carbonaceous materials, CNT-based carbocatalysts have garnered attention in the last few decades due to their large ratio of accessible mesopores and unique side curvature with π-conjugative properties. The exterior surface is the prime adsorption site for larger molecules, with small molecules getting conveniently absorbed on internal sites. A well-defined and uniform structure at the atomic scale and hexagonally arranged carbon atoms in CNTs provide exceptional interaction with dissolved pollutants. Therefore, the chapter discusses the fundamentals of CNT carbocatalysts and emphasizes the recent developments in the application of CNT-based carbocatalysts for various water treatments.

1.2 Fundamentals of carbon nanotubes

The application of CNTs has gained more attention in the research community due to their exotic electronic arrangements. Further, fascinating properties such as exceptionally robust mechanical properties and outstanding electron-transfer characteristics have encouraged their potential use in various applications. The morphology of CNT was first observed by Sumio Iijima in 1991 through transmission electron microscopy (Iijima & Ichihashi, 1993). Two years before this observation, SWCNTs were experimentally invented by Iijima at the NEC Research Laboratory in Japan and by Bethune at the INM Almaden Laboratory in California. CNTs are described as one or several concentric graphite-like layers rolled up into cylindrical shapes. Carbon hexagons are established in a concentric pattern, with both ends containing fullerene-like structures integrating pentagons. The dimension of CNTs ranges from 0.4 nm to tens of nanometers in diameter and micrometers in length. The adjacent sheets are held together by van der Waals force, and the interspacing space is 0.34 nm. CNTs are classified by their layered structure, namely SWCNT and MWCNT.

SWCNTs are single graphene layers rolled into hollow cylindrical tubes of one atom thickness incorporated with a regular hexagonal lattice structure. Due to the larger length-to-diameter ratio and appreciably small diameter, SWCNTs are classified as 1D carbonaceous materials. Based on the positioning of the carbon hexagons, SWCNTs can produce three different forms: zigzag, armchair, or chiral (Mishra & Sundaram, 2023). MWCNTs consist of two or more cylindrical-shaped graphene sheets with diameters starting from 1 nm. MWCNTs have a low pore volume and specific area due to their complicated structural features compared to SWCNTs. Further, their fundamental properties are not disrupted during surface modification by chemical treatment (Chan et al., 2022).

1.3 Characteristics and applications of carbon nanotubes

The phenomenal structural properties and electronic configurations of CNT have drawn encouraging intrigue from the nanomaterial research society. Table 1.1 explains the features and important properties of SWCNT and MWCNTs (Aslam et al., 2021; Saifuddin et al., 2013). Certainly, CNTs typically offer many benefits compared to other conventional nanomaterials. For example, the 1D buckytube-shaped CNTs, comprising a hollow internal structure, a layered outer surface, a good specific surface area, and unusual sidewall curvature, enable excellent advantages for different water purification technologies (Lee, 2018). Moreover, the presence of appropriate pore size, well-defined pore (mesopores/micropores) arrangements, hydrophobicity, and high water permeability encouraged the utilization of CNT-based substances. Internal sites in the cylindrical hollow structure of CNT, interstitial sites in the channels between the individual nanotubes, outer surface functionalities, and outer grooves contribute greatly to the application of CNTs.

To augment the industrial applications and scientific applications of CNTs, it is needed to scale up the synthesis techniques for large-scale production of defect-free nanotubes. Herein, three main synthetic approaches are outlined, namely, (i) arc-discharge, (ii) laser-ablation, and (iii) chemical vapor deposition (CVD) methods.

1.3.1 Arc-discharge method

An arc-discharge method is performed at high temperatures (> 3000°C), essentially for operating carbon atoms into plasma. This method has also been employed to

Table 1.1 Features of carbon nanotubes.

Properties	MWCNTs	SWCNTs
Nanotube diameter (nm)	5–100	1.2–2
Electron mobility (cm^2/V/second)	−105	−105
Electron conductivity (S/cm)	103–105	102–106
Thermal conductivity (W/m/K)	3000	6000
Tensile strength (GPa)	150	150
Young modulus	1200	1054
Purity level	High	Poor
Defect level	Low	High
Density (g/cm^3)	2.1	1.33–1.40
Specific gravity	<1.8	−0.8

prepare both SWCNT and MWCNT. Catalytic agents (Ni, Co, Yt, Fe, etc.) are required for the synthesis of individual SWCNTs, whereas the catalyst is not essential for the synthesis of MWCNTs. SWCNTs were successfully prepared via the arc-discharge method using a metal catalyst (Bethune et al., 1993). The Arc-discharge method typically carries out the existence of an electric arc between two closely spaced electrodes under an inert gas atmosphere (He or Ar) with reduced pressure between 50 and 700 mbar. Under high temperatures in the plasma region, carbon vaporizes, and the electrode is corroded and consumed for CNT growth. Pure graphite electrodes or graphite electrodes containing metal are also used as anode materials. In recent days, the nonvacuum medium arc-discharge method has also been adopted using ethanol or deionized water as the liquid medium (Bethune et al., 1993; Sagara et al., 2014; Yousef et al., 2013). Similarly, liquid N_2 has also been employed as the plasma medium and graphite electrodes for the CNT growth (Scalese et al., 2010a, 2010b).

1.3.2 Laser-ablation method

This method also propagates the graphite vaporization in the plasma region at 1200°C, whereas the growth of high-grade pure CNT is mostly governed by the purity of graphite. Furthermore, the CNT growth and conversion ratio is appreciable and found superior to the arc-discharge approach. For example, the bundles of uniform and small-diameter SWCNTs were fruitfully ($>70\%$ yield) prepared through a laser ablation route by Smalley and coworkers in 1996 (Thess et al., 1996). Generally, nanotube growth via laser ablation strategy is explained by the "scooter" mechanism. A single-atom catalytic material (Ni or Co) on the open edge of a nanotube with electronegativity metal NPs hampers the formation of other carbonaceous by-products (especially fullerene) and facilitates the high conversion of graphite to CNT growth. The scooter mechanism described the process of metallic agents passing around the nanotube open end, which absorbs small carbon molecules and transforms them into graphite-like sheets. The tube propagates until too many metal atoms are combined on another nanotube end. Eventually, this scooter process leads to the formation of armchair-type nanotubes with defined chirality (10, 10) and a high purity structure of CNTs (Bethune et al., 1993; José-Yacamán et al., 1993).

1.3.3 Chemical vapor deposition

Mass production of CNTs is preferentially developed via CVD techniques using transition metal (TM) catalysts (Fe, Ni, and Co) (José-Yacamán et al., 1993). Commonly, CVD techniques involve catalyst-driven decomposition of hydrocarbons (ethylene or acetylene) carried out using tube reactors at 550°C–750°C. CNT was developed on the catalyst upon decreasing the reaction temperature to cool. Metal (Fe, Ni, and Co) NPs were widely employed as catalysts. These TMs break down the hydrocarbons into hydrogen and carbon; consequently, the carbon dissolves into metal NPs until the carbon solubility reaches the saturation limit.

Depending on the contact force between the substrate material and metallic NPs, nanotubes grow into the pore embedded in the metallic NPs.

Well-defined and aligned CNTs were also synthesized at a large-scale level through CVD along with Fe NPs (Li et al., 1996). Similarly, silicon wafers patterned with micrometer-scale catalytic materials were employed in the CVD technique and used to develop high-quality SWCNTs (Letters to Nature, 1973). To achieve small-diameter SWCNTs, methane was used as a carbon source, and the reaction was carried out at high temperatures ranging between 850°C and 1000°C. Further, in the category of integrating methods, plasma-enhanced, and laser-assisted CVD techniques were identified to improve the quality and yield of CNT growth. However, they obstruct the fundamental large-scale up, resulting in the combination system correlated with the same scale-up issues, which in turn interrupts arc-discharge and laser-ablation methods. Companies like Arkema, Nanocyl™, and Bayer Material Science commercially grow MWCNTs via the CVD process (White et al., 2016). CVD is the most practically applicable option for researchers and manufacturers to mass-produce CNTs at low operation costs (Chan et al., 2022).

CNTs are used in commercial applications in the fields of anticorrosion paints, polymers, thin films and coatings, engineering plastics, displays, and transparent and nontransparent conductive electrodes. On-going research is running in fields like fuel cells, optics, batteries, solar cells, water desalination, advanced devices, and so on (Yan et al., 2015). Arc-discharge and laser vaporization synthesis are the currently used processes to obtain small quantities of high-quality CNTs. Both of these methods involve evaporating the carbon source and end with unclear decisions on how to scale up production at the industrial level. Vaporization methods grow CNTs in tangled forms with unwanted forms of carbon, which are very difficult to purify, manipulate, and assemble for building nanotube-device architectures for practical applications.

1.4 Surface engineering for catalytic active sites

1.4.1 Surface functionalization

Fig. 1.1A describes that different surface functionalization methods assist in the fabrication of versatile catalysts with optimal behavior depending upon the demand (Liu et al., 2021). The common functionalizations are doping of heteroatoms, surface oxidation, chemical or electrochemical coating of polymers or NPs, grafting of organic groups or macromolecules, plasma activation, irradiation, and so on (Yan et al., 2015). CNTs are functionalized to reduce aggregation and make them available for selected pollutants through their improved adsorption behavior. The presence of oxygen-containing functional groups enhances the adsorption performance and converts to a hydrophilic form for easy dispersion in aqueous media. Oxidation is very easy to introduce $-C=O$ and $C-OH$ to the sidewalls of CNTs (Yu et al., 2014). Functionalization methods are irradiation, mechanical, and physicochemical methods that involve π-conjugated parts (covalent bonding), noncovalent bonding

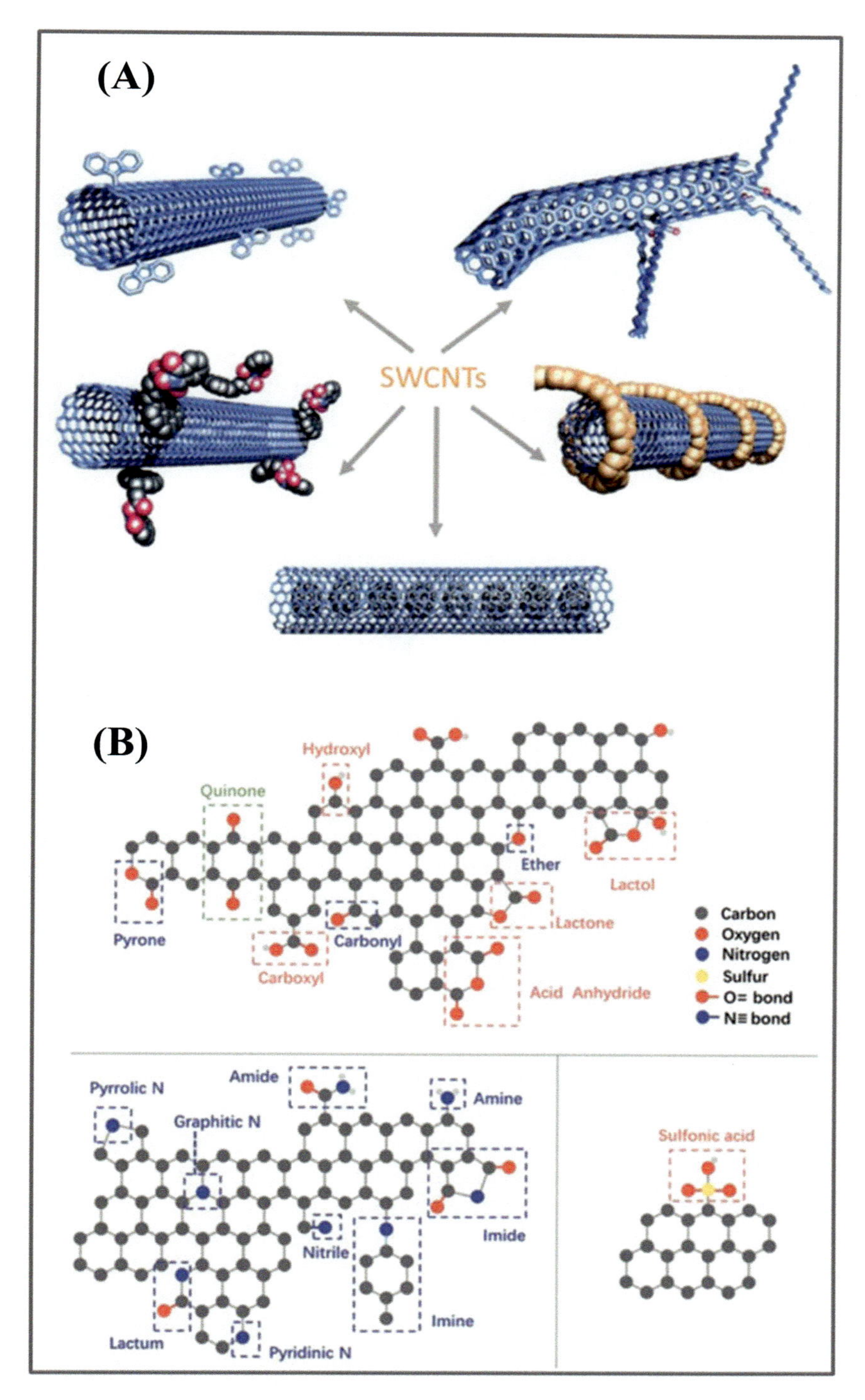

Figure 1.1 Carbon nanotube modification—surface functionalization 1. (A) Different strategies in surface functionalization of carbon nanotubes and (B) oxygenated surface functional groups and N-configurations (Mallakpour & Soltanian, 2016).

with a variety of groups, and enriching the internal hollow structure by embedding (Mallakpour & Soltanian, 2016).

Surface functionalization of CNT plays a vital role in CNT-based water treatment because of its low cost. $-C=O$, $-COOH$, and $-OH$ functional groups are immobilized into CNT surfaces by acid oxidation or air oxidation. Electrostatic interactions, covalent bonding, ion exchange, hydrophobic interactions, $\pi-\pi$ electron binding, mesopore filling, and hydrogen bonding are the key factors for the interaction between water contaminants and functionalized CNTs. These functional groups promote hydrophilicity and adsorption capacity for polar pollutants and less molecular weight atoms (e.g., phenols and 1,2-dichlorobenzene). Surface functionalization increases the number of oxygen, boron, nitrogen, or other groups on the CNT surface and enhances dispersibility and specific surface area. Carboxylate ($-COOH$), carbonyl ($-COO$), and hydroxyl ($-OH$) functional groups attached to the CNT surface as surface oxide absorb positively charged metal ions like Cu(III), Hg(II), Pb(II), and Cd(II) from wastewater. It also enables the covalent attachment of ethylenediamine and (3-mercaptopropyl) trimethoxysilane (MPTMS) for $-NH_2$ and -SH, respectively (Arora & Attri, 2020).

Oxygen functional groups and sp^2 covalent carbon network at the defective edges run the redox cycle (Duan et al., 2015). Apart from air oxidation, liquid-phase oxidation is also done using oxidants such as H_2O_2, H_2SO_4, HNO_3, and K_2MnO_4. HNO_3 adds carboxylic groups ($-COOH$), in which carbon structure is destroyed at boiling point, and H_2O_2 introduces the phenol group. Oxygen functional groups are categorized into acidic, basic, or neutral groups. Catalytic activity has been determined by regulating the quantity and positioning of the oxygen groups. Oxygen functionalities are decomposed at specific temperatures, such as carboxylic acid ($-COOH$) at 200°C–450°C, anhydrides at 450°C–650°C, lactose and lactones at 550°C–700°C, phenols at 600°C–750°C, $-C=O$ at 700°C–950°C, and quinone at 700°C–95°C.

The extreme oxygen content groups destroy the aromatic rings and conjugate $C=C$ double bonds. It occupies the defective sites and delays oxidation activation via stereo-hindrance. Thus, it leads to the disturbance of the $C=O$ group in redox reactions, the charge transport ability of the graphitic carbon network, and the electron density of carbon atoms (Chen et al., 2018). There is a gas-phase functionalization of gas-phase oxidation in carbon dioxide, oxidative plasmas, ozone, and hydrogen peroxide using nitric acid vapors. It allows selective functionalization depending on the temperature. The carboxyl ($C=O$) group is obtained at high temperatures. Oxygen is double-bound to carbon in ketones, quinones, and aldehydes, and oxygen is singly bound to $O-C$ in phenols, ethers, and hydroxyls. Decomposition of carbonyl, phenol, and ether occurs at low temperatures (Su et al., 2013).

Carbonyl groups ($-C=O$) are electron-rich, with lone-pair electrons showing better catalytic performance and acting as Lewis basic sites to activate O_3 by electron-transfer process. It has high nucleophilicity with critical active sites in persulfate activation. Carbonyl groups at the boundary have a strong affinity for peroxymonosulfate (PMS) due to the elongation of the peroxide $O-O$ bond in PMS. Carboxyl groups ($-COOH$) act as Lewis basic sites for O_3 activation with their high charge density to directly oxidize the pollutants by O_3. Deprotonated carboxyl groups react with O_3 to form electrophilic oxygen species for the radical promoters to produce HO. Hydroxyl groups (−OH) serve as electron donors and increase the electron density of the conjugated π system in the electron-transfer process. Surface functionalization engineering can also be done to promote surface oxidation and reduction modifications. Surface oxidants are HNO_3, H_2O_2, $KMnO_4$, H_2SO_4, and O_3 for introducing oxygen groups like −OH, −COOH, $-SO_3H$, and $-NO_2$ on the carbon surface to increase its acidic nature. The surface reduction process occurs at high temperatures in a reduced alkaline atmosphere to form a basic nature with −C = O, $-NH_2$, etc. Thus, fabricating optimized oxygen contents increases catalytic performance (Liu et al., 2021; Ren et al., 2020).

A moderate electron-donating capacity of the hydroxyl group increases the electron density of the adjacent carbon and promotes electrophilic interaction. For instance, ozonation could oxidize hydroxyl groups into quinone/ketonic groups. Edge sites containing ketonic and hydroxyl groups act as redox centers, and hydroxyl groups disperse the nanocarbon in water through hydrophilicity and mass transfer processes. Hydroxyl groups on the metal surface occur by surface hydroxylation because surface metal sites act as Lewis's acid sites, forming electron acceptor−donor complexes with water molecules (Wang et al., 2020). The lone-pair electrons of −C = O at the structural defect position or edges of a graphene framework play as Lewis basic sites to provide electrons. The ketonic oxygen group activates adjacent sp^2 carbons too; rGO and g-C_3N_4/O_3 with a high amount of the −C = O group (electron-rich) showed better activity for activation of O_3 via the electron-transfer process. The −C = O group was oxidized to -COOH and performed well for the persulfate activation (Duan et al., 2016). The existence of C = O in the carboxyl group (−COOH) increases the charge density of oxygen functional groups and simplifies electron transfer from the Lewis basic site. −COOH performs direct oxidation of target organics on the surface of CNTs-COOH. Carboxylates are also destroyed by O_3 to start radical chain reactions. Thus, carboxyl groups on the graphene framework possess high negative adsorption energies and are found to be more active than the ketonic group (Song et al., 2019).

1.4.2 Heteroatom doping

Typically, the CNT surface has a limited number of oxygen-related functional groups (hydroxyl group (C−OH), carbonyl group (C = O), and carboxylic acid group (COOH). The catalytic activity of the pristine CNT improved multiple times *via* the introduction of a heteroatom (B, F, N, O, P, and S) onto the graphitic network. Heteroatom doping leads toward establishing the Lewis basic sites, more oxygen functional groups, as well as defective sites. The catalytic performance of the heteroatom-doped CNT is governed by the hybrid sp^3-C/sp^2-C electronic structures, heteroatom/oxygen functionalities, and the density of the defective sites. Heteroatoms such as N, O, and S have similar atomic sizes with different electronegativities to break the delocalized π system for local charge polarization. Heteroatom doping on the carbon basal plane creates electron-rich regions for electrophilic interaction with O_3 and nucleophilic interactions between positively charged C atoms and terminal O atoms in O_3, resulting in the local density of the graphene lattice and work function, which determine the capacity and direction of electron transfer. N-doping influences the mobility of electrons and charge density in graphene, which is due to the large electronegativity difference between N (3.07) and C (2.55). The optimized electronic structures with N-doping increase the catalytic activity. Pyridinic N offers electrons from the *p*-orbital to the π-conjugation of the carbon matrix to enhance the electron-donating capacity (Pedrosa et al., 2018).

Graphene lattice doping with B or P atoms leads to positively charged B- or P-doped catalysts with polarizing paired covalent electrons with C atoms because of the smaller electronegativity. Theoretically, doped B/P atoms have high local spin density and favor π−π interactions with aromatic pollutants. A higher defective site is observed in P-doping than in B-doping due to the large atomic size. A partially localized π network and a highly defective level of P-doping are responsible for the improved catalytic activity (Yin et al., 2017). The large atomic radius of S (1.04 Å) is a greater challenge than N and B atoms for doping S atoms within the graphene lattice. S-doped atoms have developed more defective sites owing to the thermodynamic instability of the large radius in the carbon network (Wang et al., 2020). F is a highly electronegatively charged atom, which makes it quite challenging to be used as a dopant. A strong oxidative capability precursor is required for the treatment of carbon precursors. For example, the use of F-doped CNTs toward catalytic ozonation of oxalic acid is reported to form sp^3-hybridized C sites rather than sp^2. Doped F atoms maintain the nature of a delocalized π network in a graphene lattice, and positively charged C atoms induce nucleophilic activation of O_3 (Wang et al., 2018).

CNT hybrids were also established during heteroatom doping. The number of functional groups that are added will enhance the selectivity of adsorption (Fig. 1.1). For example, oxygen-containing groups with hydrophilicity support dispersion in aqueous media and impact surface reactions. N-functional groups provide a synergistic effect and do not severely affect the porosity of the adsorbent. It offers properly activated sites and properties for persulfate activation (Liang et al., 2017). Whereas S-containing groups are used for selective adsorption of some metals (Yang et al., 2019). N-doping on CNTs can establish unique catalytic active sites and good charge distribution of adjacent carbon atoms in the conjugated sp^2-carbon framework, which prompts a high chemical potential for catalytic reactions. Most recent studies also confirmed that N-doping can form electron-rich Lewis basic sites, including boundary N (pyridinic N, pyrrolic N), substitutional N (graphitic N), and oxidized N (Fig. 1.1B). Further, N-doping can also build oxygen groups (C-OH, C = O, and COOH) in the carbon network, which could also facilitate good catalytic performances.

Localized electrons in unsaturated carbon atoms in the defective structure of CNT pave the way for the O, H, and other heteroatoms to bind in evolving oxygen-containing functional groups (OFGs). It includes phenols and ethers (C−O), ketones (C = O), hydroxyls (−OH), and carboxyl groups (O−C = O). −COOH and −OH functional groups introduce $-NH_2$ and −S atoms to bind based on the cation exchange and hydrophilic properties. C = O bond has a high chemical potential to initiate the redox process for the removal of organic pollutants, e.g., directly participating in the catalytic reactions in the PS-advanced oxidation processes (AOPs) system (Duan et al., 2018). OFGs on the surface of CNT are thermally unstable due to the choice of annealing temperature, thus making it easy for C = O to get adjusted on the surface. Excessive OFGs destroy the C = C bond, conjugated double bond, and aromatic ring and end up suppressing catalytic activity. At the same time, excessively vacant sites deactivate the physical accessibility of persulfates due to electrostatic repulsion interactions (Chen et al., 2018). The advantages of heteroatom doping are introducing structure defects and new active sites, changing electron density, and breaking the inertia of the conjugate electron system.

1.4.3 Metal and composite

The introduction of metal species to the CNT structure is one of the strategic approaches to improving the efficiency of the catalytic activity due to the interfacial coupling and charge transfer between the different counterparts. Oxygen functional groups on the CNT network, heteroatom dopants, defective sites, and edges can ease the metal species due to the efficient binding strength and accelerated electron transfer process. The *d*-orbitals of the electron-donating metal species on the CNT can significantly magnify the electron conductivity and electron transfer capacity of carbocatalysts by synchronizing their electronic properties (spin density and work functions) (Peng et al., 2021). So, the effective interfacial interactions by metal doping on the CNT framework could appreciably alter the charge density and catalytic active sites, resulting in managing the carbocatalyst activity.

Metal NPs are doped at the curved carbon layer, or mesh-doped CNT. Mesh-doped CNT is the fabrication of CNT with encapsulated NPs of a single atom, metal, metal oxide, or multiple metal oxides. TMs cobalt (Co) and iron (Fe) are preferentially used for doping because of their high activity and variable valence states that offer dynamic electron transfer to persulfate activation (Boczkaj & Fernandes, 2017). Fe species are the most commonly used metals for their easy availability and high reactivity. The main disadvantage is the metal leaching and production of sludge. Thereby, interactions between dispersed carbon materials and stabilized metal NPs suppress metal leaching and induce electron transfer. Co-doped carbon composites also retard leaching and improve the activity of Co. CoO_x-doped ordered mesoporous carbon. The multivalence nature of TM oxides is active in catalytic ozonation due to the redox cycle of the metal cations (Me^{n+}/Me^{n+1}). Practically, TMs induce metal leaching because of their low activity in O_3 activation, but they need a low dosage and strong surface confinement to reduce metal leakage. For example, manganese oxides are loaded onto the nanocarbon due to their abundance on earth and lower toxicity (Yang et al., 2019). CNTs/Mn composites can be used for catalytic ozonation, during which Mn^{2+} is adsorbed/desorbed on the surface of CNTs by electrostatic forces.

Zero-valent iron, a strong reducing agent ($E^\circ(Fe^{2+}/Fe^\circ)$), and an electron donor demonstrate high reactivity in reductive reactions. But the poor stability is obtained by loading it with nanocarbons of Fe—O—C complex structure. Fe°/CNTs aid electron transfer from Fe° to CNTs with no oxidation of Fe° to form iron oxides. Partial oxidation occurs by O_3 (ozonation) and ends with Fe^{2+} with the help of H^+ and is defined as the CNTs-catalyzed ozonation system. Another mechanism with zero-valent zinc (ZVZ) and carbon matrix formed of Zn-N bonds between ZnO and N-doped CNT promotes electron transfer. ZVZ first oxidizes to ZnO to form a Zn (OH)-O_3 surface complex, and the formed Zn-N bonds regulate the electron density and electron transfer process as well as the higher density of defective sites (Wang et al., 2020).

1.4.4 Carbon structure and defects

The physicochemical properties of carbon materials are controlled by their crystal structure. It is evaluated by the intensity ratio I_D/I_G, which represents the defective degree obtained from Raman spectra. Two peaks explain the structure and defect. The G peak gives the level of graphitization, and the D peak represents the defective sites in the sp^2 lattice. The ranges are 1570—1606 cm^{-1} and 1310—1363 cm^{-1}, respectively. A high I_D/I_G value defines a greater number of defective sites, and a low I_D/I_G indicates high graphitized carbon. The overall interpretation can be varied or modulated by the thermal annealing process (Ren et al., 2020). High graphitization degree states about the sp^2 hybridized carbon bonds, which are more effective than sp^3 hybridized carbon for persulfate activation and transfer of electrons (Oyekunle et al., 2021).

Defective structures are classified into:

- Structure defects: A six-membered carbon structure is surrounded by a non-six-membered ring structure.
- Bond rotation or particle boundary: Stone—Wales defect structure formed on the surface of the graphene.
- Doping defects: N or B atoms replace C atoms in the six-membered ring to regulate electronic structure and conduction characteristics.
- Non-sp^2-carbon defects include carbon chains, edges, vacancies, suspension bonds, interstitial atoms, and adsorption atoms.
- High-strain folding of graphite layer-interlayer trap centers in the multiwall nanotube is filled with perforated defects.

Structural defects from nonhexagonal rings destroy the structural integrity of the graphene lattice and break the conjugated π network, forming electron-localized regions (Fig. 1.2).

There will be high affinities toward O_3 due to the charge distribution of C atoms near the defective site due to a biased nature with unpaired electrons (Rezaei et al., 2021). Density functional theory explains that surface-adsorbed atomic O is evolved by peroxide bond (—O—O—) breakage when placed on zigzag, vacancy, and armchair edges. It directly decomposes organics via nonradical oxidation with high charge density. The N vacancy on g-C_3N_4 is the critical defective site to regulate the electronic structure, enhance the surface adsorption capacity, and induce charge mobility

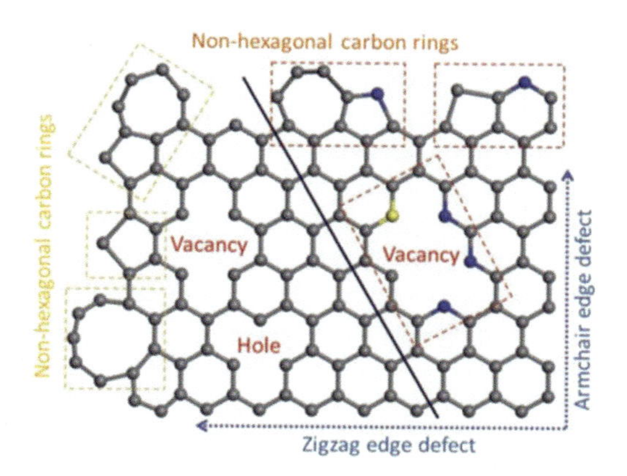

Figure 1.2 Structural defects on carbon nanotube network. Different structural defects in carbon nanotube framework (Liu et al., 2021).

between g-C_3N_4 and adsorbates. The N vacancy promotes the localization of π electrons within the heptazine units and increases the electron density of adjacent C atoms. The introduction of oxygen functionalities in g-C_3N_4 creates nitrogen vacancies for better performance in catalytic ozonation. Thus, structural defects are important to bind metal atoms or metal oxides for their charge transfer and strong binding strength. Vacancies play the role of electron acceptor for Coulomb interaction between graphene basal plane and metal oxides (Song et al., 2019; Wang et al., 2020).

Other defective structures, such as double-vacancy defects, predominate in the reactivity of CNTs. Cumulative electrons lower the electron density of neighboring carbon atoms in the structural defect (Shao et al., 2020). More structural defects on CNT can be obtained through pretreatment like potassium permanganate oxidation, acid etching, and thermal energy, and annealing is the vital step affecting defective degree. In this process, part of the carbon matrix is oxidized and converted into saturated bonds, which are very useful for PS activation. The lone-pair electrons in defective structures could attack persulfates directly to produce oxidative free radicals. On the other hand, electron transfer occurs from the pollutant (nonradical pathway) to degrade PDS (Wu et al., 2022; Yang et al., 2020).

A high annealing temperature (600°C) decreases the defective area of CNT, and a lower I_D/I_G ratio helps in the increase of complete phenol degradation (Yang et al., 2019). Defects in nanocarbons, with the help of catalysts (oxides), lead to structural rearrangement in the graphitic network. A pentagon SWCNT turns into a sharp cone, and a heptagon turns into a horn. The pentagon–heptagon pair (5/7) is the most common type of defect, depending on the formation energies of the tube diameter. The condition of CNT growth (precursor, temperature, pressure, etc.) and the rate of CNT length increase determine the defect density. Thus, defects in CNTs lead to functionalization, single-site basic heterogeneous catalyst selection, and CNT characteristics like chirality, diameter number and type of defects, functionalization, etc., determine the production of reactive oxygen species (Su et al., 2013).

Defective sites can be tailored or engineered effectively for the enhanced electronic properties of carbon catalysts. Nanocarbons inherited with topological defects and edges have unpaired electrons that interrupt the consistency of electronic structures of basal planes and carbon skeleton. Catalytic activity proceeds from dangling bonds of high-energy sites possessing high reactivity toward reactants to initiate radical chain oxidations. The unpaired electrons of carbon atoms exist in a "localized state" and strongly react with small molecules to form an attached complex. Graphene edges can transfer electrons faster than the honeycomb basal plane. Topological defects like nonhexagonal carbon rings and vacancies disrupt the conjugated π network and redistribute the electrons. Downsizing carbon creates defects in the carbon framework too. Edges and topological defects can be designed during N-doped graphene at high temperatures for the decomposition of N-rich precursors. Direct oxidation, microwave reduction, and plasma technology have been proposed for effective defect-rich electrocatalysts in controlled conditions. Defects also show poor stability owing to the unsaturated dangling bonds and long-paired electrons that induce electrophilic attack on oxidants. Hence, a balance between stability and reactivity manipulation must be taken care of for better performance (Duan et al., 2016; Wu et al., 2022).

1.5 Carbon nanotube applications in water and wastewater treatment

1.5.1 Adsorption of emerging contaminants

Toxic chemicals (inorganic and organic chemicals, heavy metals, dyes, pharmaceutical and refractory substances, and persistent organic substances) present in the water bodies end up with severe contamination, posing a serious threat to the environment. Industries (cosmetics, textiles, paper, and leather) also dispose of potential organic solutes in the water environment, causing problems in the removal of toxic components from wastewater. Industries must follow safety concerns and environmental consequences. In such cases, researchers find the problem and solve it through laboratory experiments. In this regard, effective water treatment solutions have become an important norm of the day for the world to make natural water resources safe for human consumption. Section 1.1 lists various conventional treatment methods, among which adsorption has been extensively adopted due to the easy of operation, selective capture, high efficiency, low cost, and reusability of adsorbents. The productivity of the adsorption process depends on the nature of the adsorbent and target pollutants. In-depth, pore diameter, interior and exterior morphology, surface area, and functional groups also play a crucial role in the adsorption process (Das et al., 2014).

Various adsorbents were developed for the removal of pollutants like pesticides, phenols, pharmaceuticals or drugs, dyes, and metals from water. AC is the most commonly used commercial adsorbent for its better adsorption capacity for organic contaminants. Conventional water treatment technologies face difficulties on a day-to-day basis to provide safe water and meet the WHO standard for wastewater discharge. Thus, nanomaterials, especially CNTs, took a remarkable place in wastewater treatment and were made suitable for organic, inorganic, and biological water pollution removal. The application of nanostructured material as an adsorbent focused on removing harmful and toxic organic substances from wastewater is equipped with higher removal efficiency and higher adsorption capacity. The targeted removal of pollutants has drawn the attention of researchers to experiment with the adsorption process with CNTs. The functionalization of CNTs with various functional groups proves to be the best and newest adsorption sites for the pollutants (Sajid et al., 2022).

Adsorption is the best choice for removing organic pollutants, and the commercial adsorbent AC plays the true part of adsorption in removing organic contaminants. It also has shortcomings for their recovery. So, the next phase of research brought nano-structured materials as adsorbents to remove harmful organic contaminants from wastewater (Mohmood et al., 2013). Thus came the discovery of CNTs, an excellent adsorbent with a layered and hollow structure and a high specific surface area (Aslam et al., 2021; Selvaraj et al., 2020).

The hydrophobic surface of CNTs attached to alcohols like ethanol, benzene, 2-propanol, hexane, cyclohexane, methyl ethyl ketone, and toluene has the best adsorption behavior. Further, the adsorption of phenanthrene was experimented with varying diameters of CNTs (Yin et al., 2020). SWCNTs demonstrate better adsorption properties with high specific surface area in removing toluene, benzene, and reactive blue 29 (RB29) (Jain & Kanu, 2021). Nonpolar organic compounds like nitroaromatics, nonpolar aliphatics, and nonpolar aromatics were also found to have high adsorption efficiency (Mishra & Sundaram, 2023). Sudan red I, II, III, and VI dye usage in food is prohibited due to their high level of biotoxicity, which results in potential carcinogenic and mutagenic effects in humans. Conventional and physical oxidation methods make it difficult for recalcitrant and biodegradation due to its complex aromatic structure. So, it happened to be the CNTs as adsorbents for the targeted removal of organic dyes. The absorption capacity of CNT works better with rough surface topography than with smooth surfaces. A plausible adsorption mechanism by nanocarbons is illustrated in Fig. 1.3.

Physical adsorption, electrostatic interaction, ion exchange, surface complexation, and coprecipitations were simultaneously driven by the effective adsorption process and removed the dissolved organic and inorganic pollutants (Yang et al., 2019). Better pollutant abatement means that the appreciable interactions between organic molecules and the catalyst surface and hydrophobic sites that exist on the outer surface of CNTs are the most vital criteria for adsorption. Fundamentally, the interaction between organic and carbocatalysts is mainly due to the $\pi-\pi$ interaction, Van der Waals, and electrostatic interactions. The $\pi-\pi$ interactions occur between π electrons in sp^2 hybridized orbitals of carbon atoms in CNT and delocalized π electrons of $-C=C$ in pollutants. Modified CNTs show $\pi-\pi$ interactions with delocalized π electrons of phenyl rings and carbonyl groups. Pure CNTs with sp^2 hybridization have hydrophobic or $\pi-\pi$ interactions that are strong in adsorbing

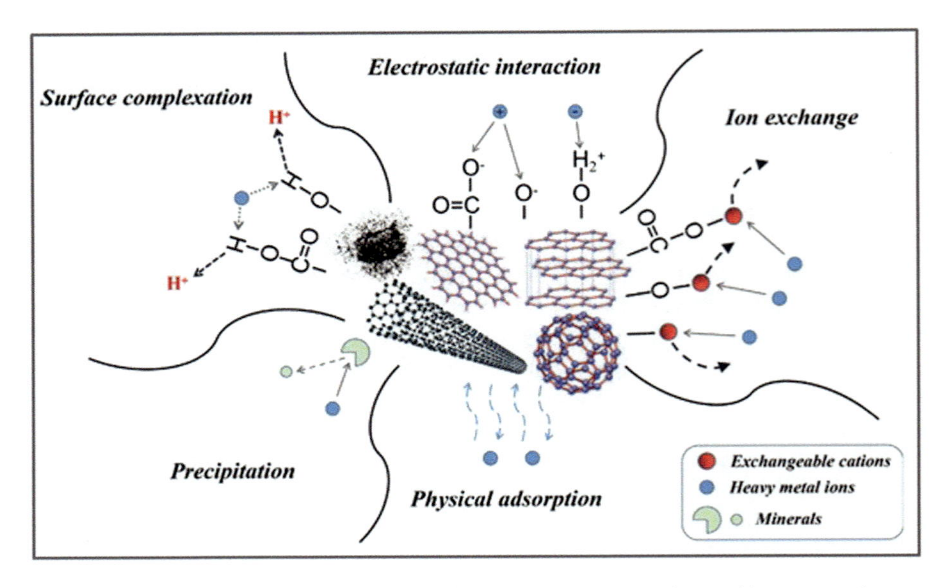

Figure 1.3 Adsorption mechanism. Plausible adsorption mechanisms of heavy metals onto carbon adsorbents (Yang et al., 2019).

organic pollutants (Sajid et al., 2022). For example, pollutants with an aniline or phenol group have a high affinity. The magnitude of adsorption is affected by the geometry of organic pollutants too, in a way that planar structures are absorbed stronger than nonplanar structures. One more example is the adsorption of PAHs on CNTs, which is associated with a volume ratio of mesopore to micropore and surface area.

Physical adsorption of metal ions can be done on the surface available to pure CNTs. Enhanced adsorption requires modifications or the composite formation of functional groups on the surface. Acid treatment or oxidizing agents develop negative charge functional groups on the surface of CNTs like ketone, hydroxyl, aldehyde, and carboxylic acid. In turn, it can be used to adsorb positive-charge metal ions (lead, chromium, and zinc), aliphatic hydrocarbons, and hydroxylated aromatic carbons through hydrogen bonding or electrostatic interactions (Rodríguez et al., 2020). Zinc, nickel, Cd(II), Pb(II), and Cr(VI) divalent metal ions have high sorption capacities on oxidized CNT (Tan & Kochmann, 2017).

The heavy metal ion removal by CNT-based adsorption has also interpreted that the metal ions in wastewater and protons in surface functional groups of CNTs exchange in the aqueous medium. The release of H^+ ions from the CNT surface happens by pH reduction in the solution, meaning that an increase in initial metal ion concentration binds to more H^+ ions in the solution. Inner areas of open-ended CNTs and exterior layers are significant locations for the adsorption of pollutants. The adsorption process occurs between functional groups and heavy metal ions through chemical bonds. Moreover, the metal ion-functional groups of CNTs are intricate through electrostatic attraction. Apart from surface functional groups, other parameters like pore density, purity, ionic strength, pH, and dispersion state influence the adsorption mechanism. The usual order of heavy metal adsorption ability is $Cd^{2+} < Cu^{2+} < Zn^{2+} < Ni^{2+} < Pb^{2+}$.

Composite formation of CNTs with metal oxides like ZnO, alumina, and SnO_2 enhances structural properties favorable for adsorption. For example, the composition of MWCNTs and iron oxide creates a magnetic material with a better adsorption capacity for metallic pollutants (Murali et al., 2020). Adsorption of metal ions from aqueous solutions to adsorbents also occurs based on ionic strength, contact time, solution pH, and CNT dosage. Intermolecular interactions between the adsorbent depend on the ionic strength of solutions, the nature of metal ions, surface chemistry, and structural features. These interactions may be surface complexation, physical, ion exchange, electrostatic, or precipitation (Yang et al., 2019). Surface complexation is caused by metal ions binding to multiatom; physical interactions usually weaken interactions between pores and the surface of CNTs and target metals. Further, the ion exchange between adsorbents functionalized with oxygen-containing groups is one of the key phenomena in the adsorption process. Electrostatic interactions between the negative-charge CNT surface and positive-charge metal ions and co-precipitation (growth of solid products either in solution or on the surface) are also recognized as the adsorption mechanism. As far as biological absorbent is concerned, microbes could change the properties of CNT, for example, hydrophobic pristine CNTs are converted to hydrophilic, degraded C-CNTs into short CNT fragments, etc. Thus, the biological method of using CNT is very harmful and must be monitored so that NT is not sheered, cut, ground, and ripped off after absorption (Jain & Kanu, 2021).

1.5.2 Carbon nanotubes in membrane process for water purification

Filtration is defined as the physical and mechanical process of separation that allows water to flow through the membrane while retaining excessive solutes. It has been the most widely used water purification method over a long period of time because of its low chemical mass, retention efficiency, automated process control, permeating flux, high stability, process strength, and good operation. Ultrafiltration (UF), nanofiltration (NF), forward osmosis (FO), reverse osmosis (RO), membrane distillation (MD), pervaporation, electro-deionization (EDI), and electrodialysis (ED) were the most commonly employed membrane-based separation methods. Each method has its own advantages and disadvantages, and the fabrication process involves interfacial polymerization, different types of coating, and composite casting. It uses more energy for construction, and the performance of the membrane declines due to fouling and high cost because there is a need for frequent cleaning, shortening the membrane life cycle, and replacement (Lee, 2018).

Conventional polymer-based membranes that have low chemical and mechanical stability have the nature to stop the filtration of biological, inorganic, and organic substances in hydrophobic membranes. Thus, the CNT membrane merits itself with comparable size of pores, unique surface chemistry, and sturdy antimicrobial activity. It has high water flux, and good electrical conductivity, and filters wastewater in an efficient system which is a vital point for separation and filtration (Jain & Kanu, 2021). CNT composite membranes outperform UF, NF, RO, and MD (Barrejón & Prato, 2022). CNT-based membrane combines with the effects of filtration membrane results in good permeability and better selectivity. CNT membrane surpasses the conventional membrane with no energy usage for the hydrophobic hollow CNT attracting polar water molecules.

1.5.3 Role of carbon nanotube in membrane separation

There are three main CNT-related membranes, namely Bucky-paper (Thamaraiselvan et al., 2018a, 2018b). Composite and vertically aligned CNT (VACNT) were substantially used in the separation methods. The hydrophilicity of functionalized CNTs enhances the overall permeability of water and hydrophobic membranes. Then, to avoid foulants, enhanced antifouling agents are added to increase membrane hydrophilicity on the surface. Negatively charged hydrophilic membranes help resist negatively charged organic molecules from the membrane surface and improve antifouling efficiency (Barrejón & Prato, 2022).

Further, CNTs and CNT functionalization alter the characteristics of the membrane with selectivity-permeability trade-offs, variation in porosity, antifouling capabilities, and surface roughness (Chufa et al., 2021). The modification of the membrane/filter using CNT-based materials substantially enhanced the pores present per unit area and influenced higher permeability. Mechanical densification of VACNTs gives more membrane pores, and the multiple CNT walls and nanometer size also help in permeability. CNTs with polymeric membrane separation have great potential with a high aspect ratio, high flexibility, effective interaction with aromatic compounds, a low friction coefficient in internal walls, and a greater number of nanochannels. CNTs have good mechanical strength, superhydrophobicity, and wetting resistance.

CNTs made of graphene sheets (rolled into cylinders) on a nanometer scale are present with hydrophobic channels that give low friction to water molecules. It is attached to a polymeric UF/NF membrane that gives excellent adsorption capacity through $C=C$ bonds from organic molecules and $\pi-\pi$ interaction on CNTs, and chemical functionalization aids in the removal of organics and heavy metal pollutants (Lee, 2018). The CNT membrane transports water through the channels through its intrinsic hydrophobic nature. Van der Waals force agglomerates in a polymer matrix; thus, functionalization makes it disperse easily on the membrane. Hydrophilic functional groups ($-OH/-COOH$) on CNTs enable the matrix to be well-dispersed on the matrix for their affinity to H_2O molecules and act as an antifouling group. Deprotonation of the $-OH/-COOH$ group removes Zn through Ar plasma treatment. The Zwitter ionic group has a positive charge in the amine group and a negative charge in the carboxylic group. There is an increased exchange between solvent and nonsolvent through an interaction between water (nonsolvent) and hydrophilic casting solution. It helps in the good dispersion of the hydrophilic nature of MWCNTs. Amine-functionalized CNTs/PES membranes enhance hydrophilicity and total flux loss through total fouling, water permeability, and decreased surface roughness (Jain & Kanu, 2021).

Sulfonated CNTs/PA membrane with 1,3-propane sulfone and pendant alkyl sulfonic acid groups helps for good CNT dispersion in water and enhances BSA fouling alleviation (Farahbaksh et al., 2017). COOH- and PEG-modified CNTs coated on PES membranes for lessening NOM fouling and enhancing HA flux due to the roughness and decreased surface charge on the modified membrane. Similarly, CNTs/ZnO/TiO_2 functionalized membranes improve water permeability, photocatalytic activity, and the self-decomposition of contaminants. Hydrophilic polyaniline and polypyrrole (PPY) coated with CNTs via in situ oxidative polymerization make the surface more hydrophilic and increase porosity and pore size to work well for the fouling resistance and water permeability. The functionalized CNTs with membranes experience hydrogen bonding between functional groups and water, creating a hydrophilic environment and thereby rejecting hydrophobic pollutants. Nanocomposite membranes like PVK-GO coated on nitrocellulose membranes go through covalent reactions or physisorption between functional groups and assist in removing microbes (e.g., *Bacillus subtilis*) from water. It also prevents biofilm and biofouling formation (Smith & Rodrigues, 2015).

1.5.4 Advanced oxidation processes

AOPs are the most efficient as well as promising techniques that involve various reactive oxidative species (ROS) for the detoxification of emerging organic contaminants (pharmaceuticals, personal care products, pesticides, endocrine-disrupting chemicals, and persistent and refractor organic substances) in the water system. The classical AOPs are mainly governed by hydroxyl radicals ($^{\bullet}OH$) to oxidize a wide range of emerging organic contaminants quickly and nonselectively. Acidic pH environments and shorter half-life periods are also major drawbacks to decelerating pollutant abatement efficiency. AOPs driven through $SO_4^{\bullet-}$ have gained more interest in this field. Owing to the advantages of $SO_4^{\bullet-}$ over $^{\bullet}OH$, it has a wider half-life period (30–40 μS) compared to $^{\bullet}OH$ (2×10^{-9} second), and $SO_4^{\bullet-}$ could propagate effective oxidation steps in a wide range of pH (2.0–8.0). Moreover, $SO_4^{\bullet-}$ has equal or even higher redox potential (2.5–3.1 V), than that of OH (1.8–2.7 V) (Hu et al., 2015).

Instead of hydrogen peroxide (classical AOPs), PMS and PDS are the most common oxidants utilized in $SO_4^{\bullet-}$ - AOPs. The $SO_4^{\bullet-}$ generates in-situ *via* breaking the peroxide bond (–O–O–) in the persulfate molecules through electron transfer and energy transfer, as described in Fig. 1.4 (Lee et al., 2020).

The molecular structure per-oxo bond distance, per-oxo bond energy, and standard redox potential are provided in Table 1.2. Technically, $SO_4^{\bullet-}$ is formed by homolytically or heterolytically dissociating the peroxide bond in the persulfate structure. The peroxide bond usually activates through a wider variety of methods, including healing, photolysis, ultrasound, TM/metal oxide catalysts, and carbocatalysts. Among these,

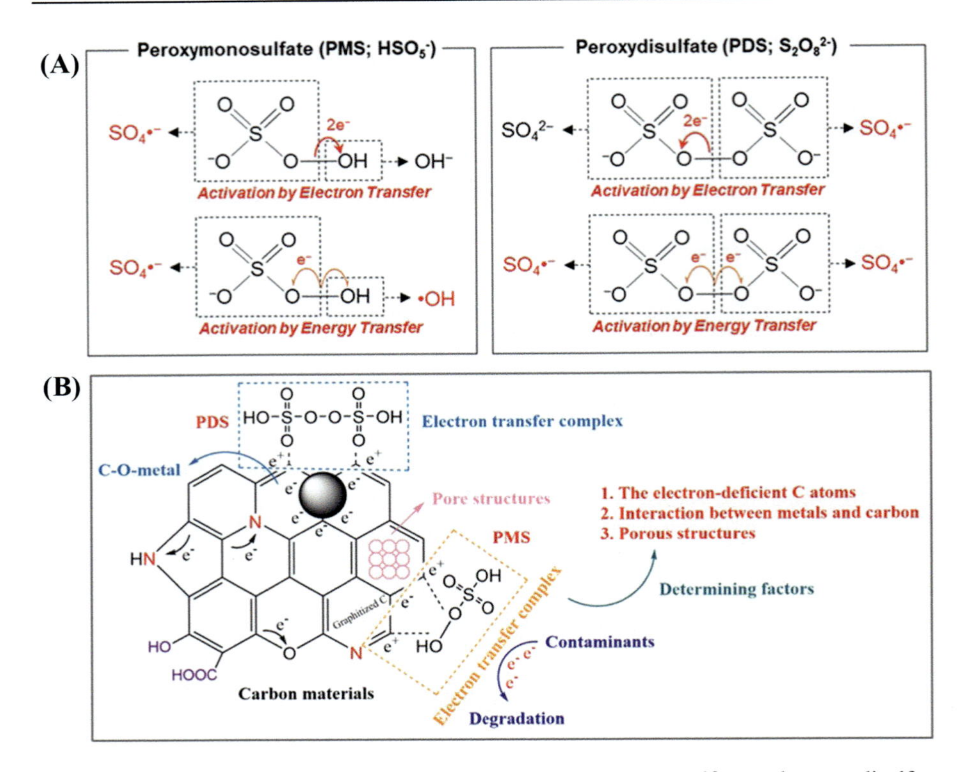

Figure 1.4 Persulfate activation phenomenon. (A) Peroxymonosulfate and peroxodisulfate activation via electron transfer and energy transfer process (Lee et al., 2020). (B) Surface-complex formation in carbon-persulfates in advanced oxidation processes (Luo et al., 2022).

metal-free carbon nanocatalysts can act as green and effective catalysts to activate persulfates (Oyekunle et al., 2021).

Owing to the peculiar sp^2 hybridized carbon network along with low-level defects (vacancies defective edges) and oxygen-containing functional groups in CNT, it could remarkably be used to activate persulfates rather than other carbon analogous. Furthermore, both SWCNT and MWCNT transfer limitations, significant electron mobility, and electrical conductivity are dominantly utilized in persulfate-based AOPs. SWCNT and MWCNT can effectively destroy phenolic compounds as well as pharmaceuticals via persulfate activation. Besides, the organic compounds (nitrobenzene and benzoic acid) could not be effectively oxidized via the persulfate/CNTs system (Cheng et al., 2017). PDS/CNT demonstrated better catalytic activity than PMS/CNT in the phenolic compound degradation process, and it was rationalized based on the specific surface area. On the other hand, the PMS/CNT system provides superior performance in the detoxification of refractor organic compounds containing electron-rich moieties. The results were justified by the fact that the formation of different ROS, radicals, and nonradical species in PMS/CNT and nonradical reactive complex formation in PDS/CNT systems. The different stereochemical structure of

Table 1.2 Physical and chemical characteristics of persulfates (Lee et al., 2020).

Properties	PMS	PDS
Molecular formula	$KHSO_5 \cdot KHSO_4 \cdot K_2SO_4$	$K_2S_2O_8$
Molecular symmetry	Asymmetric oxidant $HO{-}O{-}SO_3$ (replacement of one H atom in H_2O_2 with a SO_3 group)	Symmetric oxidant $S_3O{-}O{-}SO_3$ (replacement of two H atom in H_2O_2 with two SO_3 groups)
Peroxo-bond (−O−O−) distance (Å)	1.453	1.497
Peroxide bond dissociation energy (kj/mol)	377	92
Standard electrode potential (V_{NHE})	1.75−1.82 V	2.01−2.12 V
Reactivity toward nucleophiles	Effective oxygen atom transfer reactions to nucleophiles, such as X^- and HCO_3^- (leading to secondary oxidant formation)	Negligible (stable at excess background anions)

these persulfates (symmetric structure in PDS, asymmetric structure in PMS) was the main feature enabling dissimilar catalytic activity.

Fundamentally, the activating peroxy-bond (−O−O) stimulates electron transfer between CNTs and persulfates. The reaction between CNT and persulfate contributes both free-radical mechanisms ($SO_4^{\bullet-}$ and $^{\bullet}OH$) and nonfree radical mechanisms (electron transfer, $^1O_2$, surface active, and surface active complex formation) to oxidize the organic pollutant (Fig. 1.5A).

In most cases, mechanisms may coexist in the activation of PMS and/or PDS by CNTs. In CNT/persulfate electron conduction mechanism leads to the break of the −O−O− bond in persulfate to produce $SO_4^{\bullet-}$ over $^{\bullet}OH$, as described in Eqs. (1.1)–(1.4). $SO_4^{\bullet-}$ can also further interact with OH^- ions or H_2O to generate $^{\bullet}OH$ too.

$$HSO_5^- + e^- \rightarrow OH^- + SO_4^{\bullet-} \quad (1.1)$$

$$S_2O_8^{2-} + e^- \rightarrow SO_4^{2-} + SO_4^{\bullet-} \quad (1.2)$$

$$OH^- + SO_4^{\bullet-} \rightarrow {}^{\bullet}OH + SO_4^{2-} \quad (1.3)$$

$$SO_4^{\bullet-} + H_2O \rightarrow {}^{\bullet}OH + SO_4^{2-} + H^+ \quad (1.4)$$

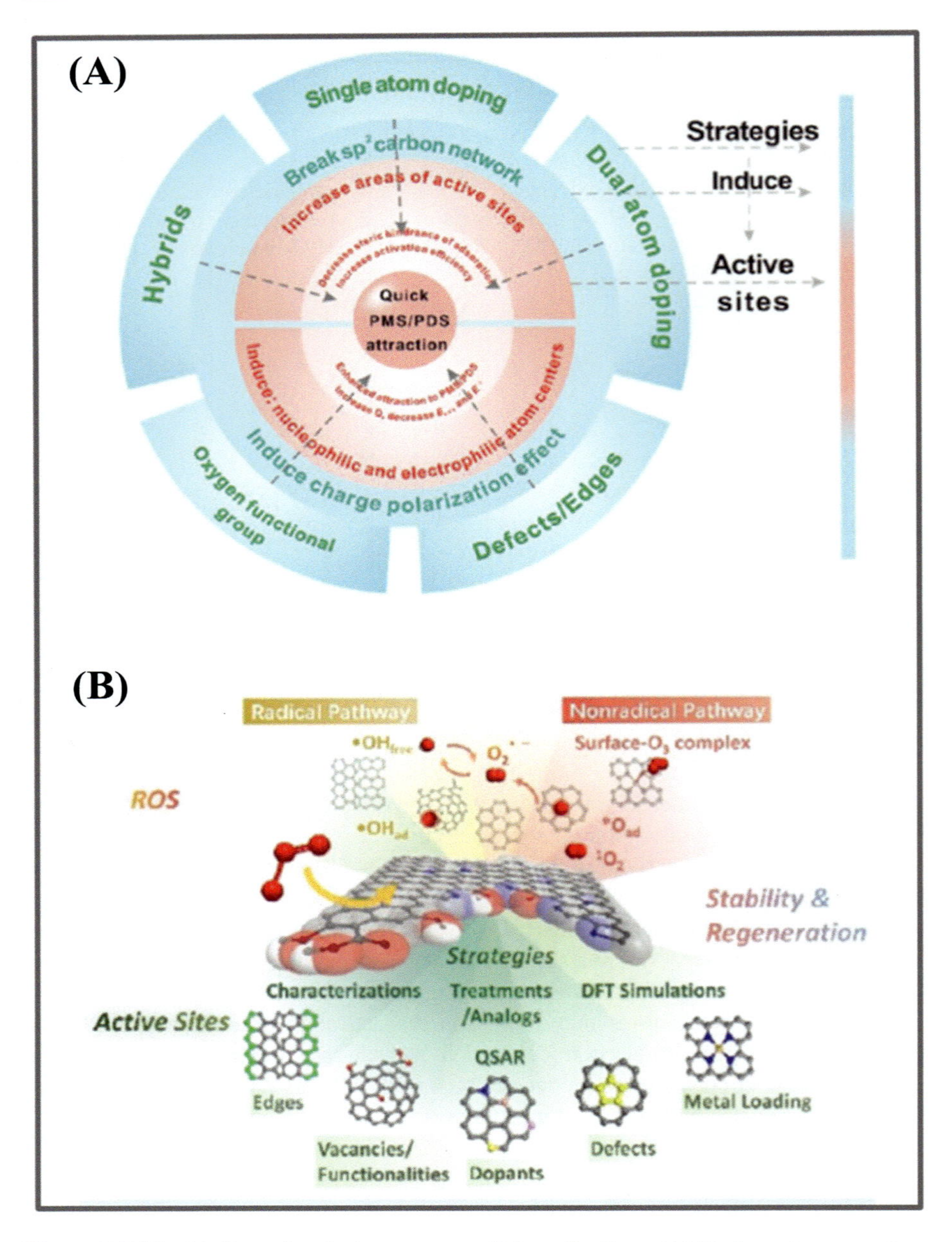

Figure 1.5 Most influencing factors on persulfate activation. (A) Various phenomenal factors influence to improve the persulfates activation by carbon materials. (B) Details strategies for promoting persulfate activation (Luo et al., 2022).

CNT catalytic activity for persulfate activation is mostly influenced by different phenomenal factors, including surface functional groups, active sites, dopant configurations, and defective edges (Wang et al., 2020) (Fig. 1.5B). The oxygen functional groups, including the ketonic group ($-C=O$), carboxyl group ($-COOH$), and hydroxyl group ($-OH$), are believed to be vital active sites for PMS/PDS activation (Duan et al., 2016, 2018). $-C=O$ is the paramount site and demonstrated the best catalytic activity to dissociate PMS into $SO_4^{\bullet-}$ and $^{\bullet}OH$, whereas the $-OH$ group could not effectively contribute to PMS activation. It attributes the highest adsorption energy of PMS to HO-CNT and oxygen groups, but the electron transfer tendency is the lowest among the three oxygen groups. The H atom in the $-OH$ could be separated from the "O" atom and bind with the PMS molecule, implying that $-OH$ is not thermodynamically feasible for PMS activation. Further, the highest reaction energy barrier (E_b) of 0.91 V for HO-CNT has a negligible impact on PMS activation. In contrast, CNT$-C=O$, which possesses the very lowest reaction energy barrier (E_b) of 0.53 V, could strongly influence $-O-O-$ bond cleavage to generate free radicals (Li et al., 2020). The study also emphasized that the limited number of oxygen functional groups was beneficial to the oxidation of organics, and excessive oxygen groups led to a determinantal impact on the persulfate/CNT catalytic activity. The excessive oxygen group not only diminishes the electron conductivity but also occupies the defective edges, which would hamper the interaction between the catalyst surface and persulfate.

$$HSO_5^- + C - \pi \rightarrow C - \pi^* + OH^- + SO_4^{\bullet-} \quad (1.5)$$

$$C - \pi^* + HSO_5^- \rightarrow C - \pi + H^+ + SO_5^{\bullet-} \quad (1.6)$$

$$-C = C = O + HSO_5^- \rightarrow SO_4^{\bullet-} + C = C = O^+ + OH^- \quad (1.7)$$

$$C = C = O^+ + HSO_5^- \rightarrow SO_5^{\bullet-} + C = C = O + H^+ \quad (1.8)$$

$$CNT_{surface} - COOH + S_2O_8^{2-} \rightarrow CNT_{surface} - COO^{\bullet} + SO_4^{\bullet-} + HSO_4^- \quad (1.9)$$

$$CNT_{surface} - OH + S_2O_8^{2-} \rightarrow CNT_{surface} - O^{\bullet} + SO_4^{\bullet-} + HSO_4^- \quad (1.10)$$

The participation of defects is generally explained by the fact that π electrons from defects can be transported to activate persulfates (PMS/PDS), facilitating $-O-O-$ bond dissociation and forming active radicals. Fundamentally, the defect levels have been determined using the integrated intensity ratio of the D band and graphite G bond (I_D/I_G) from Raman spectra. Escalating the catalytic activity of pristine CNTs through heteroatom doping (N, S, and B) on graphitic networks intrinsically alters the structural properties and chemical potential of the carbocatalyst. Whereas N-doping appreciably enhanced the CNT catalytic activity, compared to other heteroatoms, because N has five valence electrons and a similar atomic radius to carbon (0.75 vs 0.77 A for C). Thus, it can develop strong valence bonds with C atoms. N-doping could create three different N-configurations in the main graphite framework. They are, namely, pyrrolic N, pyridinic N, and graphitic N. The Lewis basic site with lone pair electrons (pyridinic N and pyrrolic N), which exhibits high redox potential and graphitic N, might play a vital role in PMS activation (Govindan et al., 2022a, 2022b; Prabavathi et al., 2019). Introduction of B (electron deficient) and N (electron-rich) into a C network can also develop the bonding configuration between B and N, which could facilitate synergistic coupling that impacts the neighboring C atoms to the oxygen reduction process and thus further enhance the catalytic activity (Oyekunle et al., 2021). Similarly, metal-doping/co-doping can also create defective edges and interrupt the uniform π-configuration, which could amend the persulfate activation.

$$HSO_5^- + M^{n+} \rightarrow M^{n+1} + OH^- + SO_4^{\bullet -} \quad (1.11)$$

$$HSO_5^- + M^{n+1} \rightarrow SO_5^{\bullet -} + H^+ + SO_5^{\bullet -} \quad (1.12)$$

$$S_2O_8^{2-} + M^{n+} \rightarrow M^{n+1} + SO_4^{2-} + SO_4^{\bullet -} \quad (1.13)$$

$$HSO_5^- + C - \pi \rightarrow C - \pi^* + OH^- + SO_4^{\bullet -} \quad (1.14)$$

1.5.5 Catalytic activity: ozone activation

The heterogeneous catalytic ozonation process has also been considered a promising remediation technology in water/wastewater treatments. The significant oxidation potentials of ozone (O_3) (2.05 V) enable the direct destruction of unsaturated organic compounds, and catalytic activation of O_3 would also aid the oxidation of refractory organics via multiple oxidative species (Liu et al., 2021). The metal-free catalytic activation of O_3 has been recognized as an alternative catalyst to conventional noble metal-based catalysts in water decontamination. Nanocarbon catalysts prevent the metal-leaching problem with their strong oxidation ability and stimulate diverse oxidative pathways, providing excellent decontamination in water treatment. Among various carbocatalysts, CNT-based catalytic ozonation gained considerable attention.

Heterogenous catalytic ozonation involves multiple reaction pathways for the detoxification of organic substances through the high oxidation potential of O_3 attributable to direct oxidation and the catalytic activation of O_3 that generates ROS. Carbocatalyst O_3 activation mechanisms are generally described into two major pathways: (i) radical pathways, relying on $^{\bullet}OH$ and/or superoxide anion radical ($O_2^{\bullet -}$) from decomposition O_3 on the active sites of carbocatalyst; and (ii) non-radical pathways, singlet oxygen ($^1O_2$), surface activated complexes, and direct electron transfer (Wang et al., 2020). The composition of oxygen functional groups (−OH, −COOH, and −C = O) on the possible active sites (basal plane, structural vacancies, and zigzag/armchair edges) of graphitic network, and defective edges are paramount active sites to activate O_3 and produce radicals ($^{\bullet}OH$ and $O_2^{\bullet -}$). On the other hand, the complete detection of nonradical active sites and their roles in organic oxidative steps might be more complicated due to their mild oxidative potential ($^1O_2$ and surface-activated complex). The coexistence of radicals and nonradicals-based oxidation with O_3/carbocatalyst has further hindered the assessment of nonradical species. Therefore, the fundamental understanding and identification of intrinsic catalytic centers in catalytic ozonation is still a major obstacle.

The well-defined graphitic carbon network (basal plane) of the carbocatalyst has one of the most important active sites for O_3 activation. The delocalized π-electrons within the sp^2-C framework lead to good affinity toward electrophilic O_3 molecules, resulting in the formation of a carbon-O_3 complex, and enrichment of O_3 on the catalyst surface tends to activate O_3 by dissociating the O_3 molecules (Zhang et al., 2017). The highly delocalized π-electrons progressively accelerate the electron conductivity to contribute to the reduction of oxidative reactions for O_3 activation (Zhang et al., 2018). In addition, the accelerated charge migration also stimulates synergistic impacts along with active sites that could escalate the catalytic activity.

The Lewis basic sites $-COOH$, and $-C=O$ groups are typically demonstrated as key active sites for ozonation activation to produce radicals ($^{\bullet}OH$ and $O_2^{\bullet -}$). Particularly, carboxylic acid-functionalized CNTs offer a higher catalytic activity than that of hydroxyl-functionalized CNTs. This was attributed to the low charge density of OH^- groups, which was unfavorable for the direction of oxidation of O_3 molecules (Zhang et al., 2017). Thus, the good charge density of the $-COOH$ group was more supportive of direct O_3 oxidation (Qu et al., 2015). The deprotonated carboxyl group has also interacted with O_3, forming electrophilic oxygen species, which could ease the generation of $^{\bullet}OH$ (de Oliveira et al., 2011; Vecitis et al., 2010). The electron-rich OH^- further reinforced the electron density of the conjugate π system and propagated the electron transfer process.

Nonradical mechanisms in catalytic ozonation fundamentally occur via electron transfer (direct oxidation) or O_3 decomposition products anchored on the catalyst surface. The nonradical mechanism is more sensitive to the adsorption behavior of O_3 and organics on the catalyst surface. Radical capacity is beneficial to the degradation of unsaturated (electron-rich) organic substances. The adsorption phenomenon of O_3 molecules on active sites leads to two nonradical pathways, including (i) degradation of organics through surface complexes due to the O_3 adsorption or organic adsorption on catalyst; and (ii) organic decomposition via $^1O_2$ from catalytic O_3 dissociation on the active sites (Fig. 1.6A) (Yu et al., 2020).

The surface complex formation tends to transfer the electrons in the dissociation of O_3 molecules through intramolecular electron transfer. Whereas O_3 directly attacks the surface organic complex without the formation of free radicals. The adsorption of O_3 on the catalytic surface is mostly governed by the pH of the catalyst, the *pKa* value of parent organic pollutants, and the solution pH (Afzal et al., 2016; Park et al., 2004; Qi et al., 2010). The MWCNT/O_3 system involved both radical and nonradical mechanisms due to the surface-adsorbed activated oxygen

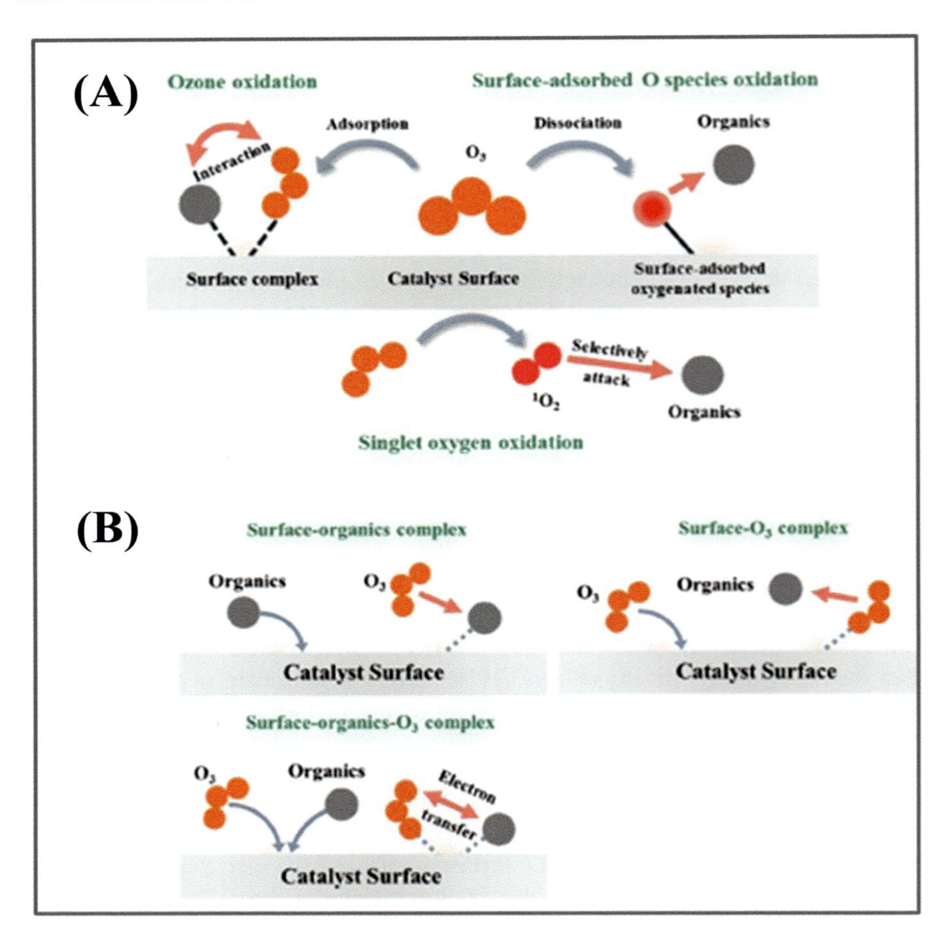

Figure 1.6 Activation of ozone. (A) Typical nonradical oxidation process. (B) Oxidation pathways for the surface−complex reaction in persulfate/O_3 system (Yu et al., 2020).

species. The good surface of the catalyst by raising the pH_{PZC} of the CNTs magnifies the adsorption, resulting in the breaking of O_3 to generate surface-adsorbed activated oxygen (Fig. 1.6B) (Yu et al., 2020).

1.5.6 Photocatalytic process

The photocatalysis process is one of the destructive processes that have been demonstrated to completely eradicate organic pollutants in water media. Nanocarbon-supported metal/metal oxide photocatalysts are widely employed to decompose a broad range of organic contaminants. Because the intrinsic properties (electronic configuration/structure, defects) and extrinsic characteristics (surface area topography, etc.) are beneficial for nanocarbon-catalyst-assisted photocatalyst catalytic activity. The development of nanocomposite delivers synergistic catalytic activity along with exceptional reusability and chemical stability. Nanocomposite photocatalysts are frequently constructed using anisotropic 1D tubular CNT or 2D planar graphene (Ahmad et al., 2023). CNT is most famous and used to synthesize nanocomposite photocatalysts as CNT has facilely tunable surface chemistry and well-organized sp^2 electronic configuration in the π-conjugations and aids for interfacial interaction between CNT and catalysts. The unique inner and outer tubular electronic configuration in CNT is also beneficial to accommodate the active catalyst via effective interfacial contact.

Commonly, two major intrinsic factors that notably impact the nanocomposite photocatalytic performance (Albero & Garcia, 2015) are the degree of graphitization and interfacial interactions. A strong interconnected electronic configuration established at high graphitization favors dynamic reaction kinetics (Ma et al., 2010). In the photocatalysis era, TiO_2 has always been the most popular and promising photocatalyst due to its photocatalytic properties. Anatase and rutile TiO_2 polymorphs were preferentially used in the photocatalytic process for the eradication of both organic and inorganic contaminants. The higher charge carrier conductivity of 80 $cm^2/V/s$ in anatase always demonstrates superior photocatalytic performances compared to rutile TiO_2 polymorphs. Besides, the mixture of these two TiO_2 phases provides even higher activity at a specific ratio (AEROXIDE TiO_2, Degussa P25) (Woan et al., 2009). This was rationalized due to the formation of special electronic states of two crystal structures (semiconductor—semiconductor junction). The combination of two semiconductor materials, semiconductors along with metal NPs or nanocarbons, significantly magnified the TiO_2 photocatalytic performances (Ahmad et al., 2023). This assumption was further explored by the fabrication of a TiO_2 photocatalyst supported by CNTs, which has sparked great interest. This mixture of CNT and TiO_2 has a larger active surface and is favorable for the efficient adsorption of water contaminants. Higher adsorption ability by TiO_2-CNT composites is the paramount phenomenon (properties) in the photocatalysis processes.

The photocatalysis process involves the irradiation of incident light (photons) on the active photocatalyst surface to produce photogenerated charge carriers, electron—hole ($e^- - h^+$) pairs. This conduction band electron and valence band hole can further react with hydroxide ions/water molecules to form basic ROS ($^\bullet OH$ and $O_2^{\bullet -}$) and $e^- - h^+$ pairs. These reactive species are responsible for the destruction of complex/refractory organic substances and their complete mineralization into small molecules, CO_2, H_2O, etc. A diverse CNT-catalytic photocatalytic mechanism is schematically represented in Fig. 1.7.

In nanocomposite, a photocatalytic (CNT-catalyst) mechanism driven by high irradiation leads to exciting the electrons from the valence band to the conduction band of TiO_2 particles. Later, the photogenerated electrons transport into the CNT configuration, and the valence band holes remain in TiO_2. On the other hand, the CNT configuration can also act as a sensitizer and transport electrons to the main TiO_2 photocatalyst. The photochemically formed electrons are inserted into the TiO_2 conduction band and produce $O_2^{\bullet -}$ through adsorbed O_2. Then the positively

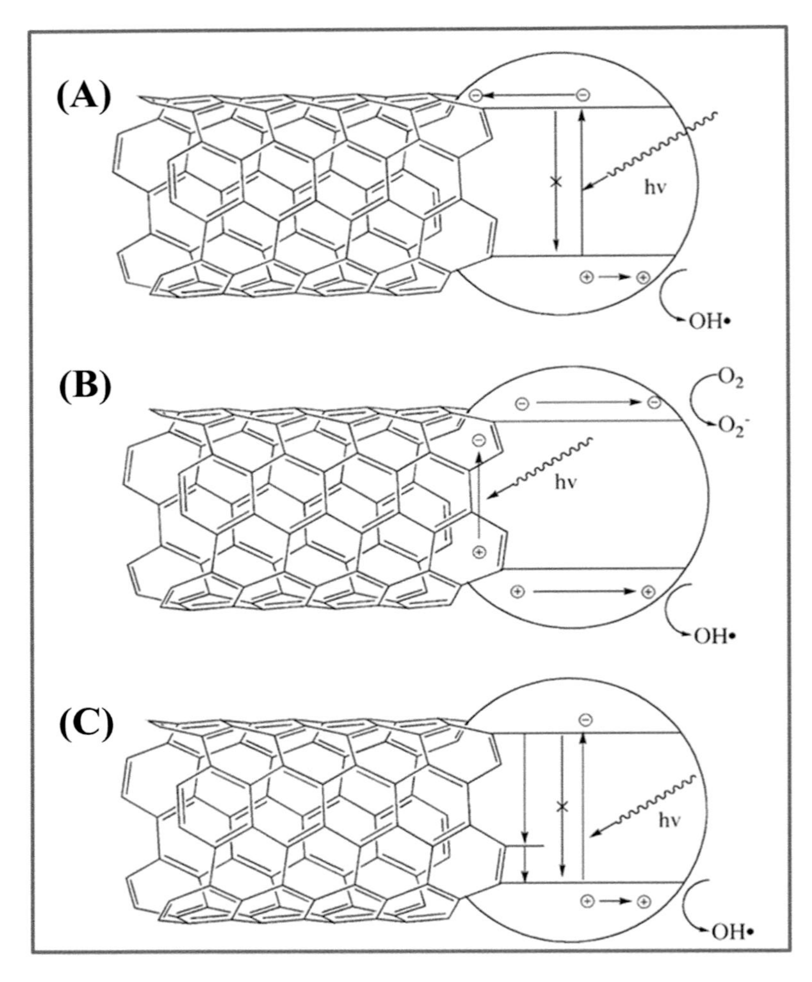

Figure 1.7 Carbon nanotube (CNT)-mediated photocatalytic mechanism. CNT-mediated photocatalytic mechanism: (A) CNT acts as an electron sinks and scavenges away the electrons hindering recombination. (B) The photon generates an electron–hole pair in the CNT. Based on the relevant positions of the bands, an electron (or the hole) is injected in the titania generating $^{\bullet}OH$ and $O_2^{\bullet-}$ species. (C) The nanotubes can act as impurity through the Ti–O–C bonds (Woan et al., 2009).

charged CNT configuration excludes an electron from the TiO_2 valence band and the hole. Consequently, the positively charged TiO_2 can react with surface-adsorbed H_2O and generate $^{\bullet}OH$. The photocatalytic performance was also enhanced due to the formation of a carbon–oxygen titanium bond that extends the light absorption to a longer wavelength (visible region). Thus, the electronic band structure between TiO_2 and CNT is also one of the key factors governing the photocatalytic activity.

1.6 Influences of water matrix and recent developments in catalytic applications

Natural wastewater is basically derived from different backgrounds of chemicals such as surfactants, dyes with complex organic structures, detergents, inorganic salt, and so on, which will hinder the catalytic conversion process by suppressing the oxidant's reactivity. It also contains ions like CO_3^{2-}, HCO^{3-}, SO_4^{2-}, and H_2PO^{4-} that generate radical and nonradical reactive species. There is also the existence of humic acid and natural organic matter in the presence of carbon-based moieties (carbonyl, phenolic, and carboxylate groups with high reactivity) and oxygen. It affects the steric and electrostatic effects and also delays the production of reactive complexes (Oyekunle et al., 2021). Some pollutants persist in the environment for long periods, viz., arsenic, mercury, lead, cadmium, chromium, iron, barium, selenium, silver, and tin, which are called inorganic pollutants. Organic pollutants, such as polycyclic aromatic hydrocarbons, nitroaromatic compounds, phthalates, polybrominated diphenyl ethers, petroleum hydrocarbons, and azo dyes, are harmful and hazardous substances to the ecosystem (Vandana et al., 2022). These pollutants cause adverse impacts on the natural biota and humans, and researchers take suitable measures like CNTs that are in the current experiment to help in the removal and decrease the efflux of the pollutant into the environment.

Carbonaceous materials applied to current research investigations are:

1. A nitrogen-sulfur co-doped industrial graphene/PMS system degrades filtered surface water and tap water spiked to remove 90% of MP.
2. The NCNT/PMS system removed 92% of BPA from real industrial wastewater.
3. Mineral–carbon material NPC-FeOOH/Mt/PDS is very good for its catalytic activity in degrading BPA from the Pearl River sample.
4. S-OMCs/PDS degrades PNP very well in complex real water (Oyekunle et al., 2021).
5. CNT-based PS-AOPs have the potential to degrade organic pollutants in soil by decomposing them into smaller molecules or converting them into CO_2 and H_2O. It avoids operation costs and energy losses. Application in soil fields needs a little research under different environmental conditions (Wu et al., 2022).
6. MWCNTs are more applicable than SWCNTs for their better performance in environmental sensing. It is less toxic to nontarget microorganisms and exhibits higher regeneration capability.

1.7 Conclusions and future perspectives

This chapter emphasizes some peculiarities of CNT-based catalytic processes for water treatment. The carbon hexagons of graphitic-like layers of CNTs, with their phenomenal structural properties and electronic configurations, help in water purification and wastewater treatment. Surface functionalization, defective degree, and heteroatom doping can be engineered according to the demands of the pollutants.

These structurally designed CNTs are applied to the water and wastewater treatment either through adsorption, membrane filtration, or AOPs. Application of CNTs at a large scale is very limited and must be investigated elaborately in terms of cost-effectiveness and toxicity assessments.

CNT-based technologies for water treatment are growing and are not yet extensive at the industrial level. A better understanding of the catalytic functionalities of CNTs at the nanoscale and their integration in the reactor design can aid in the implementation of this catalytic application on a larger scale. One key point that should be considered when constructing environmental catalysts for water treatment is the nonhazardous nature of the catalysts and their preparation through sustainable methods with respect to the principles of green chemistry. Eco-friendly strategies must be employed, which are as follows: (1) less impact chemicals; (2) green solvents; (3) biosurfactants; and (4) low energy consumption. Most recent investigations demonstrate the removal of single pollutants, while real water and wastewater encompass diverse pollutant mixtures. All existing advanced water treatments must conform to complete oxidation and mineralization of the effluent constituents, which means that degradation and transformation by-products should be characterized to investigate the kinetics of their oxidation mechanisms with respect to their potential toxicity. Moreover, the key challenge in the novel catalytic application is the complex chemical analysis of the solutions.

Molecular modeling of gas/liquid/solid interfaces and analytical characterization under working conditions are vital, which could lead to a better understanding of the complexity of the system and the coexistence of various solid–liquid–gas interfaces have been determined. Further, a rapid gain of computational and experimental data must facilitate and enhance the consistency of the discovery of novel CNT-based materials or prioritize existing findings based on statistical analysis and machine learning techniques. Despite the benefits of CNT technologies in terms of kinetics of transformation, selectivity, and energy consumption, a greater number of investigations are needed for implementation at an industrial level.

While CNTs demonstrate high treatment efficiency at the laboratory scale, still more attempts will be required related to the reuse of adsorbent, analysis of the efficiency and leaching of CNT-originated membranes, and rate of deterioration of functionalities of CNTs. Moreover, with the increasing applications of CNTs in water treatment, there are emerging apprehensions about their possible harmfulness to the environment and human health. The research analysis shows that some nanomaterials may have negative effects on human health. However, standards for measuring the harmfulness of nanomaterials are reasonably inadequate. Therefore, a complete assessment of their harmfulness is essential to safeguarding their real-time applications. As a result, future studies are necessary in large-scale continuous operations to figure out the performance evaluation of CNTs in water and wastewater treatment, as well as the assessment of their harmfulness, which is essential in real-time applications.

References

Afzal, S., Quan, X., Chen, S., Wang, J., & Muhammad, D. (2016). Synthesis of manganese incorporated hierarchical mesoporous silica nanosphere with fibrous morphology by facile one-pot approach for efficient catalytic ozonation. *Journal of Hazardous Materials*, *318*, 308–318. Available from https://doi.org/10.1016/j.jhazmat.2016.07.015, http://www.elsevier.com/locate/jhazmat.

Ahmad, A., Ali, M., Al-Sehemi, A. G., Al-Ghamdi, A. A., Park, J. W., Algarni, H., & Anwer, H. (2023). Carbon-integrated semiconductor photocatalysts for removal of volatile organic compounds in indoor environments. *Chemical Engineering Journal*, *452*. Available from https://doi.org/10.1016/j.cej.2022.139436, http://www.elsevier.com/inca/publications/store/6/0/1/2/7/3/index.htt.

Albero, J., & Garcia, H. (2015). Doped graphenes in catalysis. *Journal of Molecular Catalysis A: Chemical*, *408*, 296–309. Available from https://doi.org/10.1016/j.molcata.2015.06.011.

Arora, B., & Attri, P. (2020). Carbon nanotubes (CNTs): A potential nanomaterial for water purification. *Journal of Composites Science*, *4*(3). Available from https://doi.org/10.3390/jcs4030135, https://www.mdpi.com/2504-477X/4/3/135/pdf.

Aslam, M. M. A., Kuo, H. W., Den, W., Usman, M., Sultan, M., & Ashraf, H. (2021). Functionalized carbon nanotubes (Cnts) for water and wastewater treatment: Preparation to application. *Sustainability (Switzerland)*, *13*(10). Available from https://doi.org/10.3390/su13105717, https://www.mdpi.com/2071-1050/13/10/5717/pdf.

Barrejón, M., & Prato, M. (2022). Carbon nanotube membranes in water treatment applications. *Advanced Materials Interfaces*, *9*(1). Available from https://doi.org/10.1002/admi.202101260, http://onlinelibrary.wiley.com/journal/10.1002/(ISSN)2196-7350.

Bethune, D. S., Kiang, C. H., de Vries, M. S., Gorman, G., Savoy, R., Vazquez, J., Beyers. R. (1993). Cobalt-catalysed growth of carbon nanotubes with single-atomic-layer walls. *Nature*. *363*, 605–607, Available from https://doi.org/10.1038/363605a0.

Boczkaj, G., & Fernandes, A. (2017). Wastewater treatment by means of advanced oxidation processes at basic pH conditions: A review. *Chemical Engineering Journal*, *320*, 608–633. Available from https://doi.org/10.1016/j.cej.2017.03.084, http://www.elsevier.com/inca/publications/store/6/0/1/2/7/3/index.htt.

Chan, K. F., Maznam, N. A. M., Hazan, M. A., Ahmad, R. N. A., Sa'ari, A. S., Azman, N. F. I., Mamat, M. S., Rahman, M. A. A., Tanemura, M., & Yaakob, Y. (2022). Multi-walled carbon nanotubes growth by chemical vapour deposition: Effect of precursor flowing path and catalyst size. *Carbon Trends*, *6*, 100142. Available from https://doi.org/10.1016/j.cartre.2021.100142.

Chen, X., Oh, W. D., & Lim, T. T. (2018). Graphene- and CNTs-based carbocatalysts in persulfates activation: Material design and catalytic mechanisms. *Chemical Engineering Journal*, *354*, 941–976. Available from https://doi.org/10.1016/j.cej.2018.08.049, http://www.elsevier.com/inca/publications/store/6/0/1/2/7/3/index.htt.

Cheng, X., Guo, H., Zhang, Y., Wu, X., & Liu, Y. (2017). Non-photochemical production of singlet oxygen via activation of persulfate by carbon nanotubes. *Water Research*, *113*, 80–88. Available from https://doi.org/10.1016/j.watres.2017.02.016, http://www.elsevier.com/locate/watres.

Chufa, B. M., Murthy, H. C. A., Gonfa, B. A., & Anshebo, T. Y. (2021). Carbon nanotubes: A review on green synthesis, growth mechanism and application as a membrane filter for fluoride remediation. *Green Chemistry Letters and Reviews*, *14*(4), 640–657. Available from https://doi.org/10.1080/17518253.2021.1991484, https://www.tandfonline.com/loi/tgcl20.

Das, R., Abd Hamid, S. B., Ali, M. E., Ismail, A. F., Annuar, M. S. M., & Ramakrishna, S. (2014). Multifunctional carbon nanotubes in water treatment: The present, past and future. *Desalination*, *354*, 160–179. Available from https://doi.org/10.1016/j.desal.2014.09.032.

de Oliveira, T. F., Chedeville, O., Fauduet, H., & Cagnon, B. (2011). Use of ozone/activated carbon coupling to remove diethyl phthalate from water: Influence of activated carbon textural and chemical properties. *Desalination*, *276*(1–3), 359–365. Available from https://doi.org/10.1016/j.desal.2011.03.084.

Duan, X., Tian, W., Zhang, H., Sun, H., Ao, Z., Shao, Z., & Wang, S. (2019). sp 2/sp 3 framework from diamond nanocrystals: A key bridge of carbonaceous structure to carbocatalysis. *ACS Catalysis*, *9*(8), 7494–7519. Available from https://doi.org/10.1021/acscatal.9b01565.

Duan, X., Sun, H., Ao, Z., Zhou, L., Wang, G., & Wang, S. (2016). Unveiling the active sites of graphene-catalyzed peroxymonosulfate activation. *Carbon*, *107*, 371–378. Available from https://doi.org/10.1016/j.carbon.2016.06.016, http://www.journals.elsevier.com/carbon/.

Duan, X., Sun, H., Kang, J., Wang, Y., Indrawirawan, S., & Wang, S. (2015). Insights into heterogeneous catalysis of persulfate activation on dimensional-structured nanocarbons. *ACS Catalysis*, *5*(8), 4629–4636. Available from https://doi.org/10.1021/acscatal.5b00774, http://pubs.acs.org/page/accacs/about.html.

Duan, X., Sun, H., & Wang, S. (2018). Metal-free carbocatalysis in advanced oxidation reactions. *Accounts of Chemical Research*, *51*(3), 678–687. Available from https://doi.org/10.1021/acs.accounts.7b00535, http://pubs.acs.org/journal/achre4.

Farahbaksh, J., Delnavaz, M., & Vatanpour, V. (2017). Investigation of raw and oxidized multiwalled carbon nanotubes in fabrication of reverse osmosis polyamide membranes for improvement in desalination and antifouling properties. *Desalination*, *410*, 1–9. Available from https://doi.org/10.1016/j.desal.2017.01.031.

Govindan, K., Kim, D.-G., & Ko, S. (2022a). Role of N-doping and o-groups in unzipped N-doped CNT carbocatalyst for peroxomonosulfate activation: Quantitative structure-activity relationship. *SSRN Electronic Journal*. Available from https://doi.org/10.2139/ssrn.4102518.

Govindan, K., Kim, D.-G., & Ko, S.-O. (2022b). Catalytic oxidation of acetaminophen through pristine and surface-modified nitrogen-doped carbon-nanotube-catalyzed peroxydisulfate activation. *Journal of Environmental Chemical Engineering*, *10*(5), 108257. Available from https://doi.org/10.1016/j.jece.2022.108257.

Hu, H., Xin, J. H., Hu, H., Wang, X., & Kong, Y. (2015). Metal-free graphene-based catalyst-insight into the catalytic activity: A short review. *Applied Catalysis A: General*, *492*, 1–9. Available from https://doi.org/10.1016/j.apcata.2014.11.041, http://www.sciencedirect.com/science/journal/0926860X.

Iijima, S., & Ichihashi, T. (1993). Single-shell carbon nanotubes of 1-nm diameter. *Nature*, *363*(6430), 603–605. Available from https://doi.org/10.1038/363603a0.

Jain, N., & Kanu, N. J. (2021). The potential application of carbon nanotubes in water treatment: A state-of-the-art-review. *Materials Today: Proceedings*, *43*, 2998–3005. Available from https://doi.org/10.1016/j.matpr.2021.01.331, 22147853 Elsevier Ltd. India, https://www.sciencedirect.com/journal/materials-today-proceedings.

José-Yacamán, M., Miki-Yoshida, M., Rendón, L., & Santiesteban, J. G. (1993). Catalytic growth of carbon microtubules with fullerene structure. *Applied Physics Letters*, *62*(2), 202–204. Available from https://doi.org/10.1063/1.109315.

Lee, J. (2018). *Carbon nanotube-based membranes for water purification nanoscale materials in water purification* (pp. 309–331). Australia: Elsevier. Available from https://www.sciencedirect.com/book/9780128139264, https://doi.org/10.1016/B978-0-12-813926-4.00017-3.

Lee, J., Von Gunten, U., & Kim, J. H. (2020). Persulfate-based advanced oxidation: Critical assessment of opportunities and roadblocks. *Environmental Science and Technology, 54* (6), 3064–3081. Available from https://doi.org/10.1021/acs.est.9b07082, http://pubs.acs.org/journal/esthag.

Letters to Nature. (1973). *Nature, 246*(5429), 170. Available from https://doi.org/10.1038/246170a0.

Li, J., Li, M., Sun, H., Ao, Z., Wang, S., & Liu, S. (2020). Understanding of the oxidation behavior of benzyl alcohol by peroxymonosulfate via carbon nanotubes activation. *ACS Catalysis, 10*(6), 3516–3525. Available from https://doi.org/10.1021/acscatal.9b05273, http://pubs.acs.org/page/accacs/about.html.

Li, W. Z., Xie, S. S., Qian, L. X., Chang, B. H., Zou, B. S., Zhou, W. Y., Zhao, R. A., & Wang, G. (1996). Large-scale synthesis of aligned carbon nanotubes. *Science (New York, N.Y.), 274*(5293), 1701–1703. Available from https://doi.org/10.1126/science.274.5293.1701.

Liang, P., Zhang, C., Duan, X., Sun, H., Liu, S., Tade, M. O., & Wang, S. (2017). An insight into metal organic framework derived N-doped graphene for the oxidative degradation of persistent contaminants: Formation mechanism and generation of singlet oxygen from peroxymonosulfate. *Environmental Science: Nano, 4*(2), 315–324. Available from https://doi.org/10.1039/c6en00633g, http://www.rsc.org/publishing/journals/en/about.asp.

Liu, Y., Chen, C., Duan, X., Wang, S., & Wang, Y. (2021). Carbocatalytic ozonation toward advanced water purification. *Journal of Materials Chemistry A, 9*(35), 18994–19024. Available from https://doi.org/10.1039/d1ta02953c, http://pubs.rsc.org/en/journals/journal/ta.

Luo, H., Fu, H., Yin, H., & Lin, Q. (2022). Carbon materials in persulfate-based advanced oxidation processes: The roles and construction of active sites. *Journal of Hazardous Materials, 426*, 128044. Available from https://doi.org/10.1016/j.jhazmat.2021.128044.

Ma, C. H., Li, H. Y., Lin, G. D., & Zhang, H. B. (2010). Ni-decorated carbon nanotube-promoted Ni-Mo-K catalyst for highly efficient synthesis of higher alcohols from syngas. *Applied Catalysis B: Environmental, 100*(1–2), 245–253. Available from https://doi.org/10.1016/j.apcatb.2010.07.040.

Mallakpour, S., & Soltanian, S. (2016). Surface functionalization of carbon nanotubes: Fabrication and applications. *RSC Advances, 6*(111), 109916–109935. Available from https://doi.org/10.1039/c6ra24522f, http://pubs.rsc.org/en/journals/journalissues.

Mishra, S., & Sundaram, B. (2023). Efficacy and challenges of carbon nanotube in wastewater and water treatment. *Environmental Nanotechnology, Monitoring & Management, 19*, 100764. Available from https://doi.org/10.1016/j.enmm.2022.100764.

Mohmood, I., Lopes, C. B., Lopes, I., Ahmad, I., Duarte, A. C., & Pereira, E. (2013). Nanoscale materials and their use in water contaminants removal-A review. *Environmental Science and Pollution Research, 20*(3), 1239–1260. Available from https://doi.org/10.1007/s11356-012-1415-x, https://link.springer.com/journal/11356.

Murali, A., Sarswat, P. K., & Free, M. L. (2020). Adsorption-coupled reduction mechanism in ZnO-Functionalized MWCNTs nanocomposite for Cr (VI) removal and improved anti-photocorrosion for photocatalytic reduction. *Journal of Alloys and Compounds, 843*. Available from https://doi.org/10.1016/j.jallcom.2020.155835, https://www.journals.elsevier.com/journal-of-alloys-and-compounds.

Oyekunle, D. T., Zhou, X., Shahzad, A., & Chen, Z. (2021). Review on carbonaceous materials as persulfate activators: Structure-performance relationship, mechanism and future perspectives on water treatment. *Journal of Materials Chemistry A, 9*(13), 8012–8050. Available from https://doi.org/10.1039/d1ta00033k, http://pubs.rsc.org/en/journals/journal/ta.

Park, J. S., Choi, H., & Cho, J. (2004). Kinetic decomposition of ozone and para-chlorobenzoic acid (pCBA) during catalytic ozonation. *Water Research, 38*(9), 2285–2292. Available from https://doi.org/10.1016/j.watres.2004.01.040, http://www.elsevier.com/locate/watres.

Pedrosa, M., Pastrana-Martínez, L. M., Pereira, M. F. R., Faria, J. L., Figueiredo, J. L., & Silva, A. M. T. (2018). N/S-doped graphene derivatives and TiO_2 for catalytic ozonation and photocatalysis of water pollutants. *Chemical Engineering Journal, 348*, 888–897. Available from https://doi.org/10.1016/j.cej.2018.04.214, http://www.elsevier.com/inca/publications/store/6/0/1/2/7/3/index.htt.

Peng, J., Wang, Z., Wang, S., Liu, J., Zhang, Y., Wang, B., Gong, Z., Wang, M., Dong, H., Shi, J., Liu, H., Yan, G., Liu, G., Gao, S., & Cao, Z. (2021). Enhanced removal of methylparaben mediated by cobalt/carbon nanotubes (Co/CNTs) activated peroxymonosulfate in chloride-containing water: Reaction kinetics, mechanisms and pathways. *Chemical Engineering Journal, 409*, 128176. Available from https://doi.org/10.1016/j.cej.2020.128176.

Prabavathi, S. L., Govindan, K., Saravanakumar, K., Jang, A., & Muthuraj, V. (2019). Construction of heterostructure CoWO4/g-C_3N_4 nanocomposite as an efficient visible-light photocatalyst for norfloxacin degradation. *Journal of Industrial and Engineering Chemistry, 80*, 558–567. Available from https://doi.org/10.1016/j.jiec.2019.08.035, http://www.sciencedirect.com/science/journal/1226086X.

Qi, F., Xu, B., Chen, Z., Zhang, L., Zhang, P., & Sun, D. (2010). Mechanism investigation of catalyzed ozonation of 2-methylisoborneol in drinking water over aluminum (hydroxyl) oxides: Role of surface hydroxyl group. *Chemical Engineering Journal, 165*(2), 490–499. Available from https://doi.org/10.1016/j.cej.2010.09.047.

Qu, R., Xu, B., Meng, L., Wang, L., & Wang, Z. (2015). Ozonation of indigo enhanced by carboxylated carbon nanotubes: Performance optimization, degradation products, reaction mechanism and toxicity evaluation. *Water Research, 68*, 316–327. Available from https://doi.org/10.1016/j.watres.2014.10.017, http://www.elsevier.com/locate/watres.

Ren, W., Xiong, L., Nie, G., Zhang, H., Duan, X., & Wang, S. (2020). Insights into the electron-transfer regime of peroxydisulfate activation on carbon nanotubes: The role of oxygen functional groups. *Environmental Science and Technology, 54*(2), 1267–1275. Available from https://doi.org/10.1021/acs.est.9b06208, http://pubs.acs.org/journal/esthag.

Rezaei, P., Pfefferle, L. D., & Frey, D. D. (2021). Metal-semiconductor sorting of large-diameter single-wall carbon nanotubes by pH-dependent binding to a hydrophobic-interaction adsorbent. *Carbon, 175*, 112–123. Available from https://doi.org/10.1016/j.carbon.2020.12.078, http://www.journals.elsevier.com/carbon/.

Rodríguez, C., Briano, S., & Leiva, E. (2020). Increased adsorption of heavy metal ions. *Molecules (Basel, Switzerland), 25*.

Sagara, T., Kurumi, S., & Suzuki, K. (2014). Growth of linear Ni-filled carbon nanotubes by local arc discharge in liquid ethanol. *Applied Surface Science, 292*, 39–43. Available from https://doi.org/10.1016/j.apsusc.2013.11.056, http://www.journals.elsevier.com/applied-surface-science/.

Saifuddin, N., Raziah, A. Z., & Junizah, A. R. (2013). Carbon nanotubes: A review on structure and their interaction with proteins. *Journal of Chemistry, 2013*, 1–18. Available from https://doi.org/10.1155/2013/676815.

Sajid, M., Asif, M., Baig, N., Kabeer, M., Ihsanullah, I., & Mohammad, A. W. (2022). Carbon nanotubes-based adsorbents: Properties, functionalization, interaction mechanisms, and applications in water purification. *Journal of Water Process Engineering, 47*. Available from https://doi.org/10.1016/j.jwpe.2022.102815, http://www.journals.elsevier.com/journal-of-water-process-engineering/.

Scalese, S., Scuderi, V., Bagiante, S., Gibilisco, S., Faraci, G., & Privitera, V. (2010a). Order and disorder of carbon deposit produced by arc discharge in liquid nitrogen. *Journal of Applied Physics*, *108*(6). Available from https://doi.org/10.1063/1.3475726.

Scalese, S., Scuderi, V., Bagiante, S., Simone, F., Russo, P., D'Urso, L., Compagnini, G., & Privitera, V. (2010b). Controlled synthesis of carbon nanotubes and linear C chains by arc discharge in liquid nitrogen. *Journal of Applied Physics*, *107*(1). Available from https://doi.org/10.1063/1.3275500.

Selvaraj, M., Hai, A., Banat, F., & Haija, M. A. (2020). Application and prospects of carbon nanostructured materials in water treatment: A review. *Journal of Water Process Engineering*, *33*, 100996. Available from https://doi.org/10.1016/j.jwpe.2019.100996.

Shao, P., Yu, S., Duan, X., Yang, L., Shi, H., Ding, L., Tian, J., Yang, L., Luo, X., & Wang, S. (2020). Potential difference driving electron transfer via defective carbon nanotubes toward selective oxidation of organic micropollutants. *Environmental Science & Technology*, *54*(13), 8464–8472. Available from https://doi.org/10.1021/acs.est.0c02645.

Smith, S. C., & Rodrigues, D. F. (2015). Carbon-based nanomaterials for removal of chemical and biological contaminants from water: A review of mechanisms and applications. *Carbon*, *91*, 122–143. Available from https://doi.org/10.1016/j.carbon.2015.04.043, http://www.journals.elsevier.com/carbon/.

Song, Z., Zhang, Y., Liu, C., Xu, B., Qi, F., Yuan, D., & Pu, S. (2019). Insight into [rad]OH and O_2[rad] − formation in heterogeneous catalytic ozonation by delocalized electrons and surface oxygen-containing functional groups in layered-structure nanocarbons. *Chemical Engineering Journal*, *357*, 655–666. Available from https://doi.org/10.1016/j.cej.2018.09.182, http://www.elsevier.com/inca/publications/store/6/0/1/2/7/3/index.htt.

Su, D. S., Perathoner, S., & Centi, G. (2013). Nanocarbons for the development of advanced catalysts. *Chemical Reviews*, *113*(8), 5782–5816. Available from https://doi.org/10.1021/cr300367d.

Tan, W. L., & Kochmann, D. M. (2017). An effective constitutive model for polycrystalline ferroelectric ceramics: Theoretical framework and numerical examples. *Computational Materials Science*, *136*, 223–237. Available from https://doi.org/10.1016/j.commatsci.2017.04.032.

Thamaraiselvan, C., Lerman, S., Weinfeld-Cohen, K., & Dosoretz, C. G. (2018a). Characterization of a support-free carbon nanotube-microporous membrane for water and wastewater filtration. *Separation and Purification Technology*, *202*, 1–8. Available from https://doi.org/10.1016/j.seppur.2018.03.038, http://www.journals.elsevier.com/separation-and-purification-technology/.

Thamaraiselvan, C., Ronen, A., Lerman, S., Balaish, M., Ein-Eli, Y., & Dosoretz, C. G. (2018b). Low voltage electric potential as a driving force to hinder biofouling in self-supporting carbon nanotube membranes. *Water Research*, *129*, 143–153. Available from https://doi.org/10.1016/j.watres.2017.11.004, http://www.elsevier.com/locate/watres.

Thess, A., Lee, R., Nikolaev, P., Dai, H., Petit, P., Robert, J., Xu, C., Lee, Y. H., Kim, S. G., Rinzler, A. G., Colbert, D. T., Scuseria, G. E., Tománek, D., Fischer, J. E., & Smalley, R. E. (1996). Crystalline ropes of metallic carbon nanotubes. *Science (New York, N.Y.)*, *273*(5274), 483–487. Available from https://doi.org/10.1126/science.273.5274.483, http://www.sciencemag.org.

Vandana., Priyadarshanee, M., Mahto, U., & Das, S. (2022). *Mechanism of toxicity and adverse health effects of environmental pollutants. Microbial biodegradation and bioremediation: Techniques and case studies for environmental pollution* (pp. 33–53). India: Elsevier. Available from https://www.sciencedirect.com/book/9780323854559, 10.1016/B978-0-323-85455-9.00024-2.

Vecitis, C. D., Lesko, T., Colussi, A. J., & Hoffmann, M. R. (2010). Sonolytic decomposition of aqueous bioxalate in the presence of ozone. *Journal of Physical Chemistry A, 114* (14), 4968–4980. Available from https://doi.org/10.1021/jp9115386.

Wang, J., Chen, S., Quan, X., & Yu, H. (2018). Fluorine-doped carbon nanotubes as an efficient metal-free catalyst for destruction of organic pollutants in catalytic ozonation. *Chemosphere, 190*, 135–143. Available from https://doi.org/10.1016/j.chemosphere.2017.09.119, http://www.elsevier.com/locate/chemosphere.

Wang, Y., Duan, X., Xie, Y., Sun, H., & Wang, S. (2020). Nanocarbon-based catalytic ozonation for aqueous oxidation: Engineering defects for active sites and tunable reaction pathways. *ACS Catalysis, 10*(22), 13383–13414. Available from https://doi.org/10.1021/acscatal.0c04232, http://pubs.acs.org/page/accacs/about.html.

White C.M., Banks R., Hamerton I., Watts J.F., Characterisation of commercially CVD grown multi-walled carbon nanotubes for paint applications. Progress in Organic Coatings. 90 (2016), 44–53, http://www.elsevier.com/locate/porgcoat, https://doi.org/10.1016/j.porgcoat.2015.09.020.

Woan, K., Pyrgiotakis, G., & Sigmund, W. (2009). Photocatalytic carbon-nanotube-TiO_2 composites. *Advanced Materials, 21*(21), 2233–2239. Available from https://doi.org/10.1002/adma.200802738.

Wu, L., Wu, T., Liu, Z., Tang, W., Xiao, S., Shao, B., Liang, Q., He, Q., Pan, Y., Zhao, C., Liu, Y., & Tong, S. (2022). Carbon nanotube-based materials for persulfate activation to degrade organic contaminants: Properties, mechanisms and modification insights. *Journal of Hazardous Materials, 431*, 128536. Available from https://doi.org/10.1016/j.jhazmat.2022.128536.

Yan, Y., Miao, J., Yang, Z., Xiao, F.-X., Yang, H. B., Liu, B., & Yang, Y. (2015). Carbon nanotube catalysts: Recent advances in synthesis, characterization and applications. *Chemical Society Reviews, 44*(10), 3295–3346. Available from https://doi.org/10.1039/c4cs00492b.

Yang, Q., Chen, Y., Duan, X., Zhou, S., Niu, Y., Sun, H., Zhi, L., & Wang, S. (2020). Unzipping carbon nanotubes to nanoribbons for revealing the mechanism of nonradical oxidation by carbocatalysis. *Applied Catalysis B: Environmental, 276*, 119146. Available from https://doi.org/10.1016/j.apcatb.2020.119146.

Yang, X., Wan, Y., Zheng, Y., He, F., Yu, Z., Huang, J., Wang, H., Ok, Y. S., Jiang, Y., & Gao, B. (2019). Surface functional groups of carbon-based adsorbents and their roles in the removal of heavy metals from aqueous solutions: A critical review. *Chemical Engineering Journal, 366*, 608–621. Available from https://doi.org/10.1016/j.cej.2019.02.119, http://www.elsevier.com/inca/publications/store/6/0/1/2/7/3/index.htt.

Yin, R., Guo, W., Du, J., Zhou, X., Zheng, H., Wu, Q., Chang, J., & Ren, N. (2017). Heteroatoms doped graphene for catalytic ozonation of sulfamethoxazole by metal-free catalysis: Performances and mechanisms. *Chemical Engineering Journal, 317*, 632–639. Available from https://doi.org/10.1016/j.cej.2017.01.038, http://www.elsevier.com/inca/publications/store/6/0/1/2/7/3/index.htt.

Yin, Z., Cui, C., Chen, H., Duoni., Yu, X., & Qian, W. (2020). The application of carbon nanotube/graphene-based nanomaterials in wastewater treatment. *Small (Weinheim an der Bergstrasse, Germany), 16*(15). Available from https://doi.org/10.1002/smll.201902301, http://onlinelibrary.wiley.com/journal/10.1002/(ISSN)1613-6829.

Yousef, S., Khattab, A., Osman, T. A., & Zaki, M. (2013). Effects of increasing electrodes on CNTs yield synthesized by using arc-discharge technique. *Journal of Nanomaterials, 2013*. Available from https://doi.org/10.1155/2013/392126.

Yu, G., Wang, Y., Cao, H., Zhao, H., & Xie, Y. (2020). Reactive oxygen species and catalytic active sites in heterogeneous catalytic ozonation for water purification. *Environmental Science and Technology*, *54*(10), 5931–5946. Available from https://doi.org/10.1021/acs.est.0c00575, http://pubs.acs.org/journal/esthag.

Yu, J. G., Zhao, X. H., Yang, H., Chen, X. H., Yang, Q., Yu, L. Y., Jiang, J. H., & Chen, X. Q. (2014). Aqueous adsorption and removal of organic contaminants by carbon nanotubes. *Science of the Total Environment*, *482–483*(1), 241–251. Available from https://doi.org/10.1016/j.scitotenv.2014.02.129, http://www.elsevier.com/locate/scitotenv.

Zhang, F., Wu, K., Zhou, H., Hu, Y., Preis, S. V., Wu, H., & Wei, C. (2018). Ozonation of aqueous phenol catalyzed by biochar produced from sludge obtained in the treatment of coking wastewater. *Journal of Environmental Management*, *224*, 376–386. Available from https://doi.org/10.1016/j.jenvman.2018.07.038, http://www.elsevier.com/inca/publications/store/6/2/2/8/7/1/index.htt.

Zhang, S., Quan, X., Zheng, J. F., & Wang, D. (2017). Probing the interphase "HO[rad] zone" originated by carbon nanotube during catalytic ozonation. *Water Research*, *122*, 86–95. Available from https://doi.org/10.1016/j.watres.2017.05.063, http://www.elsevier.com/locate/watres.

Hybrid heterostructured nanocatalysts for artificial photosynthesis

2

Samidha S. Narvekar and Anjani P. Nagvenkar
School of Chemical Sciences, Goa University, Goa, India

2.1 Introduction

The process of natural photosynthesis involves harvesting solar energy to convert atmospheric carbon dioxide (CO_2) into carbohydrates and release oxygen as a by-product of water oxidation. The solar energy is stored in chemical bonds in the form of carbohydrates. In natural photosynthesis, chlorophyll in plants acts as a photosensitizer by absorbing photons and causing charge separation (electron−hole pair formation), which in turn drives the oxidation of water to generate oxygen, electrons, and protons. The oxygen evolved is released into the environment, and the electrons and protons are used in the Calvin cycle to reduce CO_2 into carbohydrates via a light-independent reaction. Emulating this photosynthesis process has been the most intriguing challenge for scientists, considering the complexity of the entire process that occurs in a single leaf equipped with the chemicals needed for the multistep activity. However, researchers have been successful in producing fuels by photochemically splitting water into hydrogen and oxygen and reducing CO_2 into industrially viable products such as methanol or methane (Barber & Tran, 2013). These solar fuels are high-energy chemicals that can be stored and release energy upon burning.

The process of partially biomimicking the natural photosynthesis process to convert and store the sunlight into chemical bonds of solar fuels is termed as artificial photosynthesis. In view of mitigating the rise in human-caused CO_2 emissions by CO_2 fixation and the need to meet the demand for energy supply through clean fuel sources such as hydrogen, the field of artificial photosynthesis is attracting researchers to develop novel strategies and design materials for high-efficiency photosynthetic processes (Zhou et al., 2016). The multistage processes of water oxidation and CO_2 reduction occurring in natural photosynthesis are accomplished by various enzymes in a single leaf with chlorophyll as a photon absorber (Miyatake & Tamiaki, 2010). However, to perform artificial photosynthesis, this multistage process must be carried out in three stages, viz., light absorption, water oxidation and hydrogen evolution, and CO_2 reduction. These steps occur with the help of a synthetic chemical agent that plays a role of catalyst at each stage by replacing chlorophyll and natural enzymes.

Nanocomposites for Environmental, Energy, and Agricultural Applications. DOI: https://doi.org/10.1016/B978-0-443-13935-2.00002-4

The major challenge among scientists is to develop novel catalysts and/or optimize the existing ones that can perform the different stages of the photosynthetic reactions under a one-pot reaction strategy (Ruan et al., 2023). On an artificial mimicking platform, this is indeed highly challenging, as three reactions, that is, water oxidation, hydrogen evolution, and CO_2 reduction, demand distinct chemical and kinetic conditions, thus posing a need for a multifunctional catalyst. To date, significant progress has been made in the development of nanomaterial-based photocatalysts ranging from noble-metal-based, transition metal oxides and sulfides, transition metal complexes, especially rhenium-based complexes, and so on (Smith et al., 2020). Furthermore, to achieve multistep processes on a single catalytic surface, there has been a surge in designing a hybrid/heterostructured photocatalyst, a catalyst that is a combination of two or more components.

The advantage of designing it in the nanoform is that it mimics the dimensions of its natural counterparts (enzymes). Each component of the hybrid nanomaterial can catalyze one type of reaction in artificial photosynthesis process. A heterostructured nanocatalyst favors different types of active sites on the same catalyst on two different surfaces, along with a high surface area. The heterojunction created by two types of components on a single catalyst offers multichannel charge transport, which facilitates complementary redox reactions on the same catalyst (Yuan et al., 2021). For artificial photosynthetic reactions, metal-semiconductor and semiconductor—semiconductor heterostructures are generally explored as catalysts. The type and functionality of the two components govern the chemical and physical properties of the hybrid catalyst and thus play a crucial role in photocatalysis and photoelectrocatalysis. A photoactive material is used as a catalyst in artificial photosynthesis based on its bandgap, surface redox potential, band energies, etc. Designing an efficient photocatalyst involves tuning these crucial parameters to promote enhanced light absorption, longer diffusion length of charge carriers, and effective charge separation (Singh et al., 2014).

Herein, we discussed basic concepts and mechanisms of light-assisted water splitting and carbon dioxide reduction processes. The chapter is primarily aimed at reviewing the role of hybrid semiconductor catalysts in improving the efficiency of artificial photosynthetic methods. The work also addresses the challenges encountered by heterostructure catalysts in mimicking the natural stages of photosynthesis. In conclusion, the unexplored areas of this research and measures to resolve the intricacies are briefly outlined.

2.2 Fundamental principles governing artificial photosynthetic systems

Artificial photosynthesis can be achieved photocatalytically or photoelectrochemically. In a photocatalytic process, a photocatalyst is dispersed in an aqueous solution and employs the simplest strategy to harvest sunlight, whereas photoelectrochemically driven reactions occur on the surface of light-absorbing photoelectrodes, which are fabricated by assembling the photocatalyst on a substrate. The two routes define artificial photosynthesis, namely, water splitting to generate H_2/O_2 and CO_2 reduction (Stolarczyk et al., 2018).

2.2.1 Photocatalytic water splitting

Photocatalytic water splitting is splitting water into H_2 and O_2 on a semiconductor photocatalyst surface, which converts the sunlight into chemical energy and stores it in the form of a carbon-neutral green fuel, H_2. Thermodynamically, the splitting of water involves a positive change in the standard Gibbs free energy and is thus an energetically uphill process. The major steps in photocatalysis involve photon absorption, exciton (electron–hole pair, e^-/h^+) generation, charge separation and migration to the catalyst surface, and water oxidation and reduction reactions on the surface. The two main approaches of one-step or two-step photocatalytic water splitting are dependent on the execution of these steps on a single or two different catalytic surfaces Fig. 2.1 (Maeda & Domen, 2010).

In the first approach, H_2 and O_2 evolution reactions are achieved by a single semiconductor absorber with or without the aid of a cocatalyst (Yang et al., 2013). Such a semiconductor photocatalyst's valence band (VB) maximum and conduction band (CB) minimum should cover the redox potential crucial for water splitting. Thus, considering the standard electrochemical potential for water oxidation to O_2 (+1.23 V) and proton reduction to H_2 (0 V), the minimum theoretical bandgap of a single photocatalyst should be 1.23 eV. However, to compensate for losses during charge separation and charge transfer between photocatalysts and water molecules, a practical bandgap of 2 eV is inevitable (Nishioka et al., 2023).

The second approach is based on a two-step photoexcitation mechanism on two different photocatalyst surfaces. This approach mimics the natural photosynthesis mechanism involving two photosystems and is termed the Z-scheme. The two photocatalysts in Z-scheme are connected in series by direct contact or in the presence of a donor/acceptor mediator called a redox mediator. The redox mediators are redox-coupled ions such as Fe^{3+} and Fe^{2+}, which facilitate the transfer of electrons between two different photocatalysts. The photoexcited electrons in the CB of one photocatalyst reduce the acceptor form of mediator and are converted to donor form, which is oxidized by photo-induced holes in the VB of another photocatalyst to regenerate the acceptor ion. These tandem photocatalysts offer advantages over one-step photocatalysis by showing a wide range of visible-light absorption and suppressing the back reaction, that is, H_2-O_2 recombination to form H_2O.

The overall process of one-step and two-step photocatalysis can be summarized as follows: photocatalyst upon irradiation gets photoexcited causing electronic transitions from VB to CB and generating e^-/h^+ pairs. The generated charge carriers separate and migrate to the surface of the photocatalyst. The holes of the VB oxidize water, and the electrons of the CB reduce the protons.

$$\text{Photocatalyst} \rightarrow \text{Photocatalyst} * \left(e^- + h^+\right)$$

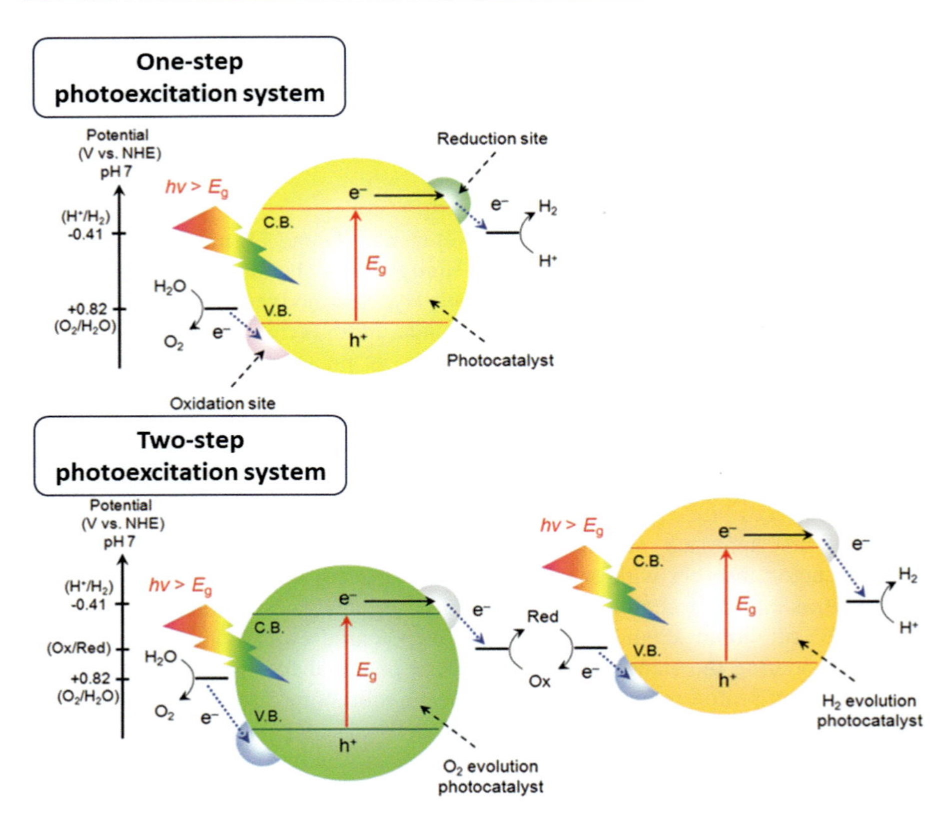

Figure 2.1 One-step and two-step photoexcitation mechanism of photocatalytic water splitting. One-step and two-step photoexcitation mechanism of photocatalytic water splitting (E_g = band gap energy).
Source: Reproduced with permission from American Chemical Society Copyright 2010. From Maeda, K. & Domen, K. (2010). *Journal of Physics & Chemistry Letters*, *1*(18), 2655–2661. https://doi.org/10.1021/jz1007966.

$$H_2O + h^+ \rightarrow \frac{1}{2}O_2 + 2H^+$$

$$2H^+ + 2e^- \rightarrow H_2$$

$$H_2O \rightarrow H_2 + \frac{1}{2}O_2$$

Considering the different sets of optoelectronic and kinetic conditions desired for the simultaneous occurrence of two types of reactions for the evolution of H_2 and O_2, it is inevitable to have a two-component photocatalyst (a hybrid photocatalyst) that can serve two different active sites for the improvement of the photocatalytic process. Engineering such heterostructured systems is crucial to tuning the overall photophysical properties of a photocatalyst. Apart from promoting charge separation by facilitating charge migration through junction/interfaces formed between the reactive surface and light absorber, such systems play a major role in providing immediate active sites on the surface, which mitigates the major challenge of charge recombination. Noble metals are used as effective candidates to fabricate hybrid systems (Liu et al., 2016). Due to their high work function compared to other semiconductors, they serve as easy electron trap sites. The metal traps electrons from light-absorbing semiconductors, and the Schottky barrier formed at the metal/semiconductor interface facilitates the charge separation. This is in addition to the proton reduction sites provided by noble metals, which lower the activation energy for the water reduction reaction.

Photocatalytic generation of H_2 and O_2 is also carried out using a sacrificial agent (Schneider & Bahnemann, 2013; Zhou et al., 2014). Sacrificial agents are electron donors (e.g., methanol, sulfides, sulfites, ethylenediaminetetraacetic acid, and triethanolamine) or electron acceptors (e.g., Ag^+ and Fe^{3+}), which minimize the recombination tendency of the electron–hole pair. If the CB of the photocatalyst is located above the reduction potential of water, the sacrificial electron donor promotes the H_2 release by CB electrons by oxidizing itself in place of H_2O from the photogenerated holes in VB. So also, the sacrificial electron acceptor irreversibly reduces itself in place of H^+ by accepting electrons from CB, thus facilitating the oxidation of H_2O at the VB.

Photocatalytic water splitting is generally performed under inert conditions, using an Ar atmosphere, or under degassed conditions for quantitative analysis of the gases evolved. Two types of reactor systems, namely, closed systems and flow systems are used for H_2/O_2 generation/collection and analysis (Gupta et al., 2022). In closed systems, the gases are generated in a closed reaction cell containing water and a photocatalyst. The reaction vessels are irradiated by an external light source (Xenon lamp or a solar simulator), and thus these vessels are made up of a material, usually Pyrex glass or quartz, that is optically transparent to the incident light.

Depending upon the method of analysis of the products formed, the closed reactor systems are further categorized into batch-type systems or closed gas circulation systems (Nishioka et al., 2023). In batch-type system, the gas generated is collected in an airtight gas syringe and analyzed by gas chromatography, whereas in a closed gas circulation system, the gas outlets from the reaction vessel are directly connected to a gas chromatograph and analyzed at regular time intervals. In the case of flow systems, the water or reaction medium is continuously purged with a carrier gas (generally Ar), and the product gases generated are steadily analyzed. The advantage of flow systems over closed systems is their suppression of the $H_2–O_2$ backward recombination reaction to produce water. In all the above-mentioned reactor systems, it is mandatory to maintain the photocatalyst in suspended form with continuous stirring for efficient diffusion of products and reactants and for optimal light absorption. For large-scale water splitting, the powder photocatalyst is fixed to a solid substrate to fabricate large-area photocatalyst sheets. In 2021, a 100-m^2 area prototype photocatalytic H_2 production system was developed using

$SrTiO_3$:Al photocatalyst sheets with Rh/Cr as H_2 evolving cocatalyst and cobalt as O_2 evolving cocatalyst (Nishiyama et al., 2021).

Various factors affect the photocatalytic activity, which include, light intensity, pH, temperature, photocatalyst structure, and band gap. The product evolution rate will be higher if high-intensity light is used. This is only true if the generated charge carriers are completely utilized in the photocatalytic reaction. The energy of the incident light should also be higher, that is, light with a shorter wavelength compared to the absorption edge of the photocatalyst can initiate the photoexcitation process. Although photocatalytic water splitting is normally carried out at a neutral pH using pure water, maintaining an optimum pH of the reaction is crucial for the chemical stability of the photocatalyst (Hameed et al., 2005). The photocatalyst band gap may vary with changes in reaction pH, so H_2 generation is more favorable at weak basic or neutral pH. An increase in reaction temperature helps to overcome the activation barrier caused by surface activities, which involve charge transfer between photocatalyst and reactant and desorption of the product molecules. However, a temperature rise may also slow down the adsorption of the water molecules.

A photocatalyst with a smaller particle size and a large surface area can enhance photocatalytic activity due to the availability of more active sites. Also, smaller particles lead to a smaller distance for the photogenerated charge carriers to travel to the catalyst surface and reach the active sites. Highly crystalline photocatalysts are also advantageous due to their lower density of defects, which in turn reduces the number of charge recombination centers. The photocatalyst bandgap should be such that the VB maximum should be more positive than the water oxidation potential, and the CB minimum should be more negative than the water reduction potential at a particular pH.

2.2.2 Photocatalytic carbon dioxide reduction

Mimicking the natural photosynthetic process of CO_2 fixation, wherein the atmospheric CO_2 is reduced into organic compounds (mainly carbohydrates), thus storing solar energy in the reduced form of CO_2, can solve most of today's global energy problems. The crisis of depleting fossil fuels and a massive rise in CO_2 emissions have compelled scientists to research the photocatalytic conversion of CO_2 as a promising route to address the challenge. Photocatalytic reduction of CO_2 is similar to photocatalytic water splitting and also involves light absorption ($h\nu = E_g$) by the photocatalyst, followed by exciton generation and separation, and finally migration of the charge on the catalyst surface for redox reactions. For a feasible CO_2 reduction process, a more negative CB edge of the photocatalyst with respect to the reduction potential of CO_2 assists in electron transfer from CB to CO_2, and a more positive VB edge compared to the water oxidation potential promotes hole transfer from VB to water, generating protons and O_2 (Karamian & Sharifnia, 2016).

The extreme thermodynamic stability of the CO_2 molecule arises from the high energy (750 kJ/mol) requirement to break the $C=O$ bond (Prabhu et al., 2020).

Similarly, single electron transfer to CO_2 to form $CO_2^{\bullet-}$ is thermodynamically unfavorable owing to its large negative reduction potential (−1.85 V vs SHE at pH 7) and very positive Gibbs free energy change (ΔG). As a result, CO_2 reduction is carried out via a proton-coupled multielectron transfer process. A wide range of reduced products of CO_2 is possible at various redox potentials (vs NHE at pH 7), as listed in the following set of equations (Lingampalli et al., 2017; Wang et al., 2022):

$$CO_2(g) + e^- \rightarrow CO_2^{\bullet-} \; E^\circ = -1.85 \text{ V} \tag{2.1}$$

$$CO_2(g) + 2H^+ + 2e^- \rightarrow HCOOH(l) \; E^\circ = -0.61 \text{ V} \tag{2.2}$$

$$CO_2(g) + 2H^+ + 2e^- \rightarrow CO(g) + H_2O(l) \; E^\circ = -0.53 \text{ V} \tag{2.3}$$

$$CO_2(g) + 4H^+ + 4e^- \rightarrow HCHO + H_2O(l) \; E^\circ = -0.48 \text{ V} \tag{2.4}$$

$$CO_2(g) + 6H^+ + 6e^- \rightarrow CH_3OH(l) + H_2O(l) \; E^\circ = -0.38 \text{ V} \tag{2.5}$$

$$CO_2(g) + 8H^+ + 8e^- \rightarrow CH_4(g) + 2H_2O(l) \; E^\circ = -0.24 \text{ V} \tag{2.6}$$

$$2H_2O + 4h^+ \rightarrow O_2(g) + 4H^+ E^\circ = +0.81 \text{ V} \tag{2.7}$$

$$2H^+ + 2e^- \rightarrow H_2(g) \; E^\circ = -0.42 \text{ V} \tag{2.8}$$

These products are obtained (via proton-coupled electron transfer) at a more positive thermodynamic potential (E° between −0.20 and −0.60 V) with relatively less positive ΔG values Fig. 2.2. However, high selectivity toward preferable products remains a major challenge due to the analogous reduction potentials of the products. During the process, H_2O, as an electron donor, is oxidized to generate O_2 and H^+. These generated protons can compete with CO_2 reduction reactions by accepting the photoexcited electrons to evolve H_2. Nevertheless, due to the lower negative potentials of CO_2 photoconversion to CH_3OH and CH_4 [Eqs. (2.5) and (2.6)], these reactions are more favorable than H_2 evolution. Further, experimentally observed redox potentials for CO_2 reduction are higher than the thermodynamically determined ones. This overpotential can be overcome by tuning the photocatalyst properties. Most of the CO_2 reduction products are one-carbon (C1) species; this is because the formation of energy-rich C2 products by C−C coupling involves complex intermediate steps to create a C−C bond and hence requires very high overpotentials. The above facts thus conclude that the electrochemical reduction potentials, complexity, and number of reduction steps together determine the yield and selectivity of the products. For instance, from Eq. (2.6), CH_4 is theoretically the most favored product considering the lowest reduction potential; however, the pathway to obtain CH_4 involves rigorous eight electron and proton transfer steps, thus lowering the yield of the product.

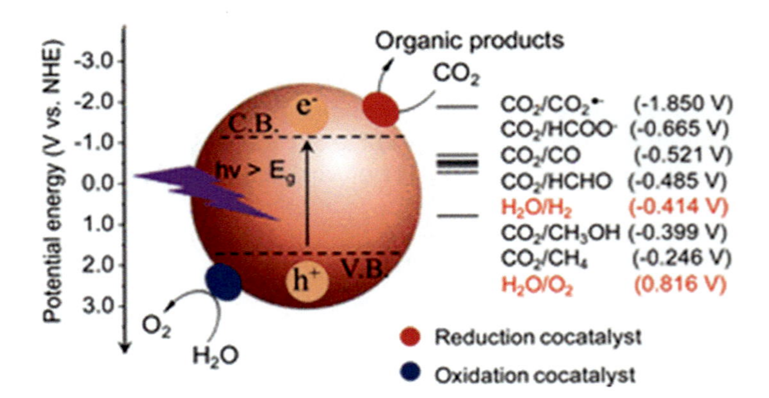

Figure 2.2 Photocatalytic CO_2 reduction pathways. Schematic illustration of photocatalytic CO_2 reduction pathways on semiconductor hybrid nanocatalyst.
Source: Reproduced with permission from American Chemical Society Copyright 2017. From Lingampalli, S. R., Ayyub, M. M. & Rao, C. N. R. (2017). *ACS Omega*, *2*(6), 2740–2748. https://doi.org/10.1021/acsomega.7b00721.

Employing a hybrid heterostructured photocatalyst for CO_2 reduction helps to overcome the limitations offered by a single photocatalyst, which include, ineffective charge separation, fewer active sites, poor selectivity, catalyst degradation, and so on. The promising approach of coupling two semiconductor catalytic systems provides the possibility to tune the band gap of the composite material for improved absorption of photons and increased production of photoinduced charge carriers. Depending on the mechanism of the charge migration pathway at the heterostructure interface, two main types of heterojunction mechanisms are reported for CO_2 reduction: type-II heterojunction and Z-scheme heterojunction. In type-II heterojunction, the photoinduced electrons in the CB band of the first semiconductor are migrated to less negative CB of the second semiconductor across the interface. The photogenerated charge carriers are thus accumulated as electrons in less negative CB of second semiconductor and holes in the less positive VB of the first semiconductor. The fast charge transfer pathway in type II heterojunction suppresses rapid charge recombination; however, by sacrificing the carriers, it compromises the strong redox abilities of the photogenerated charge carriers.

In the Z-scheme, the electrons from VB to CB are excited in both the semiconductors; however, the photogenerated electrons in CB of second semiconductor are transferred to recombine with the holes in VB of first semiconductor, which are further excited in the CB. This follows the "Z"-shape charge transfer pathway. Hence, in addition to effective charge separation, in the Z-scheme mechanism, the strong redox ability of the photogenerated charges is retained by preserving the photogenerated charges, that is, electrons in less negative CB and holes in more positive VB. Furthermore, depending on the absence and presence of an electron mediator in between the two semiconductors (as used in the case of a two-step excitation pathway in photocatalytic water splitting), the Z-scheme mechanism is classified into direct and indirect types, respectively (Zhang et al., 2020).

The photocatalytic CO_2 reduction can be performed in two types of reaction setups, namely, a gas-phase reactor or a liquid-phase reactor. In liquid-phase reactors, a saturated aqueous solution of CO_2 is used; thus, water acts both as a solvent and a reagent. However, CO_2 solubility in water is limited, causing water to be replaced with organic solvents such as isopropanol, methanol, and acetonitrile. In a gas-phase reactor, humidified CO_2 is passed through the reactor, which facilitates better adsorption of the gas molecules on the catalyst surface (Lingampalli et al., 2017). The use of gas-phase reactors is more advantageous on account of the enhanced selectivity of products achieved in gas-phase reactions. For lab-scale experiments, closed-batch reactors are commonly used. On the contrary, for large-scale output, rationally designed flow systems are employed. For effective photocatalytic reactions, the photocatalyst is usually immobilized on the substrate surface. The gas-phase products are analyzed by gas-chromatographic techniques, whereas the liquid-phase product analysis involves ^{13}C-NMR technique.

2.2.3 Photoelectrochemical artificial photosynthesis

The photoelectrochemical or photoelectrocatalytic (PEC) process combines photocatalytic and electrocatalytic approaches to carry out artificial photosynthetic reactions. The first PEC cell was fabricated by Fujishima and Honda, wherein they used TiO_2 as a photocathode and Pt as a counter electrode. Upon light absorption, VB electrons in TiO_2 were excited to CB. The photoexcited electrons under externally applied anodic potential migrated to the Pt counter electrode, reducing H^+ to H_2. Holes left at the anode got transferred to the surface of TiO_2 and oxidized water to O_2 Fig. 2.3. On the contrary, Pt/TiO_2 heterostructure nanoparticles could behave as a microintegrated PEC cell without any external bias undergoing photocatalytic water splitting Fig. 2.4. Thus, PEC systems offer several advantages over photocatalytic systems in terms of separation of gaseous products, better charge separation aided by external bias, long charge transfer pathways, and high efficiency of the process.

In photocatalytic systems, the catalyst is usually suspended in solution, whereas PEC setup demands the fabrication of photoelectrodes by deposition of the catalyst in the form of a thin film or immobilizing the catalyst on a suitable conducting substrate. The role played by hybrid nanostructures in both PEC and photocatalytic processes is identical; however, in the thin film configuration of a photoelectrode, the second component of the heterostructure can act as a passivating layer for the first component. Passivation is significant in the case of PEC to minimize photocorrosion pathways of the catalyst by externally applied bias. The passivation layer also suppresses charge recombination and assists in electron transfer by reducing the surface states. Energetic matching of the multicomponent of a hybrid photoelectrode is another factor for efficient charge transfer promotion in PEC activity. Unlike in photocatalysis, where the photocatalyst particles in suspension are assumed to be irradiated homogenously in all directions, in PEC, designing the hybrid nanomaterial photoelectrode and the direction of photoelectrode illumination can determine the photoexcitation

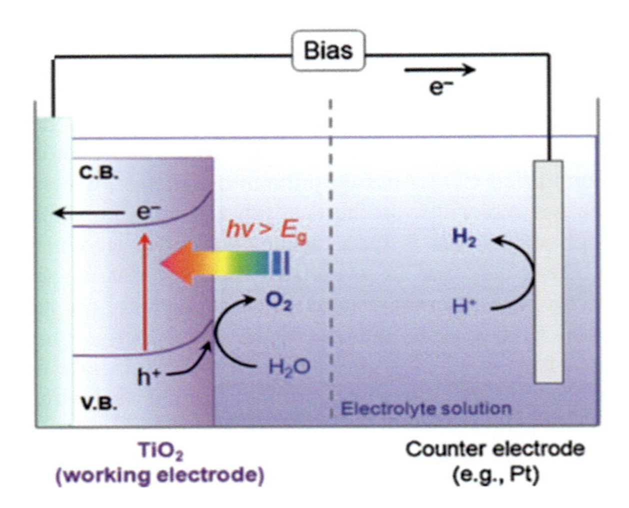

Figure 2.3 Photoelectrochemical water splitting setup. Photoelectrochemical water splitting using TiO_2 photoanode and Pt counter electrode.
Source: Reproduced with permission from Elsevier Copyright 2011. From Maeda, K. (2011). *Journal of Photochemistry and Photobiology C: Photochemistry Reviews*, *12*(4), 237–268. https://doi.org/10.1016/j.jphotochemrev.2011.07.001.

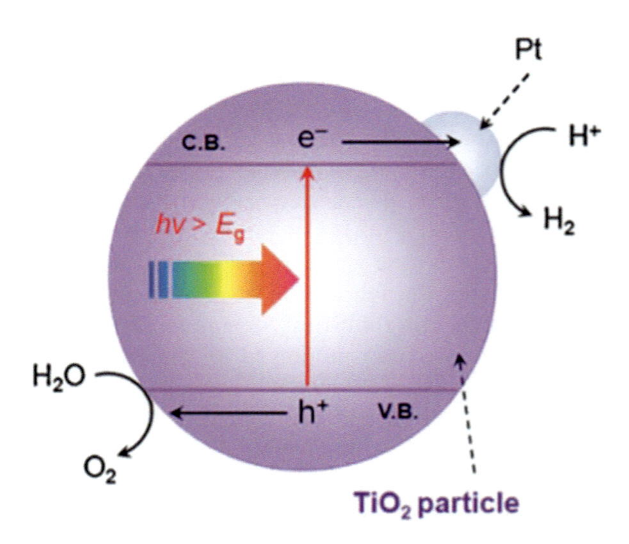

Figure 2.4 Photocatalytic water splitting (microintegrated photoelectrochemical cell). Photocatalytic water splitting using Pt/TiO_2 heterostructure photocatalyst, a short circuit model.
Source: Reproduced with permission from Elsevier Copyright 2011. From Maeda, K. (2011). *Journal of Photochemistry and Photobiology C: Photochemistry Reviews*, *12*(4), 237–268. https://doi.org/10.1016/j.jphotochemrev.2011.07.001.

efficiency. The substrate used in the photoelectrode should be a transparent/glass material, which should be irradiated from the front side of the photoelectrode.

PEC water splitting is the electrolysis of water to generate H_2 and O_2 using light-absorbing photoelectrodes. The fundamental principle involves the absorption of light by photoelectrodes (photoanode or photocathode), followed by a flow of photogenerated electrons at the surface of the cathode to reduce H^+ to H_2, and the oxidation of water by holes to evolve O_2. As depicted in Fig. 2.5, the hybrid nanostructures play a similar role both in photocatalysis and PEC water-splitting reactions. In photocatalysis, the two-component system decreases the activation energy for water oxidation. In the PEC cell, the use of two-component hybrid nanocatalysts lowers the activation overpotential, as indicated by decreased onset potential and improved photocurrent density. PEC water splitting can be carried out with or without external bias (Abe, 2010).

The configurations of biased single light absorber PEC cells are (i) n-type photoanode-counter electrode (cathode for H_2 evolution) and (ii) counter electrode (anode for O_2 evolution)-p-type photocathode. For PEC cell configurations to operate without external bias (unbiased), they are: (i) photoanode–photocathode tandem cell; and (ii) photoanode–photovoltaic tandem cell (Kim et al., 2019). Thus, PEC cells can operate with or without external bias by using appropriate n-type or p-type hybrid nanomaterials with matching band structures to function via the Z-scheme heterojunction mechanism.

In the PEC conversion process of CO_2, H_2O (electron donor) is oxidized at the anode by photogenerated holes, and electrons are transported to the cathode for reduction of CO_2 to various hydrocarbon species. As compared to the photocatalytic process, the use of externally applied bias helps in band bending of the photocatalyst at the electrode/electrolyte interface. This facilitates the oriented transfer of photoexcited electrons through the external circuit, thus reducing the recombination rate of the electron–hole pairs (Pang et al., 2018). Conversely, compared to the

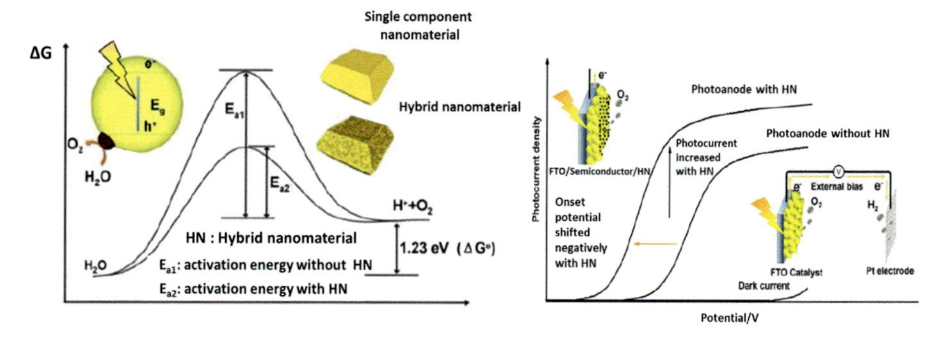

Figure 2.5 Functioning of hybrid nanostructures. Role of hybrid nanomaterial in photocatalytic water oxidation and photoelectrochemical water oxidation.
Source: Reproduced with permission from American Chemical Society Copyright 2011. From Yang, J., Wang, D., Han, H. & Li, C. (2013). *Acc. Chemistry Research*, *46*(8), 1900–1909. https://doi.org/10.1021/ar300227e.

electrocatalytic process, the use of photoelectrodes generates photovoltage, which minimizes the dependency on external bias for CO_2 reduction (Ochedi et al., 2021). The other fundamental principles governing the PEC CO_2 reduction process are similar to photocatalytic CO_2 reduction.

Unlike a photocatalytic reaction chamber, which involves the evolution of gaseous products in a single reactor, the experimental setup of a PEC cell evolves gaseous products separately at the respective electrodes, that is, O_2 at photoanode and CO_2 reduction or H_2 generation at photocathode. The two chambers (anodic and cathodic) are separated by membranes to avoid back reactions. In a single photoelectrode system, either a photoanode or photocathode is fabricated and illuminated (Wu et al., 2020). The photogenerated charges at the photoelectrode, under the influence of an external voltage, are driven to the counter electrode to induce the counter reaction. Such single photoelectrode systems are generally studied to test the activity of the designed photoelectrode. For application purposes, a dual-photoelectrode system is fabricated, wherein a p-type photocathode and an n-type photoanode are assembled. This is a self-driven bias-free PEC system wherein internally generated voltage drives the redox reactions at the respective electrodes.

2.3 Hybrid artificial photosynthetic nanosystems

As discussed, hybrid nanostructures are semiconductor materials characterized by the presence of two or more nanosized components with unique functions, exhibiting superior performance compared to their individual counterparts. This is primarily attributed to hybrid nanomaterials' ability to leverage each component's strengths while mitigating their weaknesses. Through seamlessly integrating diverse functionalities, hybrid nanomaterials unlock the potential for enhanced performance, increased efficiency, and improved overall functionality due to synergistic effects (Povolotskaya et al., 2015). Hybrid nanomaterials exhibit diverse classifications, including core-shell structures (Meng et al., 2013; Mongin et al., 2011; Yang et al., 2011; Zhang et al., 2011), Janus-type nanostructures (Menagen et al., 2009; Shaviv et al., 2011; Sotiriou et al., 2011), and nanostructures composed of nanoparticles, nanotubes, or nanoplates that contain encapsulated or surface-deposited nanocomponents (Bahns et al., 2011; Henley et al., 2008).

Particular attention is given to hybrid structures containing plasmonic nanoparticles (PNPs). PNPs are used in artificial photosynthesis to concentrate solar energy in molecular-sized regions (Giannini et al., 2011). They interact strongly with light through localized surface plasmon resonance (LSPRs), which enhances their absorption capabilities. Plasmonic materials are usually metals having negative permittivity (i.e., with a high concentration of free electrons), such as silver, gold, and copper, which can tune their plasmon resonances across the UV, visible, and near-infrared (NIR) regions of the electromagnetic spectrum (Coronado et al., 2011). They act as efficient light harvesters and can concentrate light in small volumes near their surface, making them effective nanoreactors.

The catalytic activity of plasmonic metals for fuel production is hindered by the short lifetime of photoexcited charge carriers, which generally lasts less than 1 picosecond (Zhang et al., 2018). This interval is not enough for carriers to act as adsorbate. Hence, hybrid plasmonic nanocatalysts have been developed to improve performance by combining PNPs with other nanomaterials. Unlike conventional light absorption processes by molecular sensitizers or interband excitation of metals, LSPR involves a multielectron excitation that concentrates the absorbed light energy at the nanoparticle surface through intense electric fields (Maier, 2004). The intense LSPR energy generated near the surface of the plasmonic NPs can then be transferred to a semiconductor photocatalyst (Brongersma et al., 2015) in contact with the plasmonic NPs, e.g., metal/TiO_2, metal/ZnO, metal/Fe_2O_3, and metal/CdS. Out of all semiconductors, CdS is a semiconductor material with a narrow band gap of ~2.4 eV. Variety of polymorphs of CdS, and its excellent optical/electronic properties, makes CdS an extensively studied material for solar energy utilization (Cheng et al., 2018). The inherent characteristics of CdS make it suitable for photocatalytic water splitting, reduction of CO_2, etc. However, pure CdS has certain drawbacks, such as photocorrosion, electron–hole pair recombination, and reduced surface activity for catalytic reactions (Li et al., 2018). Thus, engineering the materials to exploit light harvesting and catalytic properties is one of the greatest challenges.

2.3.1 Hybrid nanostructures for hydrogen evolution reaction

PNPs play a crucial role in the photocatalytic process of generating green H_2 from water, specifically through the hydrogen evolution reaction (HER).

$$2H^+ + 2e^-_{CB} \rightarrow H_2\,(E^\circ_{redox} = -0.41\ V_{NHE})$$

Water splitting can be achieved with purely metallic photocatalysts. Pt is commonly used in these metal hybrids due to its excellent electron sink properties and H_2 evolution capacity. This is attributed to Pt's low Fermi level and the presence of specialized catalytic sites for H_2 formation. Plasmonic metals like Au are often combined with Pt to enhance efficiency in alloys or core-shell structures. This facilitates electron transfer between the two metallic components, leading to highly efficient plasmonic photocatalysts. For example, Pt-tipped Au nanorods (NRs) have demonstrated that the surface plasmon resonance mode of Au NRs serves as the primary pathway for hot-electron transfer from Au to Pt, enabling efficient H_2 generation when exposed to visible and NIR light irradiation (Lou et al., 2016).

Plasmonic metal/semiconductor heterostructure is another most studied class of hybrid nanostructures (Gesesse et al., 2020). Metal@TiO_2core–shell nanostructures have shown superior HER compared to catalysts consisting of metal particles merely deposited onto TiO_2. The active sites in merely deposited metal particles on semiconductors diminish the active sites, and isolated Au NPs agglomerate into larger ones, reducing the photocatalysts' lifespan (Wang et al., 2015). Also, metal hybrids are often combined with semiconductors such as TiO_2 (nanotube)–Au (core)–Pt (shell) (TiO_2–Au@Pt). Further, well-defined gold–platinum core–shell nanostructures (Au@Pt) achieve a synergistic effect (Hung et al., 2016). The sequential junctions from Au to Pt induce an additional electric field, facilitating electron transport, which a random distribution counterpart cannot accomplish. Optimizing the morphology within the core–shell structure can offer further improvements in the rate of H_2 production. An intriguing approach involves utilizing Au@mesoporous TiO_2 yolk–shell hybrid nanostructures as photocatalysts for H_2 evolution. The photocatalyst has a different photoexcitation mechanism under UV light and visible light in hollow nanosphere core–shell (yolk–shell) structures, demonstrating superior performance to Au deposition on TiO_2 spheres. These yolk–shell structures achieve a higher H_2 production rate (69.77 vs 37.80 mmol/g/h) (Hung et al., 2016). The advantages of such a configuration are twofold. First, Au nanoparticles in the structure can capture electrons, facilitating charge separation within the Au@TiO_2 matrix and ultimately enhancing the overall catalytic activity. Second, the hollow cavity and mesoporous shell create extra active sites and pathways, promoting the efficient conversion of reactants into H_2. However, the current fabrication process involves multiple intricate steps and corrosive solvents, suggesting a more scalable and environmentally friendly method to produce well-defined core-shell (Au@TiO_2) hollow nanocomposites (Ngaw et al., 2014).

However, Janus metal-semiconductor structures of Au–TiO_2 offer superior performance for HER compared to their core–shell counterparts, which is attributed to the strong localization of plasmonic electric fields near the Au–TiO_2 interface (Chauhan et al., 2018). These near-fields interact closely with light and affect how electrons move in the TiO_2, enhancing the absorption of light and helping create electron–hole pairs more efficiently, which is essential for photocatalysis. A remarkable H_2 evolution activity of 56 mmol/g/h under visible light illumination is exhibited by Janus Au–TiO_2, which is approximately twice the rate achieved by the core-shell equivalents (Seh et al., 2012). CdS, as discussed earlier, is also used in H_2 evolution reactions. Au NPs with CdS have shown promising results, wherein the conduction band level allows for H^+ reduction, while the small band gap (2.4 eV) overlaps well with the LSPR of Au. To further enhance the production of H_2, a ternary system of ZnO/Au/CdS NRs is effective. The flower-like ZnO structure provides efficient charge storage and surface area, improving water-splitting performance. Au nanoparticles between ZnO and CdS increase the HER rate as electron mediators. The ZnO/Au/CdS NRs demonstrate a gas evolution rate of 502.2 μmol/g/h (Wang et al., 2015) Fig. 2.6.

Another ternary photocatalyst uses Ag nanoparticles, on which TiO_2 nanoparticles are deposited and attached to the surface of CdS nanowires. This CdSe/Ag/TiO_2 catalyst achieves a hydrogen evolution rate of 1.91 mmol/g/h, much higher than that of CdSe/TiO_2, CdSe/Ag, and pure CdS nanowires. In the CdSe/Ag/TiO_2 Z-scheme system, the photogenerated electrons on the CB of TiO_2 are rapidly transferred to the VB of CdS by Ag nanoparticles and recombined with the photogenerated holes (Zhao et al., 2018). Hence, Ag nanoparticles act as an excellent electron mediator as they also transport trapped electrons from the CB of TiO_2 to the VB of

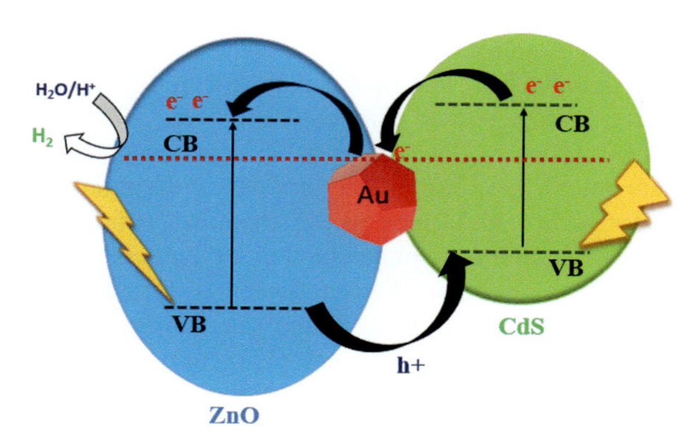

Figure 2.6 Hybrid nanorod structures for H_2 evolution. ZnO/Au/CdS hybrid nanorod structure.

CdS nanowires, and they keep the electrons and holes segregated effectively. Under UV light irradiation, the simultaneous transfer and accumulation of electrons and holes at the Z scheme system increases the separation rate of carriers and suppresses their recombination. The accumulated electrons on the CB of CdS nanowires reduce H^+ ions to H_2 (Zhao et al., 2018).

Plasmonic nanomaterials effectively decrease the overpotentials for PEC HER. The most successful plasmonic systems are metal–metal plasmonics. Pt/Fe@Au NRs, which have a unique dumbbell-like structure, enables the transfer of electrons from Au to Pt/Fe when stimulated by light, enhancing the electrocatalytic activity. These findings confirm that the light-absorbing material (Au) and the catalyst (Pt/Fe) must be properly connected and aligned. This also applies to the oxygen evolution reaction (OER), where holes contribute to the catalytic process (Guo et al., 2018).

Plasmonic metals combined with semiconductors and 2D materials also achieve low energy requirements. For instance, an overpotential of 45 mV is obtained for nanoperforated gold films placed on a nanometer-thick Ti/TiO_2 film at a current density of 10 mA/cm^2, thanks to improved electron transfer at the Au–TiO_2 interface (Pang et al., 2020).

2.3.2 Hybrid nanostructures for oxygen evolution reaction

In photocatalytic water splitting, the evolution of O_2 is a key step, as the reaction provides electrons for the reduction of H^+ to H_2. However, energetically, it is challenging to achieve the water oxidation step owing to the major task of extracting four protons and four electrons from two molecules of water. Since one photon generates one electron–hole pair at a time, to oxidize two water molecules in one step, the oxidation catalyst should be capable of accumulating and storing the charge carriers (Styring, 2012). To achieve this, in an artificial photosystem, the VB of the photocatalyst must be more positive than the water oxidation potential (1.23 V vs NHE at pH 0). Hybrid nanostructures are thus superior water oxidation photocatalysts as they offer the possibility for band structure modification, which promotes charge transfer. Also, one of the components of the heterostructure can act as a reservoir for photogenerated charges (Yang et al., 2014).

Graphitic carbon nitride (g-C_3N_4)-based hybrid nanomaterials have proven to be star photocatalysts for O_2 evolution. The limitations of pristine g-C_3N_4, such as poor charge transfer ability, and quick recombination of charge carriers, are overcome by coupling it with another semiconductor oxygen evolution catalyst, forming a hybrid assembly. In this regard, 2D nanosheets as cocatalysts were capable of suppressing electron–hole recombination and promoting charge transfer (Deng et al., 2016). In the case of InSe/g-C_3N_4 heterostructures, the VB and CB of InSe are lower than those of g-C_3N_4 (type-II band alignment), thus providing a driving force for thermodynamically permissible reactions. This band alignment in the hybrid photocatalyst also facilitated electron and hole separation (He et al., 2019). Besides, noble metal-based structures, especially Pt-based g-C_3N_4 hybrid structures, are known to accelerate the transport of generated electrons (Liu et al., 2021b; Pan et al., 2017). However, to develop a cost-effective photocatalyst, researchers attempted noble-metal-free heterostructures. In this sense, $NiCo_2O_4$ coupled g-C_3N_4 resulted in a reduced overpotential for water oxidation reaction (Cheng et al., 2021). The high efficiency of the oxygen evolution process in $NiCo_2O_4$/g-C_3N_4 was attributed to enhanced charge separation and migration due to the build-up of surface bonding states on the coupled heterostructure.

Overcoming high activation energies involved during the charge-transfer process poses a major challenge in the direct oxidation of water to diatomic oxygen. In recent years, this has been achieved by silver orthophosphate (Ag_3PO_4) photocatalyst, which can catalyze oxygen evolution by reducing the overpotential (Yang et al., 2019). Nevertheless, Ag_3PO_4 suffers from the severe issue of low photostability. The generated photoelectrons may get consumed in the self-reduction of the photocatalyst to Ag. This drawback is prevented by coupling it with a semiconductor having high electrical conductivity and a suitable band structure to quickly pick up the photogenerated electrons. g-C_3N_4 being an ideal catalyst satisfying this requirement, several hybrid heterostructures have been designed and employed by pairing Ag_3PO_4 with g-C_3N_4 (Tian et al., 2018). Such Ag_3PO_4/g-C_3N_4 hybrid systems demonstrated superior oxygen evolution ability. This enhanced activity was ascribed to the formation of in situ Z-schemes, wherein in situ generated Ag nanoparticles acted as mediators. The photogenerated electrons in CB of Ag_3PO_4 formed metallic Ag by the reduction of Ag^+ ions. Thus, after illumination, the two-component (Ag_3PO_4/g-C_3N_4) hybrid system converted into a three-component (Ag_3PO_4/Ag/g-C_3N_4) system. Ag as a mediator facilitates charge transfer between the CB of Ag_3PO_4 and the VB of g-C_3N_4, promoting additional pathways for charge migration and effective water oxidation (Yang et al., 2015). Similar Ag-mediated Z-scheme mechanisms for oxygen evolution are reported for heterostructures of Ag_3PO_4 coupled with MoS_2 (Cui et al., 2018; Wang et al., 2018), $MoSe_2$ (Li et al., 2019), and WO_3 (Pudukudy et al., 2020).

Another class of photocatalysts that attracted immense attention for water oxidation reactions are cobalt oxide-based heterostructures. Co_3O_4 has been studied as a potent catalyst for oxygen evolution, considering its anticorrosion stability in alkaline media (Zeng et al., 2020). However, the lack of active catalytic sites and poor electronic conductivity are the factors that compelled the scientists to design hybrid structures based on Co_3O_4 to nullify the limitations and improve the overall catalytic efficiency of the process. CuO/Co_3O_4 heterojunctions were designed to combine the advantage of the plasmonic effect offered by CuO with the robustness and stability of the heterostructure (Hu et al., 2018). Additionally, coupling Co_3O_4 with g-C_3N_4 resulted in the introduction of oxygen vacancies in Co_3O_4, improved surface atom distribution, and improved the electrical conductivity of the catalyst (Zeng et al., 2020). The high photocatalytic activity of such a heterojunction system was attributed to its stable crystal structure and composition. The synergistic effect of constructing a Co-based hybrid assembly was also observed in the case of Co-based metal-organic framework heterojunctions (Xu et al., 2016). For instance, Co-based 1,4-benzenedicarboxylate (BDC) and 4,4′-biphenyldicarboxylate (BPDC), a Co-BPDC/Co-BDC heterojunction, showed improved photocatalytic activity (Zha et al., 2020).

Overall, high efficiency in photocatalytic OERs has been achieved by hybrid nanostructures as compared to single-component systems. This emphasizes the significance of heterostructure systems in mechanistically boosting such complex artificial synthesis processes.

2.3.3 Hybrid nanostructures for CO_2 reduction

Owing to multiple reaction pathways, CO_2 reduction reactions are much more complex than HERs. It produces a mixture of various products, including carbon monoxide (CO), methane (CH_4), formic acid (HCOOH), methanol (CH_3OH), propane (C_3H_8), ethane (C_2H_6), and ethanol (CH_3CH_2OH), among others. Generating products with multiple carbon atoms is desirable but challenging, and the presence of byproducts such as H_2 gas reduces hydrocarbon selectivity. Therefore, it is crucial to design catalytic sites that can lower the activation barrier for CO_2 and limit the production of byproducts.

PNPs can generate high energy carriers when resonantly illuminated, which is advantageous in reducing CO_2, and one such plasmonic material is Au. However, intermediates such as CO readily detach from Au surfaces, producing CO as a major product. Unfortunately, this considerably decreases the ability to produce other hydrocarbon compounds selectively. Alloying Au with Pd atoms is an effective approach to addressing this problem. A strong binding of CO on Pd surfaces enables efficient hydrogenation of this intermediate. Also, incorporating Pd atoms in Au nanoparticles facilitates the removal of H atoms from water to form hydrocarbons, inhibiting competing HERs (Hammer et al., 1996; Vilé et al., 2015; Yu et al., 2018). For further enhancing the selectivity to form hydrocarbons, creating a heterojunction between metals and semiconductors has shown excellent results. To exemplify Au–Pd nanoparticles attached to TiO_2 nanocrystals high selectivity for hydrocarbons was attained, reaching 85% in the CO_2 reduction reaction under visible light exposure (Jiao et al., 2017). Along with CH_4, other minor products, such as ethane (C_2H_6) and ethylene (C_2H_4), confirmed the ability to facilitate C–C coupling by Au-Pd@TiO_2 catalyst. Although CH_4 production was observed when only Pd was deposited on TiO_2, the inclusion of Au resulted in a significant 5.2-fold increase (Chen et al., 2019).

AuNPs on gC_3N_4 nanosheets represent plasmonic metal 2D hybrids for CO_2 reduction. The band structure at the metal-2D interface influences the formation rates and selectivity of reduction products. The selectivity of CO_2 reduction processes can be tuned by the composition of plasmonic metal-2D nanocomposites. At the Au-MoS_2 interface, electron transfer from MoS_2 to Au was hindered, resulting in electron accumulation on MoS_2 upon light irradiation (Sun et al., 2020). Conversely, at the Ag–MoS_2 junction, electron transfer from MoS_2 to Ag was favored, leading to less electron accumulation on MoS_2. Consequently, the Ag-based system mainly produced CO, while the Au-based system favored CH_4 as the primary product. Thus, the selectivity of CO_2 reduction processes can be tuned by the composition of plasmonic metal-2D nanocomposites (Sun et al., 2020).

To promote sustainability in these processes, it is essential to seek cost-effective alternatives. In this context, inexpensive plasmonic metals like Al or Cu could potentially serve as viable candidates. Cu has emerged as a promising photocatalyst due to its conductivity, cost-effectiveness, and LSPR effect. However, charge-carrier recombination limits its activity. By forming a bimetal structure with cobalt, the Cu@Co core-shell photocatalyst demonstrates an outstanding CO generation rate and selectivity. This catalyst shows remarkable stability and performs well without needing photosensitizers or cocatalysts. CuCo activity exceeds that of other alloys (Cu@Co $>$ CuCo $>$ CuPt $>$ CuAu $>$ CuRu $>$ CuPb $>$ CuNi), which is due to the synergistic effects of Cu and Co, enabling efficient CO_2 capture and rapid charge transfer (Lai et al., 2021).

As discussed previously, CdS-based hybrids are extensively studied photocatalysts with promising potential for solar photocatalytic reduction of CO_2 into CO, CH_4, and other hydrocarbons.

Au NPs combined with CdSe and Cu_2O enable selective hydrogen or carbon product formation control. Au/CdSe hybrid nanostructures exhibit high H_2 evolution ability, while Au/CdSe-Cu_2O hybrid nanostructures demonstrate excellent CO_2 reduction (Wang et al., 2020). Without water, Au/CdSe–Cu_2O hybrid nanostructures generate CO with 100% selectivity, while in the presence of water, methane and other carbon products are obtained. Under illumination, all three components can be simultaneously excited, accumulating photogenerated electrons in Cu. The presence of oxygen vacancies in Cu_2O NPs enhances the carrier density and number of active sites. Also, Au NPs' SPR-induced light harvesting and effective charge separation and transfer in Au/CdSe–Cu_2O hybrid nanostructures contribute to improved photocatalytic H_2 evolution reaction activity (Wang et al., 2020).

Combining reduced graphene oxide (rGO) and Ag nanoparticles with CdS NRs forms an Ag-rGO-CdS hybrid. Compared to pure CdS, this hybrid shows an approximately eightfold increase in CO_2 to CO conversion yield. The enhanced efficiency is attributed to rGO acting as an electron acceptor with better conductivity, facilitating efficient electron transfer, and preventing the recombination of electron–hole pairs. Additionally, Ag nanoparticles serve as electron trappers and active sites for CO_2 reduction (Zhu et al., 2018). rGO is also incorporated in the CdS/TiO_2 core–shell structure, leading to a threefold increase in CH_4 production compared to the CdS nanoparticles alone (Kuai et al., 2015) Fig. 2.7.

The Au–CdS core–shell nanostructure supported by TiO_2 exhibits strong photocatalytic activity for converting CO_2 to CH_4. Reduction reactions are theoretically feasible as the reduction potentials for CO_2 to CO (− 0.53 V) and for CO_2 to CH_4 (− 0.24 V) are lower than those of the conduction band potentials of TiO_2 (−0.56 V) and CdS (−1.0 V) (Wei et al., 2015). The formation rates of CO and CH_4 vary depending on the nanostructure and composition of the catalysts. The vectorial electron transfer within the Z-scheme system of TiO_2→Au→CdS promotes the separation of photogenerated charge carriers. Under simulated solar irradiation, TiO_2 transfers CB-electrons to Au nanoparticles, which then transfer to the CdS shell and combine with VB-holes, facilitating the efficient separation of electron–hole pairs. The VB-holes of TiO_2 can oxidize H_2O to O_2 and H^+, while the CB-electrons of CdS are utilized for CO_2 reduction, favoring CH_4 production. Consequently, the Au@CdS/TiO_2 catalysts, utilizing the CdS–Au–TiO_2 nanojunction and the Z-scheme system, demonstrate outstanding photocatalytic performance for CO_2 reduction to CH_4 (Wei et al., 2015). Another Z-scheme hybrid catalyst, Zn-CdS/Au@g–C_3N_4, enables photocatalytic reduction of CO_2 to methanol (Madhusudan et al., 2021). Au nanoparticles were deposited on Zn–CdS, and 2D g-C_3N_4 nanosheets were wrapped over the surface. This unique configuration reduces diffusion length and improves charge separation, conductivity, and catalytic activity, resulting in a high rate of CH_3OH evolution during the photocatalytic reduction of CO_2 (Madhusudan et al., 2021).

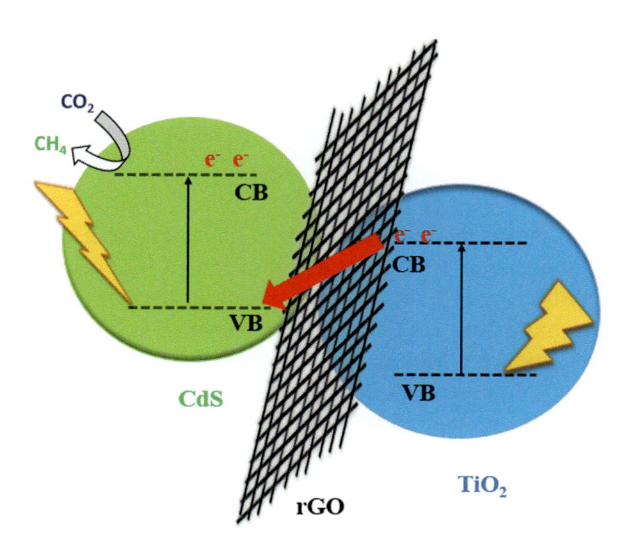

Figure 2.7 Core–shell hybrid nanostructure for CO_2 reduction. Schematic representation of rGO incorporated in CdS and TiO_2 core–shell hybrid nanostructures.

A ternary hybrid catalyst, CdS/(Cu-TNTs), based on sodium titanate nanotubes (TNTs), efficiently converts CO_2 and water into C1-C3 hydrocarbons (e.g., CH_4, C_3H_8, C_2H_4, C_2H_6, and C_3H_6) under irradiation. Photogenerated electrons are effectively transferred to elemental Cu through the titanate nanotube framework, which reduces CO_2 to C1−C3 hydrocarbons. At the same time, the photogenerated holes oxidize water (Park et al., 2015).

With the successful application of the SPR effect in photocatalysis, its utilization in photoelectrocatalysis has also become highly appealing. This approach enables the simultaneous utilization of electrical energy and solar energy, further expanding its potential applications. A notable example is the use of a metal-metal Ag/Cu hybrid plasmonic system in a PEC cell, which combines the highly plasmonic activity of Ag and the catalytic activity of nanostructured Cu, known for generating multicarbon CO_2 reduction products. At high potentials, the Cu−Ag cathode reduces CO_2 to methane, ethene, formate, and alcohol (Corson et al., 2020). Metal-semiconductor hybrid nanostructures also improve CO_2 reduction activity and selectivity through plasmon excitation. For instance, the p-type NiO film collects hot holes generated by visible-light excitation of Cu nanoparticles, while hot electrons accumulate on the Cu nanoparticles to reduce CO_2. Remarkably, Cu nanoparticles preferentially produce CO and formate ($HCOO^-$) while suppressing hydrogen evolution in aqueous electrolytes (Duchene et al., 2020).

In the nonplasmonic category involving Cu, the CuO−CdS catalyst formed by combining p-type CuO with n-type CdS (Li et al., 2018) demonstrates a significant 20.24% incident photon-to-current conversion efficiency (IPCE) at 470 nm, indicating its ability to efficiently absorb visible light. As a result, the CuO−CdS photocathode achieves a methanol yield of 35.65 μmole/L/cm^2 at an applied potential of −0.4 V versus NHE (Tarek et al., 2020).

2.4 Recent trends in artificial photosynthetic nanobiohybrid systems

One of the ways to match the maximum output capability of natural photosynthesis in an artificial system is by combining a preassembled biological apparatus with a nonbiological light-harvesting material to form a biohybrid. A nanobiohybrid involves coupling biological entities (such as enzymes and microorganisms/whole cells) with a semiconductor nanomaterial to bring about solar-to-chemical conversion. The nanomaterial component is highly configurable, engineered to nanoscale, and can pair well with the biological component due to similar dimensional scales and thus aid in efficient light absorption and charge separation, whereas the biological segment offers high product selectivity due to efficient catalytic transformations by natural enzymes. The merits of using biological species are that it aids in multielectron transfer, thus lowering the activation energy of the reactions, which is in contrast to nanomaterials, which allows sequential electron transfer and have poor selectivity, yielding only C1 reduction products of CO_2.

The chief advantage of utilizing biological components is their operation at minimal overpotential, thus achieving complex multielectron and bond formation steps at low redox potentials by channeling the energy into specific product-forming pathways. The biohybrid systems can be categorized into three types, which are as follows: (i) systems involving direct electron transport from the semiconductor to the biological component to cause CO_2 reduction; (ii) reduction of water to generate H_2, which along with electrons fed to biological component to reduce CO_2; and (iii) the reduced CO_2 products integrated with electrons and protons can be further used by biological component (especially genetically engineered microbes) to yield a wide range of $C > 1$ products (Okoro et al., 2023).

The semiconductor nanomaterial coupled with biological components in a nanobiohybrid influences the performance of the system by their surface charge (positive or negative). The cell membranes of microorganisms are generally negatively charged; thus, a nanomaterial with a positive surface charge can interact strongly with microorganisms or negatively charged cells. The nanomaterial coupled with the biological entity can be located intracellularly or extracellularly. The incorporation of a nanomaterial inside the cell leads to faster electron transfer due to the superior linking of the light-harvesting nanomaterial with the enzyme responsible for CO_2 fixation. Extracellular linking of the nanomaterial also exhibits considerable CO_2 fixing efficiency, however, with varied charge transfer pathways. A study showed the incorporation of core–shell quantum dots of different semiconductors intracellularly in *Cupriavidus necator* and *Azotobacter vinelandii* strains of bacteria (Ding et al., 2019). The quantum dots showed affinity attachment to the proteins of the bacteria via zinc linkage present on the shells of the quantum dots. The nanomaterials possessing varying band gaps when photosensitized with different light wavelengths yielded different valuable products with improved photon-to-fuel conversion.

Nanobiohybrids have been designed employing various organic and inorganic semiconductor nanomaterials as nonbiological photon absorbers. These include pristine silicon, germanium, gold, and compounds such as CdS, FeS, TiO_2, and cobalt phosphate (He et al., 2022; Liu et al., 2021a; Okoro et al., 2023). The major concern when interfacing such heavy metal semiconductors with microorganisms is their phototoxicity to the cells. Thus, choosing a biocompatible nanomaterial that does not exert a severe antibacterial effect on the cell is crucial. Also, the amount of surface interaction with the living cells is important, knowing that spherical-shaped nanomaterials have a lesser toxicity effect.

As an alternative to inorganic metal semiconductors, organic semiconductors have proven more biocompatible and exhibit good photoelectronic conversion (Zhou et al., 2022). Understanding the interaction at the nanomaterial-bio interface is fundamental to designing an efficient hybrid biocatalyst. The microorganism or proteins/enzymes take up the photoinduced electrons from the nanomaterial through the interface by which it is linked to the semiconductor. The photogenerated electrons, upon charge separation, are freed and transported to the microorganism through its cell wall. This offers various possibilities for tailoring the surface chemistry of nanoparticles using different functional groups/ligands, allowing specific interactions with the biomaterial at the bio-nano interface.

Nanobiohybrids are generally categorized into two types: nanoparticles linked to an enzyme (nanoparticle-enzyme biohybrid) and nanoparticles linked to the whole cell, that is, bacteria or microorganisms (nanoparticle-whole cell biohybrid) (Brown & King, 2020). In a nanoparticle-enzyme biohybrid, a redox enzyme is coupled to a light-sensitive nanomaterial. Generally, these are hydrogenase enzymes, which can carry out light-driven H_2 production, and dehydrogenase enzymes, which can perform CO_2 reduction. A few notable examples of hydrogen production are by Wilker et al. (2014) using CdTe, Hutton et al. (2016) using carbon dots, and Caputo et al. (2015) using TiO. CO_2 reduction to CO and formate using CO-dehydrogenase and formate-dehydrogenase enzymes coupled to TiO_2 nanoparticles is reported by Woolerton et al. (2010) and Miller et al. (2019), respectively. The advantage of using nanoparticle-whole cell biohybrids is that such biohybrids can achieve complex chemistries by simultaneously catalyzing cascades of reactions, which are possible due to channeled pathways followed by the reactants, yielding a high degree of specificity.

A report by Kim et al. (2022) studied CO_2 reduction by coupling *Sporomusa ovata* bacterial cells with silicon nanowires. Usually, nanoparticle-whole cell biohybrids utilize the Wood-Ljungdahl pathway for CO_2 fixation. Using this pathway, CO_2 is converted to acetic acid or acetate products, wherein hydrogen acts as an electron donor and CO_2 as an electron acceptor. CO_2 is converted to CO and formic acid, formate, and together they are combined to form an acetyl product with the help of Coenzyme A. In a study by Sahoo et al. (2021), CO_2 was converted to acetic acid and methanol by EAB-ZTC@TiO_2 biohybrid catalyst using the Wood-Ljungdahl pathway Fig. 2.8. The biohybrid catalyst was synthesized by coupling electroactive bacteria (EAB) strains of *Clostridium ljungdahlii* (EAB-112), *Acetobacterium woodii* (EAB-146), and *Pseudomonas aeruginosa* with zinc benzene-1,3,5 tricarboxylate@TiO_2 (ZTC@TiO_2).

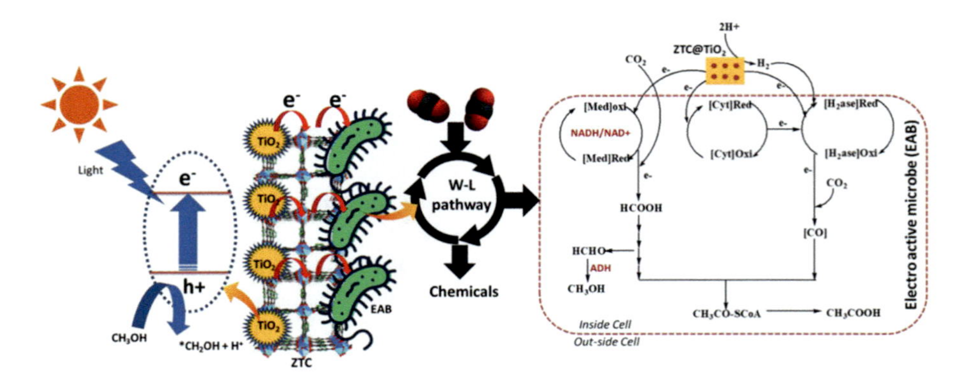

Figure 2.8 Wood–Ljungdahl pathway of CO_2 reduction. Schematic representation of nanobiohybrid catalyst (EAB- ZTC@TiO_2) demonstrating Wood–Ljungdahl pathway of CO_2 reduction.

Source: Reproduced with permission from Elsevier Copyright 2021. From Sahoo, P. C., Singh, A., Kumar, M., Gupta, R. P., Puri, S. K. & Ramakumar, S. S. V. (2021). *Molecular Catalysis*, *514*, 111845. https://doi.org/10.1016/j.mcat.2021.111845.

2.5 Conclusions and future perspectives references

In summary, the artificial photosynthesis process generally undergoes a three-stage mechanism, which involves light absorption by photo-absorber, exciton generation, and separation into electrons and holes, which can migrate to the catalyst surface, and react on the catalyst surface at the active sites. There has been a significant advancement in the design and mode of functioning of heterostructured photocatalysts for artificial photosynthesis. However, there is scope to prepare highly crystalline hybrid nanoparticles with a high aspect ratio, large-area uniformity, and precise structural control. Although hybrid nanostructures offer numerous advantages over single-component photocatalysts, there are still some challenges that need to be addressed to uplift the overall efficiency of the photosynthesis reactions. These include high overpotentials, poor absorption, and catalyst instability. Moreover, the fundamental aspects of charge kinetics at the heterojunction need to be thoroughly investigated. This is crucial to unveiling the charge transfer pathways and understanding defects and trapping sites. The design and preparation of heterostructures should take into account the active surface area, charge separation, and enhancement in the number of available active sites for redox reactions.

In the case of CO_2 reduction, the hybrid catalyst employed should be capable of enhancing the charge carrier lifetime and electron transfer to the CO_2 molecule. There is also enough scope to explore CO_2 reduction in plasmonic heterostructures. CO_2 reduction by biohybrid catalysts needs thoughtful engineering of the catalyst to incorporate specific enzymes targeting specific metabolic pathways. Further, the effect of temperature and pH on biohybrid catalyzed CO_2 reduction is yet to be meticulously studied.

Scalability is an important issue that needs to be focused on, as most of the artificial photosynthesis processes are experimentally limited to the laboratory level. Lastly, the major challenge in designing a hybrid photocatalyst that can couple proton and CO_2 reduction and water oxidation in a single system still remains unanswered.

Acknowledgments

The authors acknowledge the support of Start-up Research Grant (SRG/2022/002191) provided by Science and Engineering Research Board of Department of Science and Technology (SERB_DST), India.

References

Abe, R. (2010). Recent progress on photocatalytic and photoelectrochemical water splitting under visible light irradiation. *Journal of Photochemistry and Photobiology C: Photochemistry Reviews, 11*(4), 179–209. Available from https://doi.org/10.1016/j.jphotochemrev.2011.02.003, https://www.sciencedirect.com/science/article/pii/S1389556711000050.

Bahns, J. T., Sankaranarayanan, S. K. R. S., Gray, S. K., & Chen, L. (2011). Optically directed assembly of continuous mesoscale filaments. *Physical Review Letters*, *106*(9). Available from https://doi.org/10.1103/PhysRevLett.106.095501, http://oai.aps.org/oai?verb = GetRecord &Identifier = oai:aps.org:PhysRevLett.106.095501&metadataPrefix = oai_apsmeta_2, United States.

Barber, J., & Tran, P. D. (2013). From natural to artificial photosynthesis. *Journal of The Royal Society Interface*, *10*(81), 20120984. Available from https://doi.org/10.1098/rsif.2012.0984.

Brongersma, M. L., Halas, N. J., & Nordlander, P. (2015). Plasmon-induced hot carrier science and technology. *Nature Nanotechnology*, *10*(1), 25–34. Available from https://doi.org/10.1038/nnano.2014.311, http://www.nature.com/nnano/index.html.

Brown, K. A., & King, P. W. (2020). Coupling biology to synthetic nanomaterials for semi-artificial photosynthesis. *Photosynthesis Research*, *143*(2), 193–203. Available from https://doi.org/10.1007/s11120-019-00670-5, https://link.springer.com/journal/11120.

Caputo, C. A., Wang, L., Beranek, R., & Reisner, E. (2015). Carbon nitride-TiO_2 hybrid modified with hydrogenase for visible light driven hydrogen production. *Chemical Science*, *6*(10), 5690–5694. Available from https://doi.org/10.1039/c5sc02017d, http://pubs.rsc.org/en/Journals/JournalIssues/SC.

Chauhan, A., Rastogi, M., Scheier, P., Bowen, C., Kumar, R. V., & Vaish, R. (2018). Janus nanostructures for heterogeneous photocatalysis. *Applied Physics Reviews*, *5*(4). Available from https://doi.org/10.1063/1.5039926, http://scitation.aip.org/content/aip/journal/apr2/browse.

Cheng, C., Mao, L., Shi, J., Xue, F., Zong, S., Zheng, B., & Guo, L. (2021). $NiCo_2O_4$ nanosheets as a novel oxygen-evolution-reaction cocatalyst in situ bonded on the g-C_3N_4 photocatalyst for excellent overall water splitting. *Journal of Materials Chemistry A.*, *9*(20), 12299–12306. Available from https://doi.org/10.1039/D1TA00241D.

Cheng, L., Xiang, Q., Liao, Y., & Zhang, H. (2018). CdS-Based photocatalysts. *Energy & Environmental Science*, *11*(6), 1362–1391. Available from https://doi.org/10.1039/C7EE03640J.

Chen, Q., Chen, X., Fang, M., Chen, J., Li, Y., Xie, Z., Kuang, Q., & Zheng, L. (2019). Photo-induced Au-Pd alloying at TiO_2 {101} facets enables robust CO_2 photocatalytic reduction into hydrocarbon fuels. *Journal of Materials Chemistry A*, *7*(3), 1334–1340. Available from https://doi.org/10.1039/c8ta09412h, http://pubs.rsc.org/en/journals/journal/ta.

Coronado, E. A., Encina, E. R., & Stefani, F. D. (2011). Optical properties of metallic nanoparticles: Manipulating light, heat and forces at the nanoscale. *Nanoscale*, *3*(10), 4042–4059. Available from https://doi.org/10.1039/c1nr10788g.

Corson, E. R., Subramani, A., Cooper, J. K., Kostecki, R., Urban, J. J., & McCloskey, B. D. (2020). Reduction of carbon dioxide at a plasmonically active copper-silver cathode. *Chemical Communications*, *56*(69), 9970–9973. Available from https://doi.org/10.1039/d0cc03215h, http://pubs.rsc.org/en/journals/journal/cc.

Cui, X., Yang, X., Xian, X., Tian, L., Tang, H., & Liu, Q. (2018). Insights into highly improved solar-driven photocatalytic oxygen evolution over integrated Ag_3PO_4/MoS_2 heterostructures. *Frontiers in Chemistry*, *6*. Available from https://doi.org/10.3389/fchem.2018.00123, https://www.frontiersin.org/articles/10.3389/fchem.2018.00123/full.

Deng, D., Novoselov, K. S., Fu, Q., Zheng, N., Tian, Z., & Bao, X. (2016). Catalysis with two-dimensional materials and their heterostructures. *Nature Nanotechnology*, *11*(3), 218–230. Available from https://doi.org/10.1038/nnano.2015.340, http://www.nature.com/nnano/index.html.

Ding, Y., Bertram, J. R., Eckert, C., Bommareddy, R. R., Patel, R., Conradie, A., Bryan, S., & Nagpal, P. (2019). Nanorg microbial factories: Light-driven renewable biochemical synthesis using quantum dot-bacteria nanobiohybrids. *Journal of the American Chemical Society*, *141*(26), 10272–10282. Available from https://doi.org/10.1021/jacs.9b02549, http://pubs.acs.org/journal/jacsat.

Duchene, J. S., Tagliabue, G., Welch, A. J., Li, X., Cheng, W. H., & Atwater, H. A. (2020). Optical excitation of a nanoparticle Cu/p-NiO photocathode improves reaction selectivity for CO_2 reduction in aqueous electrolytes. *Nano Letters*, *20*(4), 2348–2358. Available from https://doi.org/10.1021/acs.nanolett.9b04895, http://pubs.acs.org/journal/nalefd.

Gesesse, G. D., Wang, C., Chang, B. K., Tai, S. H., Beaunier, P., Wojcieszak, R., Remita, H., Colbeau-Justin, C., & Ghazzal, M. N. (2020). A soft-chemistry assisted strong metal-support interaction on a designed plasmonic core-shell photocatalyst for enhanced photocatalytic hydrogen production. *Nanoscale*, *12*(13), 7011–7023. Available from https://doi.org/10.1039/c9nr09891g, http://pubs.rsc.org/en/journals/journal/nr.

Giannini, V., Fernández-Domínguez, A. I., Heck, S. C., & Maier, S. A. (2011). Plasmonic nanoantennas: Fundamentals and their use in controlling the radiative properties of nanoemitters. *Chemical Reviews*, *111*(6), 3888–3912. Available from https://doi.org/10.1021/cr1002672.

Guo, X., Li, X., Kou, S., Yang, X., Hu, X., Ling, D., & Yang, J. (2018). Plasmon-enhanced electrocatalytic hydrogen/oxygen evolution by Pt/Fe-Au nanorods. *Journal of Materials Chemistry A*, *6*(17), 7364–7369. Available from https://doi.org/10.1039/c8ta00499d, http://pubs.rsc.org/en/journals/journal/ta.

Gupta, A., Likozar, B., Jana, R., Chanu, W. C., & Singh, M. K. (2022). A review of hydrogen production processes by photocatalytic water splitting – From atomistic catalysis design to optimal reactor engineering. *International Journal of Hydrogen Energy*, *47* (78), 33282–33307. Available from https://doi.org/10.1016/j.ijhydene.2022.07.210, https://www.sciencedirect.com/science/article/pii/S036031992203261X.

Hameed, A., Gondal, M. A., Yamani, Z. H., & Yahya, A. H. (2005). Significance of pH measurements in photocatalytic splitting of water using 355nm UV laser. *Journal of Molecular Catalysis A: Chemical*, *227*(1), 241–246. Available from https://doi.org/10.1016/j.molcata.2004.10.053, https://www.sciencedirect.com/science/article/pii/S1381116904008325.

Hammer, B., Morikawa, Y., & Nørskov, J. K. (1996). CO chemisorption at metal surfaces and overlayers. *Physical Review Letters*, *76*(12), 2141–2144. Available from https://doi.org/10.1103/PhysRevLett.76.2141.

Henley, S. J., Watts, P. C. P., Mureau, N., & Silva, S. R. P. (2008). Laser-induced decoration of carbon nanotubes with metal nanoparticles. *Applied Physics A: Materials Science and Processing*, *93*(4), 875–879. Available from https://doi.org/10.1007/s00339-008-4800-z.

He, Y., Wang, S., Han, X., Shen, J., Lu, Y., Zhao, J., Shen, C., & Qiao, L. (2022). Photosynthesis of acetate by *Sporomusa ovata*-CdS biohybrid system. *ACS Applied Materials and Interfaces*, *14*(20), 23364–23374. Available from https://doi.org/10.1021/acsami.2c01918, http://pubs.acs.org/journal/aamick.

He, C., Zhang, J. H., Zhang, W. X., & Li, T. T. (2019). Type-II InSe/ g-C_3N_4 heterostructure as a high-efficiency oxygen evolution reaction catalyst for photoelectrochemical water splitting. *Journal of Physical Chemistry Letters*, *10*(11), 3122–3128. Available from https://doi.org/10.1021/acs.jpclett.9b00909, http://pubs.acs.org/journal/jpclcd.

Hu, G., Hu, C. X., Zhu, Z. Y., Zhang, L., Wang, Q., & Zhang, H. L. (2018). Construction of Au/CuO/Co_3O_4 Tricomponent Heterojunction Nanotubes for Enhanced Photocatalytic Oxygen Evolution under Visible Light Irradiation. *ACS Sustainable Chemistry & Engineering* *6*(7), 8801–8808. Available from https://doi.org/10.1021/acssuschemeng.8b01153.

Hung, S. F., Yu, Y. C., Suen, N. T., Tzeng, G. Q., Tung, C. W., Hsu, Y. Y., Hsu, C. S., Chang, C. K., Chan, T. S., Sheu, H. S., Lee, J. F., & Chen, H. M. (2016). The synergistic effect of a well-defined Au@Pt core-shell nanostructure toward photocatalytic hydrogen generation: Interface engineering to improve the Schottky barrier and hydrogen-evolved kinetics. *Chemical Communications*, *52*(8), 1567–1570. Available from https://doi.org/10.1039/c5cc08547k, http://pubs.rsc.org/en/journals/journal/cc.

Hutton, G. A. M., Reuillard, B., Martindale, B. C. M., Caputo, C. A., Lockwood, C. W. J., Butt, J. N., & Reisner, E. (2016). Carbon dots as versatile photosensitizers for solar-driven catalysis with redox enzymes. *Journal of the American Chemical Society*, *138*(51), 16722–16730. Available from https://doi.org/10.1021/jacs.6b10146, http://pubs.acs.org/journal/jacsat.

Jiao, J., Wei, Y., Zhao, Y., Zhao, Z., Duan, A., Liu, J., Pang, Y., Li, J., Jiang, G., & Wang, Y. (2017). AuPd/3DOM-TiO_2 catalysts for photocatalytic reduction of CO_2: High efficient separation of photogenerated charge carriers. *Applied Catalysis B: Environmental*, *209*, 228–239. Available from https://doi.org/10.1016/j.apcatb.2017.02.076, http://www.elsevier.com/inca/publications/store/5/2/3/0/6/6/index.htt.

Karamian, E., & Sharifnia, S. (2016). On the general mechanism of photocatalytic reduction of CO_2. *Journal of CO_2 Utilization*, *16*, 194–203. Available from https://doi.org/10.1016/j.jcou.2016.07.004, https://www.sciencedirect.com/science/article/pii/S2212982016301615.

Kim, J., Cestellos-Blanco, S., Shen, Y. X., Cai, R., & Yang, P. (2022). Enhancing biohybrid CO_2 to multicarbon reduction via adapted whole-cell catalysts. *Nano Letters*, *22*(13), 5503–5509. Available from https://doi.org/10.1021/acs.nanolett.2c01576, http://pubs.acs.org/journal/nalefd.

Kim, J. H., Hansora, D., Sharma, P., Jang, J.-W., & Lee, J. S. (2019). Toward practical solar hydrogen production – An artificial photosynthetic leaf-to-farm challenge. *Chemical Society Reviews*, *48*(7), 1908–1971. Available from https://doi.org/10.1039/C8CS00699G.

Kuai, L., Zhou, Y., Tu, W., Li, P., Li, H., Xu, Q., Tang, L., Wang, X., Xiao, M., & Zou, Z. (2015). Rational construction of a CdS/reduced graphene oxide/TiO_2 core-shell nanostructure as an all-solid-state Z-scheme system for CO_2 photoreduction into solar fuels. *RSC Advances*, *5*(107), 88409–88413. Available from https://doi.org/10.1039/c5ra14374h, http://pubs.rsc.org/en/journals/journalissues.

Lai, H., Xiao, W., Wang, Y., Song, T., Long, B., Yin, S., Ali, A., & Deng, G. J. (2021). Plasmon-induced carrier separation boosts high-selective photocatalytic CO_2 reduction on dagger-axe-like Cu@Co core–shell bimetal. *Chemical Engineering Journal*, *417*. Available from https://doi.org/10.1016/j.cej.2021.129295, http://www.elsevier.com/inca/publications/store/6/0/1/2/7/3/index.htt.

Lingampalli, S. R., Ayyub, M. M., & Rao, C. N. R. (2017). Recent progress in the photocatalytic reduction of carbon dioxide. *ACS Omega*, *2*(6), 2740–2748. Available from https://doi.org/10.1021/acsomega.7b00721.

Liu, G., Gao, F., Gao, C., & Xiong, Y. (2021a). Bioinspiration toward efficient photosynthetic systems: From biohybrids to biomimetics. *Chemistry Catalysis*, *1*(7), 1367–1377. Available from https://doi.org/10.1016/j.checat.2021.09.010, https://www.sciencedirect.com/journal/chem-catalysis.

Liu, S., Han, C., Tang, Z.-R., & Xu, Y.-J. (2016). Heterostructured semiconductor nanowire arrays for artificial photosynthesis. *Material Horizon*, *3*(4), 270–282. Available from https://doi.org/10.1039/C6MH00063K.

Liu, Y., Li, X., He, H., Yang, S., Jia, G., & Liu, S. (2021b). CoP imbedded g-C_3N_4 heterojunctions for highly efficient photo, electro and photoelectrochemical water splitting. *Journal of*

Colloid and Interface Science, *599*, 23–33. Available from https://doi.org/10.1016/j.jcis.2021.04.088, http://www.elsevier.com/inca/publications/store/6/2/2/8/6/1/index.htt.

Li, Y. H., Cheng, L., Liu, P. F., Zhang, L., Zu, M. Y., Wang, C. W., Jin, Y. H., Cao, X. M., Yang, H. G., & Li, C. (2018). Simple cadmium sulfide compound with stable 95% selectivity for carbon dioxide electroreduction in aqueous medium. *ChemSusChem*, *11*(9), 1421–1425. Available from https://doi.org/10.1002/cssc.201800372, http://www.interscience.wiley.com/jpages/1864-5631.

Li, D., Wang, H., Tang, H., Yang, X., & Liu, Q. (2019). Remarkable enhancement in solar oxygen evolution from $MoSe_2/Ag_3PO_4$ heterojunction photocatalyst via in situ constructing interfacial contact. *ACS Sustainable Chemistry and Engineering*, *7*(9), 8466–8474. Available from https://doi.org/10.1021/acssuschemeng.9b00252, http://pubs.acs.org/journal/ascecg.

Lou, Z., Fujitsuka, M., & Majima, T. (2016). Pt-Au triangular nanoprisms with strong dipole plasmon resonance for hydrogen generation studied by single-particle spectroscopy. *ACS Nano*, *10*(6), 6299–6305. Available from https://doi.org/10.1021/acsnano.6b02494, http://pubs.acs.org/journal/ancac3.

Madhusudan, P., Shi, R., Xiang, S., Jin, M., Chandrashekar, B. N., Wang, J., Wang, W., Peng, O., Amini, A., & Cheng, C. (2021). Construction of highly efficient Z-scheme ZnxCd1-xS/Au@g-C_3N_4 ternary heterojunction composite for visible-light-driven photocatalytic reduction of CO_2 to solar fuel. *Applied Catalysis B: Environmental*, *282*. Available from https://doi.org/10.1016/j.apcatb.2020.119600, http://www.elsevier.com/inca/publications/store/5/2/3/0/6/6/index.htt.

Maeda, K., & Domen, K. (2010). Photocatalytic water splitting: Recent progress and future challenges. *Journal of Physics & Chemistry Letters*, *1*(18), 2655–2661. Available from https://doi.org/10.1021/jz1007966.

Maier (2004). *Fundamentals and applications plasmonics: Fundamentals and applications* (p. 677).

Menagen, G., Macdonald, J. E., Shemesh, Y., Popov, I., & Banin, U. (2009). Au growth on semiconductor nanorods: Photoinduced versus thermal growth mechanisms. *Journal of the American Chemical Society*, *131*(47), 17406–17411. Available from https://doi.org/10.1021/ja9077733Israel, http://pubs.acs.org/doi/pdfplus/10.1021/ja9077733.

Meng, X., Fujita, K., Moriguchi, Y., Zong, Y., & Tanaka, K. (2013). Metal-dielectric core-shell nanoparticles: Advanced plasmonic architectures towards multiple control of random Lasers. *Advanced Optical Materials*, *1*(8), 573–580. Available from https://doi.org/10.1002/adom.201300153.

Miller, M., Robinson, W. E., Oliveira, A. R., Heidary, N., Kornienko, N., Warnan, J., Pereira, I. A. C., & Reisner, E. (2019). Interfacing formate dehydrogenase with metal oxides for the reversible electrocatalysis and solar-driven reduction of carbon dioxide. *Angewandte Chemie - International Edition*, *58*(14), 4601–4605. Available from https://doi.org/10.1002/anie.201814419, http://onlinelibrary.wiley.com/journal/10.1002/(ISSN)1521-3773.

Miyatake, T., & Tamiaki, H. (2010). Self-aggregates of natural chlorophylls and their synthetic analogues in aqueous media for making light-harvesting systems. *18th international symposium on the photochemistry and photophysics of coordination compounds Sapporo, 2009*, *254*(21), 2593–2602. Available from https://doi.org/10.1016/j.ccr.2009.12.027, https://www.sciencedirect.com/science/article/pii/S0010854509003555.

Mongin, D., Juvé, V., Maioli, P., Crut, A., Del Fatti, N., Vallée, F., Sánchez-Iglesias, A., Pastoriza-Santos, I., & Liz-Marzán, L. M. (2011). Acoustic vibrations of metal-dielectric core-shell nanoparticles. *Nano Letters*, *11*(7), 3016–3021. Available from https://doi.org/10.1021/nl201672k.

Ngaw, C. K., Xu, Q., Tan, T. T. Y., Hu, P., Cao, S., & Loo, J. S. C. (2014). A strategy for in-situ synthesis of well-defined core-shell AuatTiO_2 hollow spheres for enhanced photocatalytic hydrogen evolution. *Chemical Engineering Journal, 257*, 112–121. Available from https://doi.org/10.1016/j.cej.2014.07.059, http://www.elsevier.com/inca/publications/store/6/0/1/2/7/3/index.htt.

Nishioka, S., Osterloh, F. E., Wang, X., Mallouk, T. E., & Maeda, K. (2023). Photocatalytic water splitting. *Nature Reviews Methods Primers, 3*(1), 42. Available from https://doi.org/10.1038/s43586-023-00226-x.

Nishiyama, H., Yamada, T., Nakabayashi, M., Maehara, Y., Yamaguchi, M., Kuromiya, Y., Nagatsuma, Y., Tokudome, H., Akiyama, S., Watanabe, T., Narushima, R., Okunaka, S., Shibata, N., Takata, T., Hisatomi, T., & Domen, K. (2021). Photocatalytic solar hydrogen production from water on a 100-m2 scale. *Nature, 598*(7880), 304–307. Available from https://doi.org/10.1038/s41586-021-03907-3.

Ochedi, F. O., Liu, D., Yu, J., Hussain, A., & Liu, Y. (2021). Photocatalytic, electrocatalytic and photoelectrocatalytic conversion of carbon dioxide: A review. *Environmental Chemistry Letters, 19*(2), 941–967. Available from https://doi.org/10.1007/s10311-020-01131-5.

Okoro, G., Husain, S., Saukani, M., Mutalik, C., Yougbaré, S., Hsiao, Y. C., & Kuo, T. R. (2023). Emerging trends in nanomaterials for photosynthetic biohybrid systems. *ACS Materials Letters, 5*(1), 95–115. Available from https://doi.org/10.1021/acsmaterialslett.2c00752, https://pubs.acs.org/page/amlcef/about.html.

Pang, L., Barras, A., Mishyn, V., Sandu, G., Melinte, S., Subramanian, P., Boukherroub, R., & Szunerits, S. (2020). Enhanced electrocatalytic hydrogen evolution on a plasmonic electrode: The importance of the Ti/TiO_2 adhesion layer. *Journal of Materials Chemistry A, 8*(28), 13980–13986. Available from https://doi.org/10.1039/d0ta02353a, http://pubs.rsc.org/en/journals/journal/ta.

Pang, H., Masuda, T., & Ye, J. (2018). Semiconductor-based photoelectrochemical conversion of carbon dioxide: Stepping towards artificial photosynthesis. *Chemistry – An Asian Journal, 13*(2), 127–142. Available from https://doi.org/10.1002/asia.201701596.

Pan, Z., Zheng, Y., Guo, F., Niu, P., & Wang, X. (2017). Decorating CoP and Pt nanoparticles on graphitic carbon nitride nanosheets to promote overall water splitting by conjugated polymers. *ChemSusChem, 10*(1), 87–90. Available from https://doi.org/10.1002/cssc.201600850, http://onlinelibrary.wiley.com/journal/10.1002/(ISSN)1864-564X.

Park, H., Ou, H. H., Colussi, A. J., & Hoffmann, M. R. (2015). Artificial photosynthesis of C1-C3 hydrocarbons from water and CO_2 on titanate nanotubes decorated with nanoparticle elemental copper and CdS quantum dots. *Journal of Physical Chemistry A, 119*(19), 4658–4666. Available from https://doi.org/10.1021/jp511329d, http://pubs.acs.org/jpca.

Povolotskaya, A. V., Povolotskiy, A. V., & Manshina, A. A. (2015). Hybrid nanostructures: Synthesis, morphology and functional properties. *Russian Chemical Reviews, 84*(6), 579. Available from https://doi.org/10.1070/RCR4487.

Prabhu, P., Jose, V., & Lee, J. (2020). Heterostructured catalysts for electrocatalytic and photocatalytic carbon dioxide reduction. *Advanced Functional Materials, 30*(24), 1910768. Available from https://doi.org/10.1002/adfm.201910768.

Pudukudy, M., Shan, S., Miao, Y., Gu, B., & Jia, Q. (2020). WO_3 nanocrystals decorated Ag_3PO_4 tetrapods as an efficient visible-light responsive Z-scheme photocatalyst for the enhanced degradation of tetracycline in aqueous medium. *Colloids and Surfaces A: Physicochemical and Engineering Aspects, 589*. Available from https://doi.org/10.1016/j.colsurfa.2020.124457, http://www.elsevier.com/locate/colsurfa.

Ruan, X., Li, S., Huang, C., Zheng, W., Cui, X., & Ravi, S. K. (2023). Catalyzing artificial photosynthesis with TiO_2 heterostructures and hybrids – Emerging trends in a classical yet contemporary photocatalyst. *Advanced Materials*, *n/a*(n/a) 2305285. Available from https://doi.org/10.1002/adma.202305285.

Sahoo, P. C., Singh, A., Kumar, M., Gupta, R. P., Puri, S. K., & Ramakumar, S. S. V. (2021). Light augmented CO_2 conversion by metal organic framework sensitized electroactive microbes. *Molecular Catalysis*, *514*. Available from https://doi.org/10.1016/j.mcat.2021.111845, http://www.elsevier.com/locate/issn/24688231.

Schneider, J., & Bahnemann, D. W. (2013). Undesired role of sacrificial reagents in photocatalysis. *Journal of Physics & Chemistry Letters*, *4*(20), 3479–3483. Available from https://doi.org/10.1021/jz4018199.

Seh, Z. W., Liu, S., Low, M., Zhang, S. Y., Liu, Z., Mlayah, A., & Han, M. Y. (2012). Janus Au-TiO_2 photocatalysts with strong localization of plasmonic near-fields for efficient visible-light hydrogen generation. *Advanced Materials*, *24*(17), 2310–2314. Available from https://doi.org/10.1002/adma.201104241.

Shaviv, E., Schubert, O., Alves-Santos, M., Goldoni, G., Di Felice, R., Vallée, F., Del Fatti, N., Banin, U., & Sönnichsen, C. (2011). Absorption properties of metal-semiconductor hybrid nanoparticles. *ACS Nano*, *5*(6), 4712–4719. Available from https://doi.org/10.1021/nn200645h, http://pubs.acs.org/journal/ancac3.

Singh, V., Beltran, I. J. C., Ribot, J. C., & Nagpal, P. (2014). Photocatalysis deconstructed: Design of a new selective catalyst for artificial photosynthesis. *Nano Letters*, *14*(2), 597–603. Available from https://doi.org/10.1021/nl403783d.

Smith, P. T., Nichols, E. M., Cao, Z., & Chang, C. J. (2020). Hybrid catalysts for artificial photosynthesis: Merging approaches from molecular, materials, and biological catalysis. *Accounts of Chemical Research*, *53*(3), 575–587. Available from https://doi.org/10.1021/acs.accounts.9b00619.

Sotiriou, G. A., Hirt, A. M., Lozach, P. Y., Teleki, A., Krumeich, F., & Pratsinis, S. E. (2011). Hybrid, silica-coated, Janus-like plasmonic-magnetic nanoparticles. *Chemistry of Materials*, *23*(7), 1985–1992. Available from https://doi.org/10.1021/cm200399t.

Stolarczyk, J. K., Bhattacharyya, S., Polavarapu, L., & Feldmann, J. (2018). Challenges and prospects in solar water splitting and CO_2 reduction with inorganic and hybrid nanostructures. *ACS Catal*, *8*(4), 3602–3635. Available from https://doi.org/10.1021/acscatal.8b00791.

Styring, S. (2012). Artificial photosynthesis for solar fuels. *Faraday Discussions*, *155*, 357–376. Available from https://doi.org/10.1039/C1FD00113B.

Sun, S., An, Q., Watanabe, M., Cheng, J., Ho Kim, H., Akbay, T., Takagaki, A., & Ishihara, T. (2020). Highly correlation of CO_2 reduction selectivity and surface electron accumulation: A case study of Au-MoS_2 and Ag-MoS_2 catalyst. *Applied Catalysis B: Environmental*, *271*. Available from https://doi.org/10.1016/j.apcatb.2020.118931, http://www.elsevier.com/inca/publications/store/5/2/3/0/6/6/index.htt.

Tarek, M., Rezaul Karim, K. M., Sarmin, S., Ong, H. R., Abdullah, H., Cheng, C. K., & Rahman Khan, M. M. (2020). Photoelectrochemical activity of CuO-CdS heterostructured catalyst for CO_2 reduction. *IOP conference series: Materials science and engineering*, *736*(4), 1757899X. Available from https://doi.org/10.1088/1757-899X/736/4/042023, Institute of Physics Publishing Malaysia, https://iopscience.iop.org/journal/1757-899X.

Tian, L., Xian, X., Cui, X., Tang, H., & Yang, X. (2018). Fabrication of modified g-C_3N_4 nanorod/Ag_3PO_4 nanocomposites for solar-driven photocatalytic oxygen evolution from water splitting. *Applied Surface Science*, *430*, 301–308. Available from https://doi.org/10.1016/j.apsusc.2017.07.185, http://www.journals.elsevier.com/applied-surface-science/.

Vilé, G., Albani, D., Nachtegaal, M., Chen, Z., Dontsova, D., Antonietti, M., López, N., & Pérez-Ramírez, J. (2015). A stable single-site palladium catalyst for hydrogenations. *Angewandte Chemie - International Edition*, *54*(38), 11265–11269. Available from https://doi.org/10.1002/anie.201505073, http://onlinelibrary.wiley.com/journal/10.1002/(ISSN)1521-3773.

Wang, Y., Chen, E., & Tang, J. (2022). Insight on reaction pathways of photocatalytic CO_2 conversion. *ACS Catalysts*, *12*(12), 7300–7316. Available from https://doi.org/10.1021/acscatal.2c01012.

Wang, F., Jiang, Y., Lawes, D. J., Ball, G. E., Zhou, C., Liu, Z., & Amal, R. (2015). Analysis of the promoted activity and molecular mechanism of hydrogen production over fine Au-Pt Alloyed TiO_2 photocatalysts. *ACS Catalysis*, *5*(7), 3924–3931. Available from https://doi.org/10.1021/acscatal.5b00623, http://pubs.acs.org/page/accacs/about.html.

Wang, H., Rong, H., Wang, D., Li, X., Zhang, E., Wan, X., Bai, B., Xu, M., Liu, J., Liu, J., Chen, W., & Zhang, J. (2020). Highly selective photoreduction of CO_2 with suppressing H2 evolution by plasmonic Au/CdSe–Cu_2O hierarchical nanostructures under visible light. *Small (Weinheim an der Bergstrasse, Germany)*, *16*(18). Available from https://doi.org/10.1002/smll.202000426, http://onlinelibrary.wiley.com/journal/10.1002/(ISSN)1613-6829.

Wang, Z., Xu, X., Si, Z., Liu, L., Liu, Y., He, Y., Ran, R., & Weng, D. (2018). In situ synthesized MoS_2/Ag dots/Ag_3 PO_4 Z-scheme photocatalysts with ultrahigh activity for oxygen evolution under visible light irradiation. *Applied Surface Science*, *450*, 441–450. Available from https://doi.org/10.1016/j.apsusc.2018.04.149, http://www.journals.elsevier.com/applied-surface-science/.

Wei, Y., Jiao, J., Zhao, Z., Liu, J., Li, J., Jiang, G., Wang, Y., & Duan, A. (2015). Fabrication of inverse opal TiO_2-supported Au@CdS core-shell nanoparticles for efficient photocatalytic CO_2 conversion. *Applied Catalysis B: Environmental*, *179*, 422–432. Available from https://doi.org/10.1016/j.apcatb.2015.05.041, http://www.elsevier.com/inca/publications/store/5/2/3/0/6/6/index.htt.

Wilker, M. B., Shinopoulos, K. E., Brown, K. A., Mulder, D. W., King, P. W., & Dukovic, G. (2014). Electron transfer kinetics in CdS nanorod-[FeFe]-hydrogenase complexes and implications for photochemical H_2 generation. *Journal of the American Chemical Society*, *136*(11), 4316–4324. Available from https://doi.org/10.1021/ja413001p, http://pubs.acs.org/journal/jacsat.

Woolerton, T. W., Sheard, S., Reisner, E., Pierce, E., Ragsdale, S. W., & Armstrong, F. A. (2010). Efficient and clean photoreduction of CO_2 to CO by enzyme-modified TiO_2 nanoparticles using visible light. *Journal of the American Chemical Society*, *132*(7), 2132–2133. Available from https://doi.org/10.1021/ja910091z.

Wu, H., Tan, H. L., Toe, C. Y., Scott, J., Wang, L., Amal, R., & Ng, Y. H. (2020). Photocatalytic and photoelectrochemical systems: Similarities and differences. *Advanced Materials*, *32*(18), 1904717. Available from https://doi.org/10.1002/adma.201904717.

Xu, J., Wang, Z., Yu, W., Sun, D., Zhang, Q., Tung, C. H., & Wang, W. (2016). Kagóme cobalt(II)-organic layers as robust Scaffolds for highly efficient photocatalytic oxygen evolution. *ChemSusChem*, *9*(10), 1146–1152. Available from https://doi.org/10.1002/cssc.201600101, http://www.interscience.wiley.com/jpages/1864-5631.

Yang, X., Tang, H., Xu, J., Antonietti, M., & Shalom, M. (2015). Silver phosphate/graphitic carbon nitride as an efficient photocatalytic tandem system for oxygen evolution. *ChemSusChem*, *8*(8), 1350–1358. Available from https://doi.org/10.1002/cssc.201403168, http://www.interscience.wiley.com/jpages/1864-5631.

Yang, X., Tian, L., Zhao, X., Tang, H., Liu, Q., & Li, G. (2019). Interfacial optimization of g-C_3N_4-based Z-scheme heterojunction toward synergistic enhancement of solar-driven photocatalytic oxygen evolution. *Applied Catalysis B: Environmental*, *244*, 240–249. Available from https://doi.org/10.1016/j.apcatb.2018.11.056, http://www.elsevier.com/inca/publications/store/5/2/3/0/6/6/index.htt.

Yang, J., Wang, D., Han, H., & Li, C. (2013). Roles of cocatalysts in photocatalysis and photoelectrocatalysis. *Accounts of Chemical Research*, *46*(8), 1900–1909. Available from https://doi.org/10.1021/ar300227e.

Yang, J., Zhang, F., Chen, Y., Qian, S., Hu, P., Li, W., Deng, Y., Fang, Y., Han, L., Luqman, M., & Zhao, D. (2011). Core-shell Ag@SiO_2@mSiO_2 mesoporous nanocarriers for metal-enhanced fluorescence. *Chemical Communications*, *47*(42), 11618–11620. Available from https://doi.org/10.1039/c1cc15304h, http://pubs.rsc.org/en/journals/journal/cc.

Yang, L., Zhou, H., Fan, T., & Zhang, D. (2014). Semiconductor photocatalysts for water oxidation: Current status and challenges. *Physical Chemistry Chemical Physics: PCCP*, *16*(15), 6810–6826. Available from https://doi.org/10.1039/C4CP00246F.

Yuan, L., Geng, Z., Xu, J., Guo, F., & Han, C. (2021). Metal-semiconductor heterostructures for photoredox catalysis: Where are we now and where do we go? *Advanced Functional Materials*, *31*(27), 2101103. Available from https://doi.org/10.1002/adfm.202101103.

Yu, S., Wilson, A. J., Heo, J., & Jain, P. K. (2018). Plasmonic control of multi-electron transfer and C-C coupling in visible-light-driven CO_2 reduction on Au nanoparticles. *Nano Letters*, *18*(4), 2189–2194. Available from https://doi.org/10.1021/acs.nanolett.7b05410, http://pubs.acs.org/journal/nalefd.

Zeng, Y., Li, H., Xia, Y., Wang, L., Yin, K., Wei, Y., Liu, X., & Luo, S. (2020). Co_3O_4 nanocrystals with an oxygen vacancy-rich and highly reactive (222) facet on carbon nitride scaffolds for efficient photocatalytic oxygen evolution. *ACS Applied Materials and Interfaces*, *12*(40), 44608–44616. Available from https://doi.org/10.1021/acsami.0c09761, http://pubs.acs.org/journal/aamick.

Zhang, L., Blom, D. A., & Wang, H. (2011). Au-cu_2o core-shell nanoparticles: A hybrid metal-semiconductor heteronanostructure with geometrically tunable optical properties. *Chemistry of Materials*, *23*(20), 4587–4598. Available from https://doi.org/10.1021/cm202078t.

Zhang, Y., He, S., Guo, W., Hu, Y., Huang, J., Mulcahy, J. R., & Wei, W. D. (2018). Surface-plasmon-driven hot electron photochemistry. *Chemical Reviews*, *118*(6), 2927–2954. Available from https://doi.org/10.1021/acs.chemrev.7b00430, http://pubs.acs.org/journal/chreay.

Zhang, W., Mohamed, A. R., & Ong, W.-J. (2020). Z-Scheme photocatalytic systems for carbon dioxide reduction: Where are we now? *Angewandte Chemie International Edition*, *59*(51), 22894–22915. Available from https://doi.org/10.1002/anie.201914925.

Zhao, W., Liu, J., Deng, Z., Zhang, J., Ding, Z., & Fang, Y. (2018). Facile preparation of Z-scheme CdS–Ag–TiO_2 composite for the improved photocatalytic hydrogen generation activity. *International Journal of Hydrogen Energy*, *43*(39), 18232–18241. Available from https://doi.org/10.1016/j.ijhydene.2018.08.026, http://www.journals.elsevier.com/international-journal-of-hydrogen-energy/.

Zha, Q., Yuan, F., Qin, G., & Ni, Y. (2020). Cobalt-based MOF-on-MOF two-dimensional heterojunction nanostructures for enhanced oxygen evolution reaction electrocatalytic activity. *Inorganic Chemistry*, *59*(2), 1295–1305. Available from https://doi.org/10.1021/acs.inorgchem.9b03011, http://pubs.acs.org/journal/inocaj.

Zhou, H., Yan, R., Zhang, D., & Fan, T. (2016). Challenges and perspectives in designing artificial photosynthetic systems. *Chemistry – A European Journal*, *22*(29), 9870–9885. Available from https://doi.org/10.1002/chem.201600289.

Zhou, P., Yu, J., & Jaroniec, M. (2014). All-solid-state Z-scheme photocatalytic systems. *Advanced Materials*, *26*(29), 4920–4935. Available from https://doi.org/10.1002/adma.201400288.

Zhou, X., Zeng, Y., Lv, F., Bai, H., & Wang, S. (2022). Organic semiconductor-organism interfaces for augmenting natural and artificial photosynthesis. *Accounts of Chemical Research*, *55*(2), 156–170. Available from https://doi.org/10.1021/acs.accounts.1c00580, http://pubs.acs.org/journal/achre4.

Zhu, Z., Han, Y., Chen, C., Ding, Z., Long, J., & Hou, Y. (2018). Reduced graphene oxide-cadmium sulfide nanorods decorated with silver nanoparticles for efficient photocatalytic reduction carbon dioxide under visible light. *ChemCatChem*, *10*(7), 1627–1634. Available from https://doi.org/10.1002/cctc.201701573, http://onlinelibrary.wiley.com/journal/10.1002/(ISSN)1867-3899.

Piezoelectric-semiconductor hybrids as next generation nanostructures for water remediation

Sakthivel Thangavel[1], *Abiyazhini Rajendran*[2], *Nivea Raghavan*[3] *and Ji Woong Chang*[1]

[1]Department of Chemical Engineering, Kumoh National Institute of Technology, Gumi-si, Gyeongsangbuk-do, Republic of Korea, [2]Department of Physics, Bharathiar University, Coimbatore, Tamil Nadu, India, [3]Electronics and Communication Department, Saveetha Institute of Medical Sciences and Technology, Chennai, Tamil Nadu, India

3.1 Introduction

Water, the essence of life, finds itself besieged. In the unending struggle against contamination, our aquatic realms confront an unrelenting deluge of pollutants, casting dire threats upon both ecosystems and human well-being. Water pollution embodies the clandestine tainting of water bodies, encompassing rivers, lakes, oceans, and subterranean reservoirs, imperiling the fundamental life-sustaining resource on our planet. This ever-escalating predicament emanates from a myriad of. Industrial discharges, for instance, release noxious chemicals and heavy metals into aquatic realms (Kordbacheh & Heidari, 2023). Similarly, agricultural pursuits contribute to the contamination through runoff laden with pesticides and nutrients. Meanwhile, the inadequate treatment of sewage and wastewater compounds the dilemma. Furthermore, catastrophic oil spills stand as stark reminders of the far-reaching consequences of environmental negligence, while even urban locales remain vulnerable, with urban runoff introducing contaminants into our water-courses (Kataki et al., 2021). The underlying factors propelling water pollution are rooted in human activities, predominantly stemming from industrial, agricultural, and domestic practices. The issue is further exacerbated by a deficiency in crucial infrastructure, including sewage treatment facilities and waste management. Furthermore, natural phenomena such as volcanic eruptions and algal blooms can unexpectedly contribute to water pollution, adding an additional layer of complexity.

The adverse repercussions stemming from water pollution are far-reaching, permeating multiple aspects of both society and the environment. Contaminated water sources stand as breeding grounds for waterborne diseases, engendering

Nanocomposites for Environmental, Energy, and Agricultural Applications. DOI: https://doi.org/10.1016/B978-0-443-13935-2.00003-6

illnesses and fatalities, thereby posing substantial health risks. Ecological systems bear the brunt of pollution, enduring the disruption of food chains and the jeopardy of aquatic species, resulting in notable biodiversity losses (Stadler et al., 2018). Additionally, the economic fallout is considerable, impacting industries reliant on pristine water sources, which must bear the burdensome costs of water treatment and pollution mitigation (Xiao et al., 2021). Water has been polluted largely due to the repaid development of various sectors, including industrialization, pharmaceuticals, and agriculture. Some of the contaminants that are discharged into the water bodies rapidly include drugs, pesticides, dyes, and organic compounds. Before using this water in households, industry, and agriculture, it should be purified. Different technologies, such as precipitation, filtration, and biological methods, have been applied to treat the wastewater (Boinpally et al., 2023). However, most of the methods suffer from the economy; some of them are not efficient and produce secondary products that are more harmful than primary ones. Photocatalysts are environmentally friendly and are driven by semiconductors and natural sunlight. Hence, it is used in different application sectors such as water splitting, chemical synthesis, environmental remediation (water and air purification), and self-cleaning applications.

Semiconductors play a significant role in photocatalysts. Semiconductors have an optical band gap greater than that of metal and less than that of insulators. The energy gap between the top of the valence band (VB) and the bottom of the conduction band (CB) is typically called band gap energy (E_g) (Justinabraham et al., 2023). Based on the carrier transition between bands, they are classified as direct and indirect band gap semiconductors. Further, semiconductors are also classified as N-type and P-type, depending on the majority charge carrier. It tends to absorb light energy (photons) based on the band gap in the range of visible and ultraviolet frequencies and produces charge carriers. Due to their fascinating optical properties, semiconductors have been used in various devices, such as electronic chips, displays, and energy conversion devices, and in many fields, including material science, environmental science, and electronics (Bai et al., 2023).

In photocatalysts, semiconductors play a significant role in improving the reaction kinetics by using light energy. Photocatalysts absorb light photons and create charge carriers in their respective bands. It is important to note that the photon should have more energy. The produced electrons move to the CB, leaving a hole in the VB. Depending on the band potential, redox reactions occur with the charge carriers (Li et al., 2021). Particularly, the CB electron reacts with oxygen to produce superoxide radicals, and the VB hole reacts with water to produce hydroxyl radicals. These free radicals are highly energetic, and they can degrade or oxidize contaminants. The photocatalyst efficiency strongly depends on the type of materials, morphology, light absorption efficiency, active site, electron–hole pair generation, carrier separation, and active species production. In this context, semiconductor photocatalysts like titanium dioxide (TiO_2) are harnessed to disintegrate organic pollutants, such as dyes and pharmaceutical residues, by utilizing the charge carriers generated upon exposure to light (Raghavan et al., 2018). Moreover, semiconductor materials also facilitate redox reactions, wherein electrons are transferred between molecules, leading to the degradation of contaminants. This capability plays a crucial role in processes like water disinfection and the removal of heavy metals.

The potential of semiconductor materials in water treatment is vast, extending to various facets. For instance, zinc oxide (ZnO) finds application in water disinfection by producing reactive oxygen species that efficiently eradicate bacteria and viruses (Raghavan et al., 2015). Furthermore, specific semiconductor materials, like iron-doped TiO_2, demonstrate proficiency in removing heavy metals from water through a combination of adsorption and photoreduction. Despite their enormous promise, challenges persist, such as the need to maintain the stability of semiconductor materials in harsh water treatment conditions and the ongoing quest to design new materials with enhanced photocatalytic properties. Additionally, scaling up these technologies for practical implementation on a larger scale remains a priority to address the pressing global water pollution crisis. In conclusion, semiconductor materials occupy a leading role in pioneering innovative water treatment solutions (Thangavel et al., 2016). They offer a promising avenue for combating water pollution and scarcity, with their remarkable properties and adaptability driving ongoing research and development efforts in the field.

As advancements in material science and engineering continue to unfold, the future holds immense potential for semiconductor-based water treatment methods, contributing significantly to a cleaner and safer global water supply. Among other factors, the electron–hole back recombination process significantly affects the photocatalyst efficiency (Durairaj et al., 2021). In semiconductor photocatalysts, free electron–hole recombines occurs before they can participate in chemical reactions. This results in diminished efficiency for the system. Several methods have been suggested to minimize the recombination in photocatalysts. The core idea of these methods is to separate the free electron and hole recombination. This can be done in several ways, such as defect creation, fact engineering, metal–nonmetal doping, heterojunction modification using metal, and semiconductors. Photocatalysts produce electrons accepted by defects or doped metal, thereby causing charge carrier separation to occur. In recent times, photoinduced carriers have been separated by another semiconductor using a heterojunction design (Venugopal et al., 2020). This is because the difference between the semiconductor work function and fermi level can lead to different charge carrier separation mechanisms, such as *p–n*, Z-scheme, and S-scheme. These methods have their pros and cons.

In this chapter, we provide a brief overview of the piezo-semiconductor photocatalyst for water remediation. First, we delve into the piezoelectric effect and its principles. Next, we explore the synthesis methods and characterization techniques. Furthermore, we present up-to-date information gathered from reported work by various research groups. Finally, the chapter concludes with a summary, addressing the challenges and prospects of piezophotocatalysis.

3.2 Principles of piezoelectric effect and synthesis methods

Ferroelectric materials have spontaneous polarization due to the relative displacement of positive and negative charges. Under mechanical strain, the noncentrosymmetrically structured materials produced piezoelectric potential caused by charge displacement.

Among the crystal structures, 20 of the 32 crystal point groups ensured piezoelectric properties (Dong et al., 2017). Piezoelectric materials are used to convert electrical energy to mechanical energy below the Curie temperature (*Tc*). The piezoelectric coefficient is an important mark for the efficiency of piezoelectric material. Four types of piezoelectric coefficients are available, such as piezoelectric strain coefficient, piezoelectric voltage coefficient, piezoelectric stress coefficient, and piezoelectric stiffness coefficient, due to the antistrophic nature of piezoelectric material. This piezoelectric nature has been used in various fields, such as nanogenerators, self-powered sensors, self-charging power cells, piezophotocatalysts, and piezocatalysts.

In piezophotocatalysts, mechanical and light energy are applied simultaneously to the material, which produces photogenerated carriers and piezoelectric charges (Sheng et al., 2023). Piezocatalysts can be explained by two theories; energy band alignment and the screening charge effect.

Under mechanical stimulation, local temperature or local energy is produced, which induces electron and hole in VB and CB. While pressure is increasing, the electron and hole separation occurs, and the VB and CB tilting are also observed. The tilting gradient is directly proposed to apply mechanical energy. Based on the band bending theory, the fowling conclusion, the redox potential of the band is improved in piezoelectric material. The piezo potential can attract nearby charges, which is typically called the screening effect. Due to the screening effect, the piezo potential becomes weaker. Therefore, periodic polarization changes are needed for piezoelectric potential. The external mechanical force deforms the original structure, resulting in the displacement of piezoelectric charges. This is because the external force produces tensile strain and compressive strain. The tensile strain produces the positive polarization, while the compressive strain generates the negative polarization.

Research on piezocatalytic degradation of organic dyes is still emerging (Sun et al., 2023), the process can be understood through the interplay of mechanical force and charge effects. When piezoelectric materials are subjected to external pressure or vibration, their internal crystal structure generates a voltage (piezo potential) that pushes positive or negative charges to opposite surfaces. This attracts an equal and opposite charge from the surrounding environment. The continuous changes in strain and stress caused by the vibration facilitate the adsorption and desorption of molecules on the material's surface, which is crucial for piezocatalytic reactions. The positive charge is attached to the surface; the proton drives the reaction. On the other hand, an electron participates in a degradation reaction when the negative charge molecules are attached to the surface. Various piezoelectric materials have been synthesized using different methods, each offering unique advantages for specific applications.

In the realm of ceramic solid-state synthesis, lead zirconate titanate (PZT) stands as a prominent example, renowned for its robust piezoelectric properties (R Damjanovic et al., 1998). PZT is fabricated by blending high-purity metal oxides such as lead oxide (PbO), zirconium oxide (ZrO_2), and TiO_2 and subjecting them to high-temperature sintering processes, resulting in a well-defined crystalline structure. Sol-gel processing has been instrumental in crafting materials like barium titanate, offering precise compositional control at lower temperatures. Hydrothermal synthesis is employed for the creation of piezoelectric nanomaterials like zinc oxide, benefiting from its ability to produce well-defined structures under high-temperature and high-pressure aqueous conditions (Dawason, 1993). Solvothermal synthesis finds utility in lead-free piezoelectric materials such as bismuth ferrite, offering a promising alternative to traditional materials. Electrospinning techniques yield nanofibers, such as polyvinylidene fluoride (PVDF) with embedded $BaTiO_3$ nanoparticles, designed for energy harvesting and sensing applications (Chang et al., 2012). Emerging methods explore biological synthesis using bacterial nanowires, showing promise for eco-friendly material production (Xu et al., 2021).

For thin-film applications, chemical vapor deposition facilitates the creation of thin films, exemplified by PZT thin films, offering precise control over film thickness and composition. Energy harvesting technologies have been a focus of research, aiming to harness the piezoelectric properties of these materials for various applications. Lead-free ferroelectric ceramics have also gained attention, with researchers working on their synthesis and piezoelectric properties (Lin et al., 2004). Progress in piezoelectric materials for clean energy harvesting has been recently explored, highlighting their potential for sustainable energy generation (Mahapatra et al., 2021). Porwal et al. (2023) studied the piezocatalytic properties of Bismuth Zinc Borat with respect to their size. First, the $Bi_2ZnB_2O_7$ was synthesized by solid-state approach, which has a large particle size. To reduce the particle size, synthesized Bi_2ZnB_2O is subjected to ball milling. The particle size was reduced from 5 mm to 470 nm. The photocatalytic properties of methylene blue (MB) degradation under ultrasonication were studied. Solid-state synthesized $Bi_2ZnB_2O_7$ exhibits 16% MB degradation at 180 minutes. On the contrary, 60% of MB degradation was observed in ball-milled $Bi_2ZnB_2O_7$. Size reduction is an important factor for enhanced piezocatalytic performance in ball-milled $Bi_2ZnB_2O_7$. As shown in Fig. 3.1,

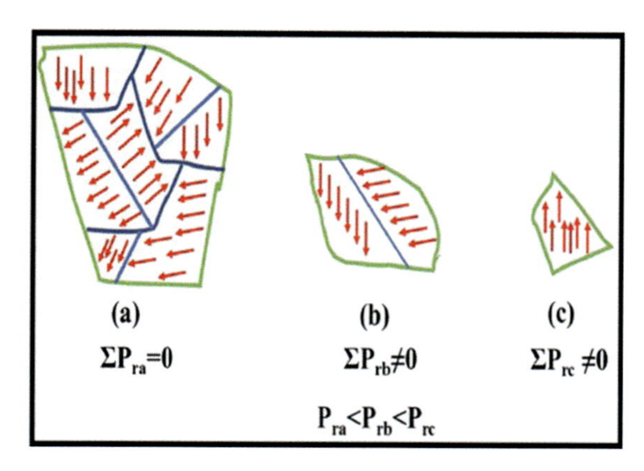

Figure 3.1 (A) A schematic summary of piezocatalysis mechanisms: energy band theory (*left*) and screening charge effect (*right*). (B) Polarization dependency on particle size. From Wang, K., han, C., Li, J., Qiu, P., Sunarso, D. & Liu, Prof. S. (2023). The mechanism of piezocatalysis: Energy band theory or screening charge effect? In: C. Porwal, A. Gaur, V.S. Chauhan, R.J.S. Vaish (eds.), *Interfaces, Enhancing piezocatalytic dye degradation through ball milling-induced polarization in nano bismuth zinc borate*, 103391.

the ferroelectric domains are reduced when reducing the particle size. A large-size catalyst has zero ferroelectric domains due to its random orientation. Furthermore, there is a size reduction in the ferroelectric domains, but it is not at net zero. The size reduction creates more active sites and increases internal polarization, which improves the catalytic properties (Porwal et al., 2023).

3.3 Influencing factors for piezocatalytic activity

There are few modification strategies reported to improve the piezocatalytic properties, for example, chemical element doping, defects/vacancies manipulation, crystal plane control, phase structure engineering, surface modification, morphology adjustment, co-modification, and construction of composite systems. Metal doping is employed to adjust the charge distribution, phase transition, and change lattice arrangement to improve the polarization. Vacancy is an important property in catalysts to change the electronic structure, improve optical absorption, and improve the active site.

In piezoelectric materials, there's a debate regarding the impact of vacancies on their properties. Some studies suggest that vacancies enhance the internal potential, while others indicate a decrease in piezoelectricity (Xiao et al., 2023). For instance, Xiao et al. (2023) investigated the influence of photoactive surface oxygen vacancies on $BaTiO_3$ piezo-photocatalytic activities. Their findings revealed a 1.5-fold decrease in piezoelectric catalytic performance but a 3.5-fold enhancement in photocatalytic efficiency. This contradictory effect results in an overall improvement in coupled piezo-photocatalytic performance due to the interplay of these factors. Similarly Liu et al. (2022) observed a decrease in piezoelectric properties in $(Bi_{1/2}Na_{1/2})TiO_3$ due to oxygen vacancies. Tang et al. (2023) reported improved piezo-photocatalytic activity in $Bi_4Ti_3O_{12}$ nanosheets with oxygen vacancies, although they didn't delve into the impact on piezoelectric properties. In the case of MoS_2, S vacancies were found to enhance the piezoelectric response in nanosheets. Li et al. (2023) investigated the synthesis of tunable S vacancies in MoS_2 through various parameters and observed improved Orange II degradation, highlighting the beneficial impact of vacancies on its properties. Additionally, element doping can alter charge distribution, phase transitions, and lattice distortion, consequently improving polarization. For instance, Ti-doped BiOCl nanowires exhibited significantly higher RhB degradation compared to undoped ones, demonstrating the effectiveness of doping as a strategy for enhancing piezophotocatalytic properties. Furthermore, Zhou et al. (2020) discovered that Rh doping induces a new built-in electric field in $SrTiO_3$, leading to improved charge carrier separation and thus enhancing piezophotocatalytic performance.

Meng et al. (2023) synthesized Co-doped ZnO lattices, which improved the electrical properties of ZnO. The d_{33} test shows that the Co-doped ZnO piezoelectric coefficient is 6.5 times. $Zn_{0.8}Co_{0.2}O$@PVDF membrane coefficient increased from 27.3 to 44.7 pC/N. Phase junction is another important strategy to improve the piezo-photocatalytic properties. The phase junction is possible in polymorphic systems, namely CdS, ZnS, etc. In this polymorphic system, certain phases possess piezoelectric properties. The wurtzite phase of ZnS and CdS exhibits piezoelectric behavior, whereas cubic does not show net zero piezo. Zhao et al. (2022) designed the phase junction to improve the piezopotential via stress created by the cubic phase. In addition, the authors synthesized phase junctions via the hydrothermal method. The CdS mixed-phase MO degradation efficiency considerably increased. The improved catalytic activity mainly improved the separation efficiency of the charge carrier. Crystal facet exposures are one of the important properties of piezophotocatalysis. BiOCl was synthesized with different crystal facts of 102 and 200, respectively. The first BiOCl exhibits piezophotodegradation of RhB of about 96.9%, and the later BiOCl shows 62.60% degradation, respectively. Likewise, Wang et al. studied the crystal-face-dependent piezophotocatalytic properties of $BiVO_4$. From HRTEM analysis, it was inferred that $BiVO_4$ lattice planes are 040 and 110, respectively. The 110 $BiVO_4$ exhibits a stronger piezoelectric (105 pm/V) coefficient than the 040 crystal (75 pm/V), respectively (Wang et al., 2022a, 2022b).

In other strategies, the catalyst surface can be modified, which separates the electron hole and enhances the piezophotocatalyst performance. Jiang et al. (2022) modified the $SrTiO_3$ surface with OH via the low-temperature sol–gel method. Similarly, the Bi_2WO_6 nanosheet surface is modified with NaOH etching. Furthermore, the piezocoefficient was significantly higher than that of pure Bi_2WO_3. The degradation analysis showed that the rate constant was improved in NaOH-etched Bi_2WO_6 nanosheets compared with bare Bi_2WO_6 nanosheets, which is about 33.9 times higher than untreated catalysts. Morphology is an important property to change the built-in electric field in piezoelectric materials. Some materials in nanorods exhibit high piezophtocatalytic decomposition properties compared with nanospheres. For example, flowers like MoS_2 and WS2 exhibit higher piezophotocatalytic activity than other morphologies. General catalytic quantum efficiency highly restricted the catalytic efficiency. The piezo-phototronic effect could be improved with the assistance of the plasmonic effect. The CdS–Au composite was synthesized by water bath heating and reflux condensation methods for the degradation of organic pollutants. Combined effects such as piezopotential and plasmonic effects improved the charge carrier separation and enhanced the quantum efficiency. The composite photocatalytic efficiency of MO degradation is improved 2.3 times when adding 4 mol% of Au. The next important plasmonic metal is Pt, which has used different photocatalytic materials to increase the efficiency of the system.

Meng et al. (2023) designed the $BaTiO_3$ nanocubes decorated with Pt heterojunctions via the photoreduction method. The piezophotocatalytic efficiency strongly depends on the percentage of PtNPs. The $Pt_{0.25}/BaTiO_3$ with 0.25 wt.% PtNPs showed excellent catalytic activity. Another important Ag is also used as plasmonic material. Ag-doped $Na_{0.5}Bi_{0.5}TiO_3$ nanospheres are prepared by hydrothermal and chemical solution approaches. The composite exhibited 98.6% RhB degradation, which was attributed to the local surface plasmon resonance effect. Single-crystalline ferroelectrics exhibit more efficiency than polycrystalline ferroelectrics. Polycrystalline ferroelectrics have randomly oriented grains; hence, the ferroelectric domains cannot be used efficiently. However, single-crystalline piezocatalyst synthesis is more complex. Moon and his research group recently synthesized single-crystalline (K, Na)NbO_3.

The single crystal KNN is combined with Cu_2O, and the catalyst efficiency is measured with RhB degradation as proof of concept. The RhB removal efficiency is 28.9% and 38.3% for polycrystalline and single-crystalline, respectively. Further, the reaction kinetics were calculated to be 0.063 and 0.089/minute polycrystalline and single crystalline, respectively (Im et al., 2023).

3.4 Types of external forces

The external force plays a vital role in piezocatalysts such as ultrasonication, water flow, and simple mechanical stirring. Ultrasound needs more power consumption, resulting in a high cost. The naturally available mechanical energy is tiny, scattered, low frequency, and low power. Wind energy, water flow energy, and noise can also be used to trigger the piezoelectric catalyst.

Prasanna et al. (2023) studied their piezocatalyst activity under simple mechanical stirring using $BaTiO_3$ as a catalyst. The RhB was chosen as a pollutants and the PTFE magnetic bar was used for magnetic stirring. Further, it has been demonstrated that the liquid height influences the piezocatalyst efficiency. When the size of the liquid is low, tangential velocity is increased, which improves the larger stress on the catalyst. Also, there is more chance of entre the air resulting. Similarly, Li et al. (2023) also studied the piezophotocatalytic efficiency related to water level. In this work, the authors performed four different water levels in the ultrasonic tank. They obtained maximum degradation efficiency at water level three. However, the authors did not explain the in-depth mechanism of piezoelectric enhancement at certain water levels. Further, heavy metal degradation was also studied by CuS/MoS_2 nanocomposites piezocatalyst, which is driven by mechanical energy (Li et al., 2022). The piezocatalytic activity is analyzed using piezoelectric current analysis. While increasing the stirring speeds, the corresponding piezoelectric current is increased. These results ensure that water stirring produced the mechanical force and deformed the MoS_2 sheet, resulting in a piezoelectric current.

Ball milling is a simple and widely used technique for mixing, grinding, and synthesizing nanomaterials at wet chemical routes. Recently, a few researchers studied the piezocatalystic properties using ball milling as an external force, which is discussed below. The screening effect is strongly mitigated by a built-in electric field, which is formed under ball milling X. A fan research group studied the piezoelectric properties of Pb_2BO_3X (X = Cl, Br, and I) for the degradation of organic pollutants such as rhodamine B, ciprofloxacin, and norfloxacin (Tang, Chen et al., 2022; Tang, Gong et al., 2022). It is important to note that two kinds of balls, namely ZrO_2 and SiO_2, are at different rpm. The former one produces 64.7 N force, which is higher than 2.2 times. Continuous collisions may produce heating. In this reaction, there was no significant rise in temperature.

PVDF is a well-known piezoelectric material known for its flexibility. A PVDF-assisted semiconductor hybrid has been reported as a piezophotocatalytic system. Our research group analyzed the flexibility and efficiency of a PVDF-TiO_2 hybrid as a piezocatalyst (Durairaj et al., 2019). The efficiency of the

PVDF-TiO_2 hybrid was evaluated by degrading MB under UV light. The overall reactor setup is presented in Fig. 3.2A and B. In this study, air bubbles were utilized as the mechanical force to induce the piezoelectric effect. Different reaction conditions were tested using various catalysts, such as PVDF-TiO_2/UV, PVDF-TiO_2/air bubble, and PVDF-TiO_2/UV/air bubble. Additionally, the concentration of air bubbles was varied, including 25%, 50%, and 100%. The reaction mechanism was systematically analyzed using radical analysis techniques. Similarly, E-MoS_2/PVDF piezoelectric microcapsules are synthesized from expanded MoS_2 with PVDF. The bubble is used as an external force for piezocatalysis for the degradation of tetracycline (TC), chlortetracycline (CH), and ofloxacin (OFL). The obtained efficiency is 91.88%, 85.48%, and 96.84% in 60 minutes, respectively. Compared with ultrasonic processes, bubbles-driven processes need 76.00 kw/h^3/order, which is only 6.43% of ultrasonic processes. In addition, the bubble-driven process generates the $^1O_2$ and O_2^-. The piezoreactor is shown in Fig. 3.2C. The reactor tube is 50 cm in length and 4 cm in diameter. Based on ROS and dissolved oxygen analysis, the mechanism is derived as described as follows. Dispersed bubbles induce localized shear and turbulence near the microcapsules, which generate strain (Huo et al., 2023).

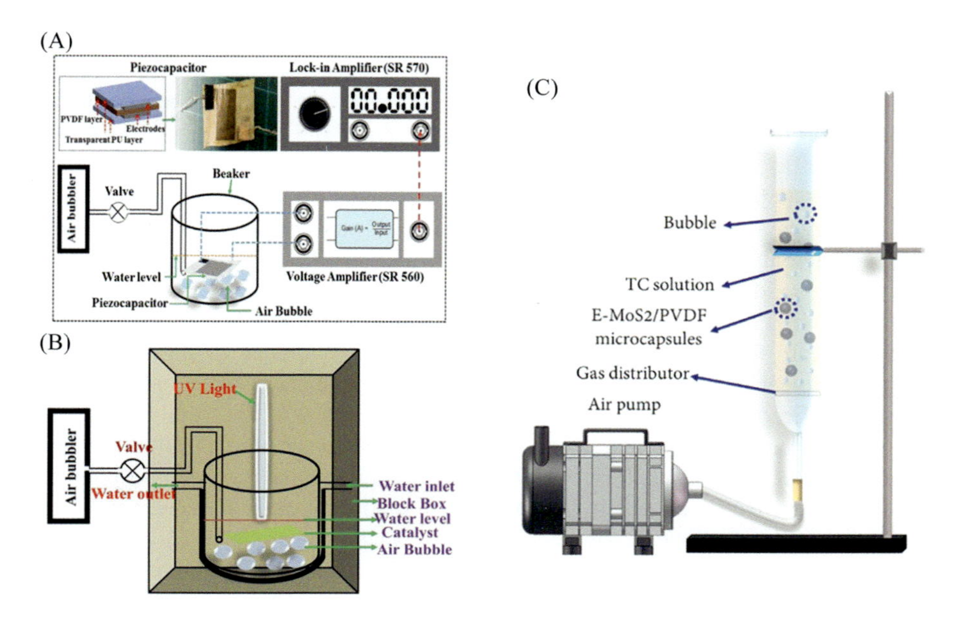

Figure 3.2 Schematic experimental setup used for (A) piezopotential and (B) piezophotocatalytic measurements. (C) Device diagram of piezocatalytic degradation by the bubble-driven method.
From Huo, B., Wang, J., Wang, Z., Zhang, X., Yang, J., Wang, Y., Qi, J., Ma, W. & Meng, F. (2023). Bubble-driven piezo-activation of E-MoS_2/PVDF piezoelectric microcapsule for antibiotic degradation with ultralow energy consumption. *Journal of Cleaner Production, 419*. https://doi.org/10.1016/j.jclepro.2023.138333.

3.5 Materials used in piezo catalyst

Piezocatalyst is an impressive technique that has promising applications in mechanical energy and electrochemical energy. The fascinating piezoelectric phenomenon has been used in different fields and sectors. In semiconductor photocatalysis, water treatment has received significant attention due to the spatial advantages that piezo potential offers in these areas. The piezo potential can separate photocatalytically synthesized electron−hole pairs, which will be separated in opposite directions by piezo potential. The population of free-charge carriers improves reaction efficiency. This type of separation could be done through two types of systems, such as a single material and a combination of piezocatalyst and semiconductor.

Only a few materials exhibit both piezoelectric and semiconductor characteristics, making them desirable for various applications. However, not all semiconductors possess piezoelectric properties. Therefore, combining piezoelectric materials with semiconductors can enhance carrier separation and improve their overall performance. In the field of photocatalysis, TiO_2 is a traditional semiconductor that has been extensively studied since 1972. TiO_2 has numerous advantages, including suitable redox potential, chemical stability, and simple synthesis. However, its efficiency is often limited by rapid electron−hole recombination, which hinders large-scale applications.

To enhance the efficiency of TiO_2, researchers have explored hybrid materials by combining them with piezoelectric materials such as $BaTiO_3$, Quartz (SiO_2), PAN, Bi_2WO_6, $MoSe_2$, MoS_2, $SrTiO_3$, $PbTiO_3$, ZnO, and polytetrafluoroethylene. Under external mechanical force, those materials generate a built-in electric field, which paves the way for charge separation macroscopically for catalysis. For example, $BaTiO_3$ is a well-known piezoelectric material, particularly when it is in the β phase. In a study, Ag_2O was successfully grown on the $BaTiO_3$ surface, creating a hybrid semiconductor-piezocatalyst (Li et al., 2015). The photocatalytic properties of the prepared hybrids ($BaTiO_3-Ag_2O$) were analyzed through RhB degradation. To investigate the piezoelectric effect, the catalytic efficiency of the hybrid was compared under different conditions, namely, light irradiation (photocatalytic) and ultrasound irradiation (sonocatalytic). It was observed that the catalytic activity improved when the reaction was conducted with both ultrasound and light simultaneously. Under light irradiation, Ag_2O produces charge carriers through a photocatalytic process. However, due to the fast recombination of these charge carriers, the efficiency of Ag_2O alone is low. By applying mechanical force and light simultaneously, $BaTiO_3$ generates a piezoelectric charge that attracts the opposite photogenerated charge in Ag_2O. This results in a decrease in the rate of charge carrier recombination, thereby increasing the efficiency of $BaTiO_3-Ag_2O$. A schematic representation of charge carrier separation is shown in Fig. 3.3.

Besides, ZnO exhibits piezoelectric properties when it undergoes one-dimensional growth under mechanical stimulation. Our group analyzed a piezoelectric-semiconductor ($ZnO-Ag_8S$) hybrid for water remediation (Venugopal et al., 2020). In this work, the $ZnO-Ag_8S$ hybrid was used as a model system to investigate the enhancement of

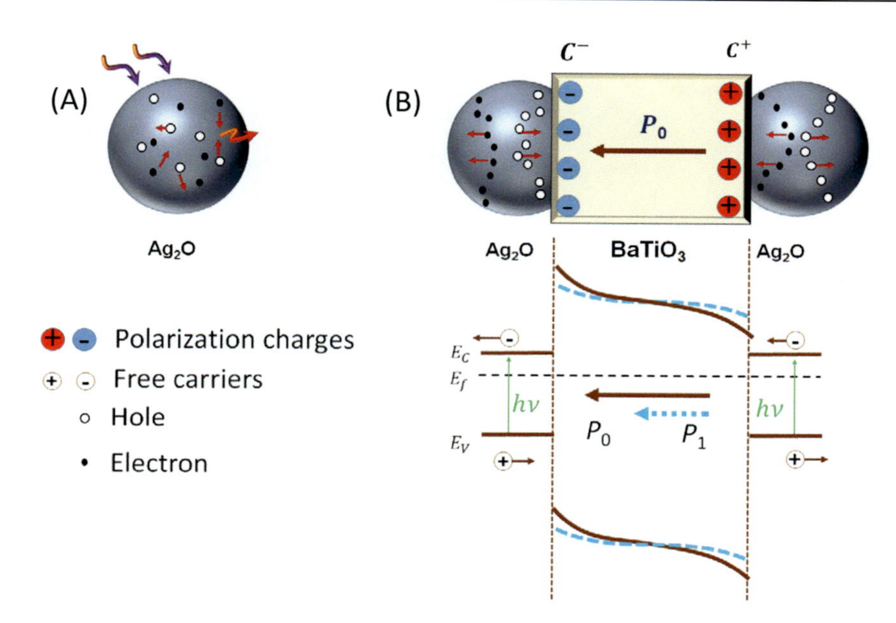

Figure 3.3 Schematic illustration of (A) charge carrier generation in a Ag_2O nanoparticle when excited by a photon. (B) The separation of electrons and holes in Ag_2O nanoparticles attached to the two opposite surfaces of a $BaTiO_3$ nanocube that have opposite polarization charges due to piezoelectric effect and the corresponding tilting in the bands. The solid line is the bands with the presence of spontaneous polarization charges at the two surfaces of the $BaTiO_3$ nanocube; the dashed line indicates the decrease of the piezoelectric polarization with a mechanical strain. From Li, H., Sang, Y., Chang, S., Huang, X., Zhang, Y., Yang, R., Jiang, H., Liu, H. & Wang, Z. L. (2015). Enhanced ferroelectric-nanocrystal-based hybrid photocatalysis by ultrasonic-wave-generated piezophototronic effect. *Nano Letters*, *15*(4), 2372–2379. https://doi.org/10.1021/nl504630j.

photocatalysis induced by piezopotential. The catalytic efficiency of the system was evaluated through the degradation of RhB. First, the ZnO tetrapods are synthesized on a substrate using the hydrothermal method. Subsequently, Ag_8S nanoparticles were coated onto the ZnO tetrapods via the SILAR method. SEM and TEM images, as shown in Fig. 3.4, confirm the successful formation of the ZnO−Ag_8S hybrid.

As prepared, the hybrid exhibits high RhB degradation efficiency, further increasing the presence of mechanical stress. Following this work, the piezopotential improved PMS activation is studied for water remediation using the piezoelectric-semiconductor hybrid $BaTiO_3$−Ag_2O (Thangavel et al., 2022). The activity of the system is analyzed by the degradation of RhB and phenol. The influence of piezoelectric properties on degradation efficiency is analyzed using single systems $BaTiO_3$, Ag_2O, physically mixed $BaTiO_3$−Ag_2O, and $BaTiO_3$−Ag_2O nanohybrid with different conditions like photocatalysis, sonocatalysis, and sonophotocatalysis. Further, the radical analysis confirms that PMS is successfully activated by the $BaTiO_3$−Ag_2O nanohybrid. It is important to note that the degradation efficiency is

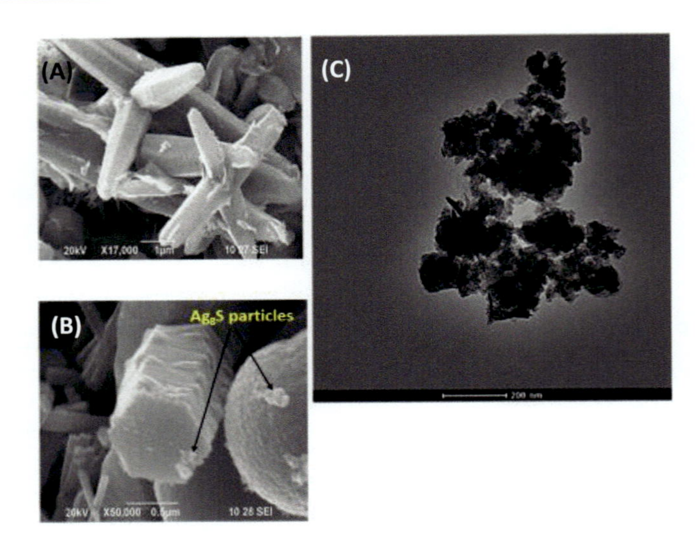

Figure 3.4 SEM images of (A) ZnO tetrapods (scale bar is 1 μm) and (b) hexagonal structure of ZnO and it is seen that Ag_8S nanoparticles coated on ZnO tetrapod (scale bar is 0.5 μm). (C) TEM image of ZnO–Ag_8S hybrid where Ag_8S particles are fully covered the ZnO tetrapods (scale bar is 200 nm).
From Venugopal, G., Thangavel, S., Vasudevan, V. & Zoltán, K. J. J. o. P. (2020). *Solids, Efficient visible-light piezophototronic activity of ZnO-Ag8S hybrid for degradation of organic dye molecule* (Vol. 143).

purely attributed to the sulfate radicals formed from PMS. Without $BaTiO_3$, Ag_2O alone does not effectively activate the PMS, resulting in low photocatalytic degradation efficiency compared to piezophotocatalysis. This study confirms that piezopotential not only enhances photocatalysis but can also be utilized for PMS activation. In comparison to $BaTiO_3$ and ZnO, SiO_2 is an abundant earth material that also possesses piezoelectric properties, further enhancing photocatalytic efficiency.

Yang et al. (2022) developed a ZnO/piezoelectric quartz core–shell structure for the decontamination of ibuprofen (IBF). The absorption efficiency of ZnO for light is unaffected when combined with quartz in a hybrid configuration. The piezophotocatalytic degradation of IBF was observed to be enhanced compared to photocatalytic and piezocatalytic processes. Photoinduced charge carriers are spatially separated by the piezoelectric field, causing electrons in the ZnO to migrate toward the negatively charged polarization regions. This spatial charge carrier separation occurs throughout the system.

Two-dimensional materials have recently emerged as potential candidates for water remediation, including MoS_2, g-C_3N_4, WS_2, $MoSe_2$, and WSe_2. Many of these 2D materials come under the category of transition metal chalcogenides (TMDs), which exhibit piezopotential properties. Generally, these TMDs are layered materials with the chemical configuration of MX_2. Here M denotes transition metal and X represents chalcogenides, including S, Se, and Te. The piezoelectricity of these materials is strongly observed at the edge, whereas the plane has zero piezopotential. For example, MoS_2 exhibits piezoelectric properties due to the asymmetric distribution of Mo and two S atoms in the unit cell. Strain-free state MoS_2 has shown spontaneous polarization because the 2S-Mo dipole enlarged while stretched. MoS_2 has various crystal forms, such as 1T, 2H, and 3R, based on the stacking type of layer. Various research groups have studied their piezo potential to improve photocatalysis for water remediation. The Ag_2O nanoparticles were modified with MoS_2 nanoflowers to efficiently utilize the full solar spectrum. The catalyst efficiency was evaluated by methyl orange (MO) degradation. External mechanical forces, such as magnetic stirring and ultrasonic waves, are applied. Fig. 3.5 represents the MO degradation by the MoS_2-Ag_2O hybrid under different experimental conditions. The results indicate that the degradation efficiency of MO increases in the presence of external mechanical forces (Li et al., 2019).

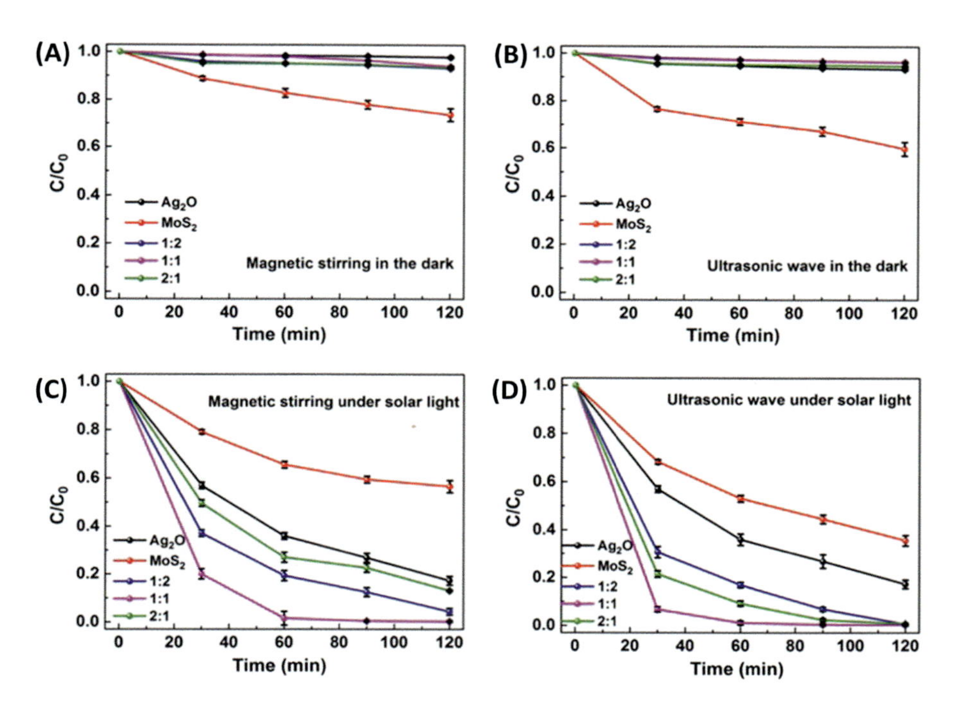

Figure 3.5 The degradation ratio of MO using Ag_2O NPs, MoS_2 NFs, and MoS_2@Ag_2O heterostructures (mole ratios 1:2, 1:1, and 2:1) under different conditions: (A) magnetic stirring in the dark, (B) ultrasonic wave in the dark, (C) magnetic stirring under full solar light illumination, and (D) ultrasonic wave under full solar light illumination.
From Li, Y., Wang, Q., Wang, H., Tian, J. & Cui, H. (2019). Novel Ag_2O nanoparticles modified MoS_2 nanoflowers for piezoelectric-assisted full solar spectrum photocatalysis. *Journal of Colloid and Interface Science*, *537*, 206–214. https://doi.org/10.1016/j.jcis.2018.11.013.

This is because the mechanical force induces the piezopotential of MoS_2, which in turn reduces the recombination of photoinduced electron−hole pairs in Ag_2O. Another important 2D material exhibiting piezoelectric properties is graphitic carbon nitride (g-C_3N_4). It also has two frameworks, such as the s-triazine and tri-s-triazine/heptazine subunits. The single-layer tri-s-triazine-based structure has strong in-plane piezoelectric.

Tang, Gong et al. (2022) and Tang, Chen et al. (2022) synthesized a g-C_3N_4/PDI−g-C_3N_4 homojunction and analyzed its piezoelectric and photocatalytic properties. The degradation efficiency was assessed using atrazine as the target pollutant, with an external force applied through ultrasonication. High-performance liquid chromatography analysis was conducted to evaluate the degradation of atrazine. Fig. 3.6 illustrates the piezophotocatalytic mechanisms of

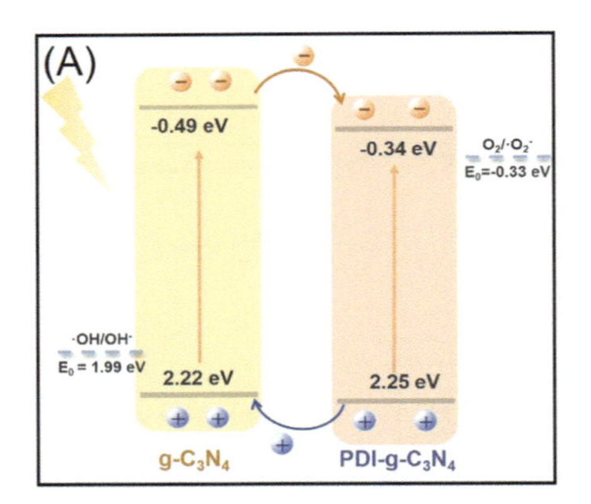

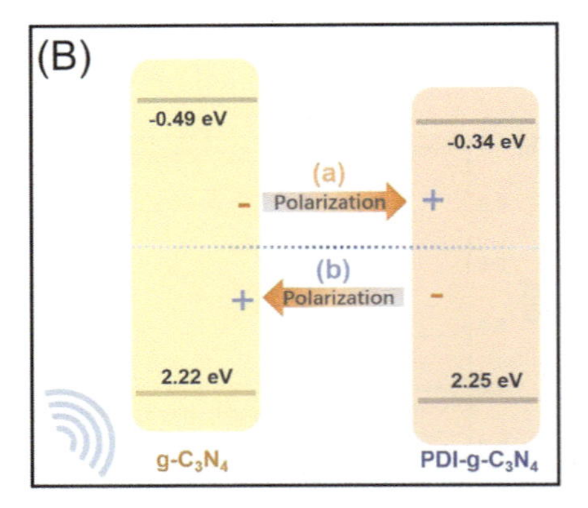

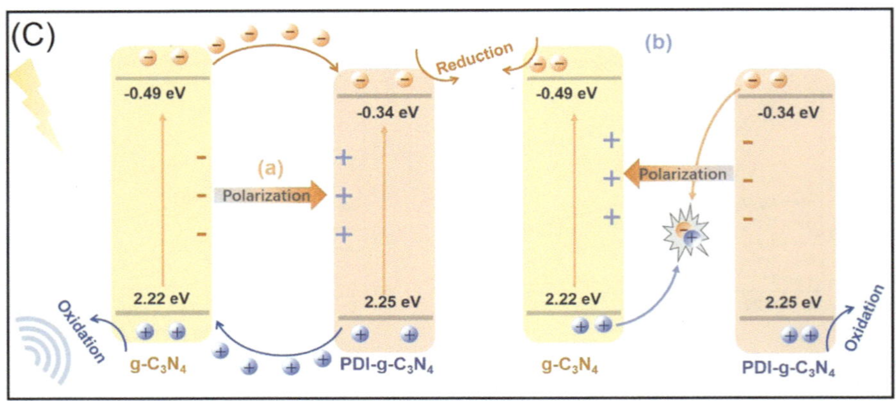

Figure 3.6 Schematic diagram showing the mechanism of the individual (A) photocatalysis, (B) piezocatalysis, and (C) synergetic piezo-photocatalysis under different polarization directions. From Tang, R., Gong, D., Zhou, Y., Deng, Y., Feng, C., Xiong, S., Huang, Y., Peng, G., Li, L. & Zhou, Z. (2022). Unique g-C_3N_4/PDI-g-C_3N_4 homojunction with synergistic piezo-photocatalytic effect for aquatic contaminant control and H_2O_2 generation under visible light. *Applied Catalysis B: Environmental, 303*. https://doi.org/10.1016/j.apcatb.2021.120929.

g-C_3N_4 and PDI-g-C_3N_4. The direction of piezopotential plays a crucial role in the charge separation mechanism. In the case of polarization from g-C_3N_4 to PDI-g-C_3N_4 (on the left side of the figure), the direction of the electron flow is the same.

On the other hand, the direction of the electric field from PDI-g-C_3N_4 to g-C_3N_4 is opposite to the direction of the photogenerated electrons, thus forming a Z-scheme mechanism. Another important piezoelectric material is SrTiO3, although it exhibits a relatively low piezoelectric potential compared to other piezoelectric materials. To improve the properties of SrTiO3, metal doping is employed.

Li and his research group synthesized an Al-doped $SrTiO_3/TiO_2$ heterostructure nanorod to improve its catalytic properties via piezo-photocatalysis (Chu et al., 2022). The catalytic performance was evaluated by analyzing the degradation of RhB. Improved RhB degradation is analyzed with the COMSOL Multiphysics software. Simulation studies confirmed that the introduction of aluminum doping creates crystalline distortion, thereby increasing the piezoelectric potential in $SrTiO_3$. In addition to solid piezoelectrics, flexible polymers also exhibit piezoelectric properties.

3.6 Performance evaluation of piezophotocatalytic water treatment

Piezophotocatalysis is an emerging technique in wastewater treatment in recent times. To successfully apply this technology to real-time water treatment, it must be analyzed well. Hence, the developed efficiencies that improved the existing system should be analyzed carefully. First, the developed system should contain piezoelectric materials that respond immediately to external forces. For example, the piezoelectric properties of $BaTiO_3$ and PVDF strongly depend on the phase of the material. Therefore, the phase of the material must be identified before analyzing the system's degradation efficiency. On the other hand, the morphology of the materials highly influences the piezo potential. For example, ZnO nanorods exhibit higher piezoelectricity than other morphologies. After confirming the phase and morphology of the materials, we should analyze the piezo potential of the material. This can be done in different ways, but piezoresponsive force microscopy (PFM) is an important tool for measuring the piezoelectric property. Fig. 3.7 shows the PFM results of cobalt-doped ZnAl-LDH (Lu et al., 2023). As seen in the figure, the maximum phase angle deformation amplitude and phase lag loop area reveal that the system has more piezoelectric properties.

Further, the piezoelectric coefficient is calculated as 7.63 and 10.56 pm/V. COMSOL Multiphysics software can be used to calculate piezoelectric properties and morphology-dependent piezoelectric potential. Fig. 3.8 shows the one-dimensional tubular g-C_3N_4 and $BaTiO_3$ distribution of strain and piezoelectric potential on the surface. Based on the strain distribution, the one-dimensional tubular g-C_3N_4 calculated about 9.63×10^3 V.

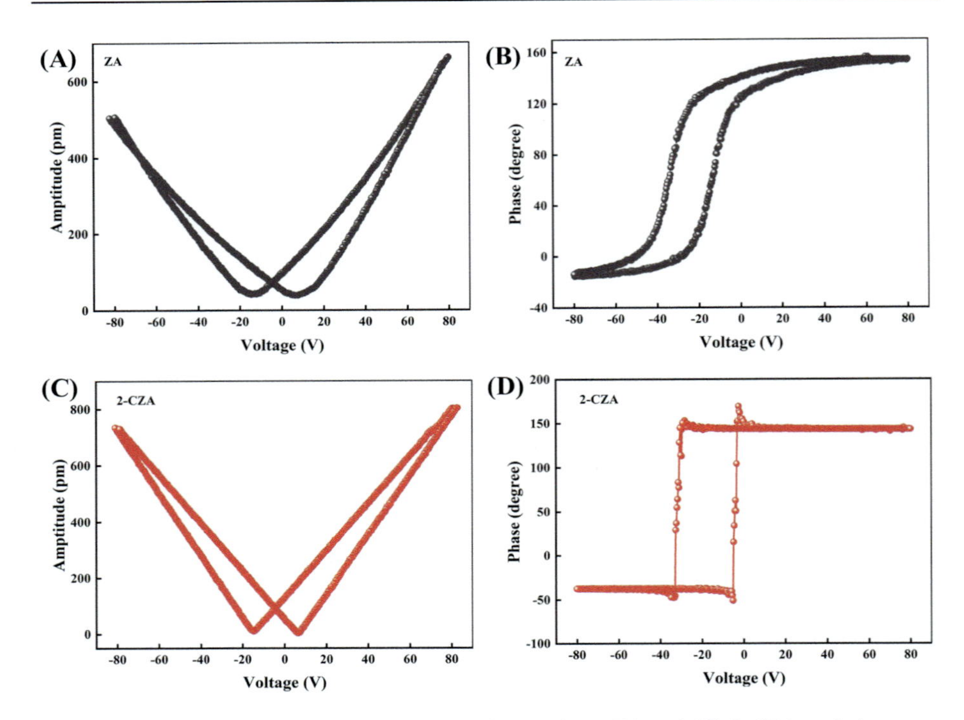

Figure 3.7 PFM amplitude–voltage butterfly loops of (A) ZA and (C) 2-CZA and phase-voltage hysteresis loops of (B) ZA and (D) 2-CZA.
From Lu, Y., Ding, C., Guo, J., Gan, W., Chen, P., Chen, R., Ling, Q., Zhang, M., Wang, P. & Sun, Z. (2023). Cobalt-doped ZnAl-LDH nanosheet arrays as recyclable piezo-catalysts for effective activation of peroxymonosulfate to degrade norfloxacin: non-radical pathways and theoretical calculation studies. *Nano Energy*, *112*. https://doi.org/10.1016/j.nanoen.2023.108515.

On the other hand, the one-dimensional tubular g-$C_3N_4/BaTiO_3$ exhibited 9.91×10^3 V. The improved piezo potential of hybrid material will improve the high photocurrent and increase the photoinduced charge carrier. Here, the dimension of the material is important during the computational calculation. To confirm the piezopotential of materials, it is important to maximize the interaction between piezoelectric materials and semiconductors during hybrid manufacturing. If the interaction is not good, the piezo force will be unable to pull in photoinduced charge carriers effectively.

Additionally, the piezo potential interacts with pollutants. The hybrid catalyst efficiency is analyzed using different modal pollutants to understand the improved efficiency of the piezo force. To comprehend the influence of the piezoelectric effect, the system is often analyzed with and without mechanical force, specifically with light only. During the efficiency analysis, the catalyst needs to separately analyze the photocatalyst, piezocatalyst, and piezophotocatalyst to understand their individual influences on the piezoelectric effect. Furthermore, the magnitude of the

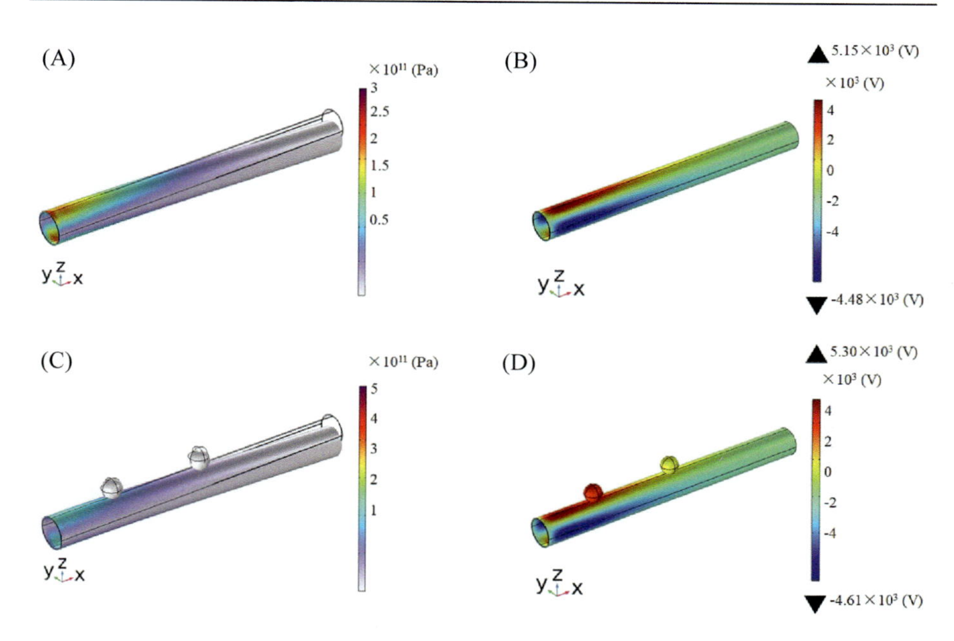

Figure 3.8 Theoretical calculation of volume stress distribution and corresponding voltage potential distribution of tubular *g*-C_3N_4 (TCN) (A, B) and $BaTiO_3$/TCN (C, D) under an external load of 100 MPa.
From Gong, S., Zhang, W., Liang, Z., Zhang, Y., Gan, T., Hu, H. & Huang, Z. (n.d.). Construction of a $BaTiO_3$/tubular *g*-C_3N_4 dual piezoelectric photocatalyst with enhanced carrier separation for efficient degradation of tetracycline.

external mechanical force can be adjusted to explore new characteristics of mechanical force. Depending on the model pollutant, different methods are used to analyze the pollutant concentration and degradation. The concentration of color-based pollutants can be measured using UV absorption spectroscopy.

Additionally, total organic carbon analysis can be utilized to measure pollutant concentration and toxicity. To analyze the piezophotocatalytic degradation pathway and identify major intermediates, iquid chromatography-mass spectrometry (LC-MS) is commonly employed. It is also important for the catalyst to undergo physical examination after repeated catalytic cycles to understand changes in the piezo potential. In recent years, in-situ analysis has rapidly grown in different fields, making it necessary to analyze the behavior of piezomaterials under mechanical stress. This can help gain a better understanding of this technology. After water remediation, the quality of the purified water should be assessed for phytotoxicity to ensure its suitability for discharge. For example, if the treated water is used for irrigation to grow plants, plant toxicity can be analyzed and compared with plants grown using pure water. Alternatively, some studies have reported the use of treated water for fish habitats, and in such cases, the toxicity of the water is analyzed after a certain period to assess its purity.

3.7 Applications of piezophotocatalytic water treatment

Piezo photocatalysis was successfully applied for different harmful pollutants in wastewater treatment sectors on a lab scale. The piezophotocatalyst produces strong radicals such as OH and O_2 that are powerful to degrade harmful chemicals such as dye molecules, colorless chemicals, antibiotics, and so on. Some of the systems are not effectively decomposed by the above radicals; hence, sulfate radicals are used to degrade the pollutant. Among the pollutants, pathogenic bacteria are one of them.

Piezophotocatalyst Au-MoS_2@CF is used as an antibacterial piezophotocatalyst by Chou et al. Few drugs are ecological risks when it comes to the environment that should be degraded. Piezophotocatalyst is considered a low-cost approach to removing novel pollutants (Chou et al., 2019). Phosphodiesterase-5 inhibitors are one of the pollutants that emerged during COVID-19. Zeng et al. (2023) studied the degradation of the drugs by 1T−2H MoS_2/$NaBi(MoO_4)_2$ piezoelectric. Degradation reached sildenafil of 99.8% at 60 minutes. To understand this best catalytic efficiency, the piezoelectric response is analyzed. 1T−2H MoS_2/$NaBi(MoO_4)_2$ showed 233.3 pm/V, which is 8.1 times higher than $NaBi(MoO_4)_2$ (28.89 pm/V). The 99.999% bacteria was eliminated in 30 minutes by piezophotocatalytic conditions. Similarly, *Escherichia coli* was also successfully cleared by a pieophotocatalyt, $NaNbO_3$ nanorods (Khosya et al., 2022). As generated, radicals attack the bacterial cell wall, and H_2O_2 directly attacks the bacterial cell. The next important source of water pollution is heavy metal, namely Cr(VI). This easily enters the cells to damage them, and on the other hand, reduced form of Cr(III) is an essential nutrient for organisms. A vast amount of research is ongoing on hazardous Cr(VI) to Cr(III) in wastewater. Piezophotocatalytic techniques are used to remove the heavy metal Cr (VI). $CaBiO_3$ nanosheet piezophotocatalytic removal efficiency analysis showed 94% removal within 120 minutes.

3.8 Challenges and prospects

Semiconductor-piezoelectric hybrids have emerged as effective environmental photocatalysts for water purification applications. When an external force is applied, piezoelectric materials generate an internal electric field that facilitates the separation of photocatalytic charge carriers. This piezo-induced spatial separation of charge carriers also helps to minimize back reactions. It is crucial during piezophotocatalytic reactions to periodically change the external field. Failure to do so may result in the screening effect, which neutralizes the piezo charge and hinders the separation of photogenerated charges. Compared to polycrystalline piezo materials, single-crystalline materials exhibit higher piezo potential. This is because polycrystalline materials have different grain boundaries, each with distinct domains and orientations, leading to an overall decrease in polarization. To achieve maximum efficiency, single-crystalline piezoelectric materials are preferred.

In piezophotocatalytic reactions, many studies have utilized ultrasonic waves as an external force, which can introduce defects in the catalyst. These defects may act as reactive sites and reduce charge carrier recombination. Therefore, it is necessary to analyze the efficiency enhancements to determine whether they arise from the piezopotential or other factors. However, it is important to note that ultrasonic irradiation at high frequencies can potentially damage the piezo material. Alternatively, renewable energy sources such as wind, tidal, stirring, and geothermal energy can be utilized as external forces. In addition, it is crucial to periodically change the external force to activate the piezoelectric field.

A study conducted by Fan and their research group explored the piezocatalytic activity of $Cu_3B_2O_6$ in the degradation of RhB dye. They employed ball milling as the mechanical force and compared it with ultrasonication. Under the external force of ball milling, they achieved 99.9% degradation of RhB, and the rate constant was found to be 10.4 times higher compared to ultrasonication. This significant enhancement in performance can be attributed to the strong force generated during ball milling, resulting in a more pronounced polarization electric field (Ren et al., 2023). Fig. 3.9A–C shows the sharing force increase in the piezoelectric increase. Likewise, Gaur et al. (2023) synthesized $BaTiO_3$, and their piezoelectric properties were analyzed with MB degradation using ball milling. Furthermore, it is necessary to analyze the electronic structure of the semiconductor through theoretical simulations in the absence and presence of external forces, along with light irradiation. Additionally, the kinetics of charge transfer and catalytic reactions need to be analyzed through in situ characterization.

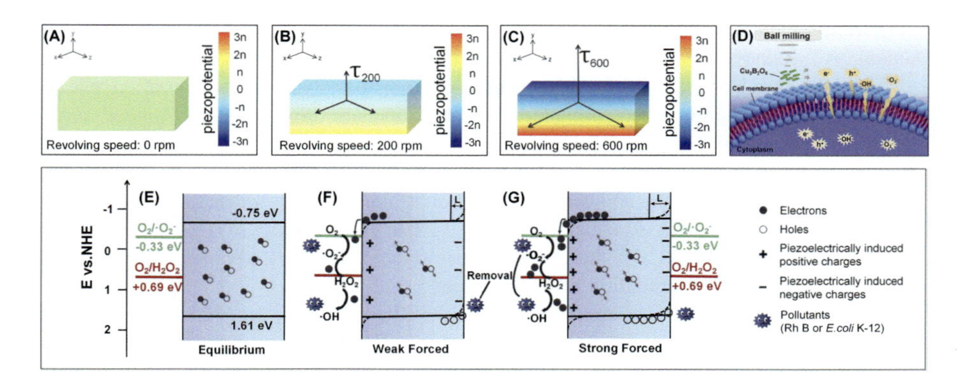

Figure 3.9 (A)–(C) The diagrams of the relationship between revolving speed and piezopotential on the basis of Eqs. (3.6) and (3.7) in which τ is the shearing force. (D) Schematic illustration of highly reactive species produced by CBO attack *E. coli* K-12 under BM. (E) Band structure of CBO when no BM is applied to the surface of the material. (F) Weak forced and (G) Strong forced induce the bands bending, forming the space charge region (L).
From Liao, X., Xie, H., Liao, B., Hou, S., Yu, Y. & Fan, X. (n.d.). Ball milling induced strong polarization electric fields in $Cu_3B_2O_6$ crystals for high efficiency piezocatalysis.

References

Bai, B., Zhang, C., Dou, Y., Kong, L., Wang, L., Wang, S., Li, J., Zhou, Y., Liu, L., Liu, B., Zhang, X., Hadar, I., Bekenstein, Y., Wang, A., Yin, Z., Turyanska, L., Feldmann, J., Yang, X., & Jia, G. (2023). Atomically flat semiconductor nanoplatelets for light-emitting applications. *Chemical Society Reviews*, *52*(1), 318–360. Available from https://doi.org/10.1039/d2cs00130f.

Boinpally, S., Kolla, A., Kainthola, J., Kodali, R., & Vemuri, J. (2023). A state-of-the-art review of the electrocoagulation technology for wastewater treatment. *Water Cycle*, *4*, 26–36. Available from https://doi.org/10.1016/j.watcyc.2023.01.001.

Chang, J., Dommer, M., Chang, C., & Lin, L. (2012). Piezoelectric nanofibers for energy scavenging applications. *Nano Energy*, *1*(3), 356–371. Available from https://doi.org/10.1016/j.nanoen.2012.02.003.

Chou, T.-M., Chan, S.-W., Yu-Jiung, L., Yang, P.-K., Liu, C.-C., Lin, Y.-J., Wu, J.-M., Lee, J.-T., & Lin, Z.-H. (2019). A highly efficient Au-MoS_2 nanocatalyst for tunable piezocatalytic and photocatalytic water disinfection. *Nanoenergy*, 57. Available from https://doi.org/10.1016/j.nanoen.2018.12.006.

Chu, X., Jiang, X., Zhang, H., Wang, C., Huang, F., Sun, X., & Li, S. (2022). Microstructure Engineering of Al doped $SrTiO_3/TiO_2$ heterostructure nanorod arrays boosting piezo-photocatalytic performances. *Advanced Materials Technologies*, *7*(12). Available from https://doi.org/10.1002/admt.202200390.

Dawason, W. (1993). *Hydrothermal synthesis of ferroelectric materials*, the Fourth US Japan Seminar in Hawaii, October.

Dong, L., Lou, J., & Shenoy, V. B. (2017). Large in-plane and vertical piezoelectricity in janus transition metal dichalchogenides. *ACS Nano*, *11*(8), 8242–8248. Available from https://doi.org/10.1021/acsnano.7b03313.

Durairaj, A., Liu, J., & Lv, X. (2021). In Vasanthkumar, S., & Sakthivel T. (eds.), *Facile synthesis of waste-derived carbon/MoS_2 composite for energy storage and water purification applications*.

Durairaj, A., Ramasundaram, S., Sakthivel, T., Ramanathan, S., Rahaman, A., Kim, B., & Vasanthkumar, S. (2019). Air bubbles induced piezophotocatalytic degradation of organic pollutants using nanofibrous poly(vinylidene fluoride)-titanium dioxide hybrid. *Applied Surface Science*, *493*, 1268–1277. Available from https://doi.org/10.1016/j.apsusc.2019.07.127.

Gaur, A., Chauhan, V. S., & Vaish, R. (2023). Planetary ball milling induced piezocatalysis for dye degradation using BaTiO3 ceramics. *Environmental Science: Advances*, *2*, 462–472.

Gong, S., Zhang, W., Liang, Z., Zhang, Y., Gan, T., Hu, H., & Huang, Z. (n.d.). Construction of a $BaTiO_3$/tubular *g*-C_3N_4 dual piezoelectric photocatalyst with enhanced carrier separation for efficient degradation of tetracycline.

Huo, B., Wang, J., Wang, Z., Zhang, X., Yang, J., Wang, Y., Qi, J., Ma, W., & Meng, F. (2023). Bubble-driven piezo-activation of E-MoS_2/PVDF piezoelectric microcapsule for antibiotic degradation with ultralow energy consumption. *Journal of Cleaner Production*, 419. Available from https://doi.org/10.1016/j.jclepro.2023.138333.

Im, E., Park, S., Hwang, G. T., Hyun, D. C., Min, Y., & Moon, G. D. (2023). Single-crystal ferroelectric-based (K, Na) NbO_3 microcuboid/CuO nanodot heterostructures with enhanced photo–piezocatalytic activity.

Jiang, Y., Xie, J., Lu, Z., Hu, J., Hao, A., & Cao, Y. (2022). Insight into the effect of OH modification on the piezo-photocatalytic hydrogen production activity of SrTiO3. *Journal of Colloid and Interface Science*, *612*, 111–120.

Justinabraham, R., Sowmya, S., Durairaj, A., Sakthivel, T., Wesley, R. J., Vijaikanth, V., & Vasanthkumar S. (2023). In Vasanthkumar, A. (ed.), *Compounds, Synthesis and characterization of SbSI modified g-C_3N_4 composite for photocatalytic and energy storage applications* (Vol. 935).

Kataki, S., Chatterjee, S., Vairale, M. G., Dwivedi, S. K., & Gupta, D. K. (2021). Constructed wetland, an eco-technology for wastewater treatment: A review on types of wastewater treated and components of the technology (macrophyte, biolfilm and substrate. *Journal of Environmental Management*, 283. Available from https://doi.org/10.1016/j.jenvman.2021.111986.

Khosya, M., Faraz, M., & Khare, N. (2022). Enhanced photocatalytic reduction of hexavalent chromium by using piezo-photo active calcium bismuth oxide ferroelectric nanoflakes. *New Journal of Chemistry*, *46*(25), 12244–12251. Available from https://doi.org/10.1039/d2nj01005d.

Kordbacheh, F., & Heidari, G .J. M. C. H. (2023). *Water pollutants and approaches for their removal* (Vol. 2, pp. 139–153).

Li, R., Liang, S., Aihemaiti, A., Li, S., & Zhang, Z. (2023). Effectively enhanced piezocatalytic activity in flower-like 2h-MoS2 with tunable S vacancy towards organic pollutant degradation. *Applied Surface Science*, *631*, 157461.

Li, H., Sang, Y., Chang, S., Huang, X., Zhang, Y., Yang, R., Jiang, H., Liu, H., & Wang, Z. L. (2015). Enhanced ferroelectric-nanocrystal-based hybrid photocatalysis by ultrasonic-wave-generated piezophototronic effect. *Nano Letters*, *15*(4), 2372–2379. Available from https://doi.org/10.1021/nl504630j.

Li, H., Xiong, Y., Wang, Y., Ma, W., Fang, J., Li, X., Han, Q., Liu, Y., He, C., & Fang, P. (2022). High piezocatalytic capability in CuS/MoS_2 nanocomposites using mechanical energy for degrading pollutants. *Journal of Colloid and Interface Science*, *609*, 657–666. Available from https://doi.org/10.1016/j.jcis.2021.11.070.

Li, M., Ramachandran, R., Sakthivel, T., Wang, F., & Xu, Z. X. (2021). Siloxene: An advanced metal-free catalyst for efficient photocatalytic reduction of aqueous Cr(VI) under visible light. *Chemical Engineering Journal*, 421. Available from https://doi.org/10.1016/j.cej.2021.129728.

Li, R., Liang, S., Aihemaiti, A., Li, S., & Zhang, Z. (2023). Effectively enhanced piezocatalytic activity in flower-like 2h-MoS_2 with tunable S vacancy towards organic pollutant degradation. *SSRN*. Available from https://doi.org/10.2139/ssrn.4414692.

Li, Y., Wang, Q., Wang, H., Tian, J., & Cui, H. (2019). Novel Ag_2O nanoparticles modified MoS_2 nanoflowers for piezoelectric-assisted full solar spectrum photocatalysis. *Journal of Colloid and Interface Science*, *537*, 206–214. Available from https://doi.org/10.1016/j.jcis.2018.11.013.

Liao, X., Xie, H., Liao, B., Hou, S., Yu, Y., & Fan, X. (n.d.). Ball milling induced strong polarization electric fields in $Cu_3B_2O_6$ crystals for high efficiency piezocatalysis.

Lin, D., Xiao, D., Zhu, J., Yu, P., Yan, H., & Li, L. (2004). Synthesis and piezoelectric properties of lead-free piezoelectric $[Bi_{0.5}(Na_{1-x-y}K_xLi_y)_{0.5}]TiO_3$ ceramics. *Materials Letters*, *58*(5), 615–618. Available from https://doi.org/10.1016/S0167-577X(03)00580-9.

Liu, D. M., Zhang, J. T., Jin, C. C., Chen, B. B., Hu, J., Zhu, R., & Wang, F. (2022). Insight into oxygen-vacancy regulation on piezocatalytic activity of $(Bi_{1/2}Na_{1/2})TiO_3$ crystallites: Experiments and first-principles calculations. *Nano Energy*, 95. Available from https://doi.org/10.1016/j.nanoen.2022.106975.

Lu, Y., Ding, C., Guo, J., Gan, W., Chen, P., Chen, R., Ling, Q., Zhang, M., Wang, P., & Sun, Z. (2023). Cobalt-doped ZnAl-LDH nanosheet arrays as recyclable piezo-catalysts for effective activation of peroxymonosulfate to degrade norfloxacin: non-radical pathways and theoretical calculation studies. *Nano Energy*, 112. Available from https://doi.org/10.1016/j.nanoen.2023.108515.

Mahapatra, S. D., Mohapatra, P. C., Aria, A. I., Christie, G., Mishra, Y. K., Hofmann, S., & Thakur, V. K. (2021). Piezoelectric materials for energy harvesting and sensingapplications: roadmap for future smart materials. *Advanced Science*, 8. Available from https://doi.org/10.1002/advs.202100864.

Meng, H., Chen, Z., Lu, Z., & Wang, X. (2023). Piezoelectric effect enhanced plasmonic photocatalysis in the Pt/$BaTiO_3$ heterojunctions. *Journal of Molecular Liquids*, *369*, 120846. Available from https://doi.org/10.1016/j.molliq.2022.120846.

Porwal, C., Gaur, A., Chauhan, V. S., & Vaish, R. J. S. (2023). Interfaces, enhancing piezo-catalytic dye degradation through ball milling-induced polarization in nano bismuth zinc borate.

Prasanna, G., Nguyen, H. D. P., Dunn, S., Karunakaran, A., Marken, F., Bowen, C. R., Le, B. N. T., Nguyen, H. D., & Pham, T. P. T. (2023). Impact of stirring regime on piezocatalytic dye degradation using $BaTiO_3$ nanoparticles. *Nano Energy*, 116. Available from https://doi.org/10.1016/j.nanoen.2023.108794.

R, D. J. (1998). *Damjanovic, ferroelectric, dielectric and piezoelectric properties of ferroelectric thin films and ceramics*.

Raghavan, N., Thangavel, S., Sivalingam, Y., & Venugopal, G. (2018). Investigation of photocatalytic performances of sulfur based reduced graphene oxide-TiO_2 nanohybrids. *Applied Surface Science*, *449*, 712–718. Available from https://doi.org/10.1016/j.apsusc.2018.01.043.

Raghavan, N., Thangavel, S., & Venugopal, G. (2015). Enhanced photocatalytic degradation of methylene blue by reduced graphene-oxide/titanium dioxide/zinc oxide ternary nanocomposites. *Materials Science in Semiconductor Processing*, *30*, 321–329. Available from https://doi.org/10.1016/j.mssp.2014.09.019.

Ren, Z., Peng, Y., He, H., Ding, C., Wang, J., Wang, Z., & Zhang, Z. (2023). Piezoelectrically mediated reactions: From catalytic reactions to organic transformations. *Chinese Journal of Chemistry*, *41*(1), 111–128. Available from https://doi.org/10.1002/cjoc.202200443.

Sheng, T., He, Q., Cao, Y., Dong, Z., Gai, Y., Zhang, W., Zhang, D., Chen, H., & Jiang, Y.J. A.A. M. (2023). Interfaces, fish-wearable piezoelectric nanogenerator for dual-modal energy scavenging from fish-tailing.

Stadler, L. B., Delgado Vela, J., Jain, S., Dick, G. J., & Love, N. G. (2018). Elucidating the impact of microbial community biodiversity on pharmaceutical biotransformation during wastewater treatment. *Microbial Biotechnology*, *11*(6), 995–1007. Available from https://doi.org/10.1111/1751-7915.12870.

Sun, Y., Shen, S., Deng, W., Tian, G., Xiong, D., Zhang, H., Yang, T., Wang, S., Chen, J., & Yang, W. (2023). Suppressing piezoelectric screening effect at atomic scale for enhanced piezoelectricity. *Nano Energy*, 105. Available from https://doi.org/10.1016/j.nanoen.2022.108024.

Tang, Q., Wu, J., Chen, X.-Z., Gual, R. S., Veciana, A., Franco, C., Kim, D., Surin, I., Ramírez, J. P., Mattera, M., Terzopoulou, A., Qin, N., Vukomanovic, M., Nelson, B. J., Luis, J. P., & Pané a, S. (2023). Tuning oxygen vacancies in $Bi4Ti_3O_{12}$ nanosheets to boost piezo-photocatalytic activity. *Nano Energy*, 108. Available from https://doi.org/10.1016/j.nanoen.2023.108202.

Tang, R., Gong, D., Zhou, Y., Deng, Y., Feng, C., Xiong, S., Huang, Y., Peng, G., Li, L., & Zhou, Z. (2022). Unique g-C3N4/PDI-g-C3N4 homojunction with synergistic piezo-photocatalytic effect for aquatic contaminant control and H_2O_2 generation under visible light. *Applied Catalysis B: Environmental*, 303. Available from https://doi.org/10.1016/j.apcatb.2021.120929.

Tang, Y., Chen, X., Zhu, M., Liao, X., Hou, S., Yu, Y., & Fan, X. (2022). The strong alternating built-in electric field sourced by ball milling on Pb_2BO_3X (X Cl, Br, I) piezoelectric materials contributes to high catalytic activity. *Nano Energy, 101*, 107545. Available from https://doi.org/10.1016/j.nanoen.2022.107545.

Thangavel, S., Pazhamalai, P., Krishnamoorthy, K., Sivalingam, Y., Arulappan, D., Mohan, V., Kim, S.-J., & Venugopal, G.J. C. (2022). *Ferroelectric-semiconductor $BaTiO_3-Ag_2O$ nanohybrid as an efficient piezo-photocatalytic material* (Vol. 292).

Thangavel, S., Thangavel, S., Raghavan, N., Krishnamoorthy, K., & Venugopal, G. (2016). Visible-light driven photocatalytic degradation of methylene-violet by rGO/Fe_3O_4/ZnO ternary nanohybrid structures. *Journal of Alloys and Compounds, 665*, 107–112. Available from https://doi.org/10.1016/j.jallcom.2015.12.192.

Venugopal, G., Thangavel, S., Vasudevan, V., & Zoltán, K. J. (2020). Solids, efficient visible-light piezophototronic activity of ZnO–Ag8S hybrid for degradation of organic dye molecule. *J. o. P., 143.*

Wang, F., Zhang, J., Jin, C. C., Ke, X., Wang, F., & Liu, D. (2022a). Unveiling the effect of crystal facets on piezo-photocatalytic activity of $BiVo_4$. *SSRN*. Available from https://www.ssrn.com/index.cfm/en/.

Wang, K., han, C., Li, J., Qiu, P., Sunarso, D., & Liu Prof. S. (2022b). the mechanism of piezocatalysis: Energy band theory or screening charge effect?.

Xiao, L., Liu, J., & Ge, J. (2021). Dynamic game in agriculture and industry cross-sectoral water pollution governance in developing countries. *Agricultural Water Management*, 243. Available from https://doi.org/10.1016/j.agwat.2020.106417.

Xiao, Q., Chen, L., Xu, Y., Feng, W., & Qiu, X. (2023). Impact of oxygen vacancy on piezo-photocatalytic catalytic activity of barium titanate. *Applied Surface Science, 619*, 156794. Available from https://doi.org/10.1016/j.apsusc.2023.156794.

Xu, Q., Gao, X., Zhao, S., Liu, Y. N., Zhang, D., Zhou, K., Khanbareh, H., Chen, W., Zhang, Y., & Bowen, C. (2021). Construction of bio-piezoelectric platforms: From structures and synthesis to applications. *Advanced Materials, 33*(27). Available from https://doi.org/10.1002/adma.202008452.

Yang, X., Wang, J., El-Sherbeeny, A. M., AlHammadi, A. A., Park, W.-H., & Abukhadra, M. R. (2022). Insight into the adsorption and oxidation activity of a ZnO/piezoelectric quartz core-shell for enhanced decontamination of ibuprofen: Steric, energetic, and oxidation studies. *Chemical Engineering Journal*, 431. Available from https://doi.org/10.1016/j.cej.2021.134312.

Zeng, Q., Jia, Z., Liu, X., & Cheng, J.-pS. (2023). Protection, a noval 1T-2H MoS2/NaBi (MoO4)2 alternating-phase piezoelectric composites for high-efficient ultrasound-drived piezoelectric catalytic removal of Sildenafil.

Zhao, Q., Zou, Y., & Liu, Z. (2022). The spontaneous polarization in CdS to enhance the piezo-PEC performance via phase transition stress engineering. *Journal of Catalysis, 416*, 398–409. Available from https://doi.org/10.1016/j.jcat.2022.11.024.

Zhou, Q., Shi, Q., Li, N., Chen, D., Xu, Q., Li, H., Hea, J., & Lu, J. (2020). Rh-Doped SrTiO 3 inverse opal with piezoelectric effect for enhanced visible-light-driven photodegradation of bisphenol A. *Environmental Science: Nano, 7*, 2267–2277.

Advances in nanocomposite membranes for CO_2 removal

4

Fauziah Marpani[1], *Nur Hidayati Othman*[1], *Nur Hashimah Alias*[1], *Muhammad Shafiq Mat Shayuti*[1] and *Sacide Alsoy Altinkaya*[2]
[1]School of Chemical Engineering, College of Engineering, Universiti Teknologi MARA, Shah Alam, Selangor Darul Ehsan, Malaysia, [2]Department of Chemical Engineering, Izmir Institute of Technology, İzmir, Turkey

4.1 Introduction

Carbon dioxide (CO_2) levels in May 2023 averaged 424.0 parts per million (ppm), the fourth-largest annual increase since the measurements began 65 years ago at the NOAA observatory in Mauna Loa, Hawaii (Ajasa, 2023). The main industrial CO_2 sources are power plants, petroleum refineries, and manufacturing plants of cement, iron, steel, and ammonia (Cuéllar-Franca & Azapagic, 2015). It is estimated that 12 gigatonnes or 40% of emissions around the globe are from electricity generation by fossil fuel power plants alone (IEA, 2023; Voumik et al., 2023). Carbon capture is one method that could mitigate the climate change caused by the abundant release of CO_2 in the atmosphere. The conventional carbon capture techniques include solvent absorption, adsorption by solid sorbents, cryogenic distillation, and pressure swing adsorption. These techniques are known to be energy-intensive and can use up to 4–6 GJ/t of CO_2 captured (Goh & Ismail, 2020).

Currently, the most adopted technology on an industrial scale is solvent absorption by monoethanolamine (MEA) due to its technological maturity, which can reach more than 90% efficiency (Goh & Ismail, 2020). Alternatively, membrane separation is more environmentally friendly as no chemicals are involved during the operation compared to the amine absorption process. Only a small fraction of solvent is used during the membrane fabrication. The efficiency of the membrane separation process is more than 80%, which is slightly lower than the MEA absorption. The cost of operation is estimated to be US\$20.5 ton^{-1} of CO_2 captured, and only about 20% of energy is required from the MEA absorption process (1 GJ/t CO_2) (Goh & Ismail, 2020). Recently, a solvent-free UV polymerization method was used to crosslink poly(ethylene glycol) methyl ether acrylate (PEGMEA) with a large molecular-size bisphenol A ethoxylate diacrylate (Sun et al., 2023). The crosslinked PEO membranes can be readily used without any further treatment, signifying the rapid, green, and material-efficient features of the technique. The optimum membrane successfully addresses the trade-off dilemma where dense microstructures reduce gas permeability and loose microstructures compromise high-pressure resistance. As a result, it achieves a remarkable CO_2 permeability of 1711 Barrer and maintains stable performance for 100 hours at 15 atm.

Nanocomposites for Environmental, Energy, and Agricultural Applications. DOI: https://doi.org/10.1016/B978-0-443-13935-2.00004-8

Among the various approaches, nanocomposite membranes have emerged as a promising solution for the efficient and selective removal of CO_2 from industrial emissions (Rezakazemi et al., 2014) and natural gas streams (Alqaheem et al., 2017). Nanocomposite membranes combine the advantages of both polymeric matrices and nanofillers, resulting in improved separation performance and enhanced selectivity. The integration of nanofillers, such as nanoparticles, nanotubes, and porous materials, into the polymer matrix enables precise control over the transport properties and interaction with target gases, such as CO_2 (Chen et al., 2021; Li et al., 2022). For example, recently, a nanocomposite membrane fabrication from pyrazole-based metal-organic framework (MOF) has shown a high CO_2 permeability of 7528.2 Barrer (Chen et al., 2023). Pebax, carboxymethyl cellulose, and MXene-based mixed matrix membranes showed an excellent CO_2 selectivity of 40.4 and enhanced mechanical stability in a continuous operation of 60 hours (Luo et al., 2022). These studies have shown that a synergistic effect arising from the interactions between the nanofillers and polymer matrix results in efficient gas separation, making nanocomposite membranes a promising technology for large-scale CO_2 removal applications.

In this chapter, we explore the recent advances in nanocomposite membranes for CO_2 removal, focusing on the key nanofillers utilized, their impact on membrane performance, and the underlying mechanisms governing the gas separation process. Additionally, we discuss the challenges and opportunities in the development of nanocomposite membranes, along with potential strategies for further improvement. With the increasing demand for sustainable solutions to mitigate CO_2 emissions, the continued progress in nanocomposite membrane technology holds the potential to revolutionize carbon capture and storage practices, contributing significantly to a cleaner and greener future.

4.2 Nanocomposite membrane configurations

The nanocomposite membrane is a novel filtration material that combines polymeric and inorganic components in a single entity, with nanosized fillers commonly dispersed in a continuous polymer matrix or embedded in a polymeric or inorganic oxide matrix, functionalizing the membrane. The primary objective behind the design of the nanocomposite membrane is to combine the superior intrinsic separation capabilities of inorganic nanomaterials with the durability and mechanical stability of polymers in a synergistic manner.

The nanocomposite membrane configuration for gas separation can be a flat sheet or hollow fiber. In large-scale applications, flat sheet membranes are assembled in spiral wound modules. These modules are pressurized with the feed gas, allowing the more permeable gas to pass through the membrane and exit through the central perforated tube. Also known as plate and frame membranes, the advantage of flat sheets is that they offer reduced pressure drops and enhanced mass transfer facilitated by the presence of a feed spacer (Rivero et al., 2023; Xu et al., 2019). The hollow fiber

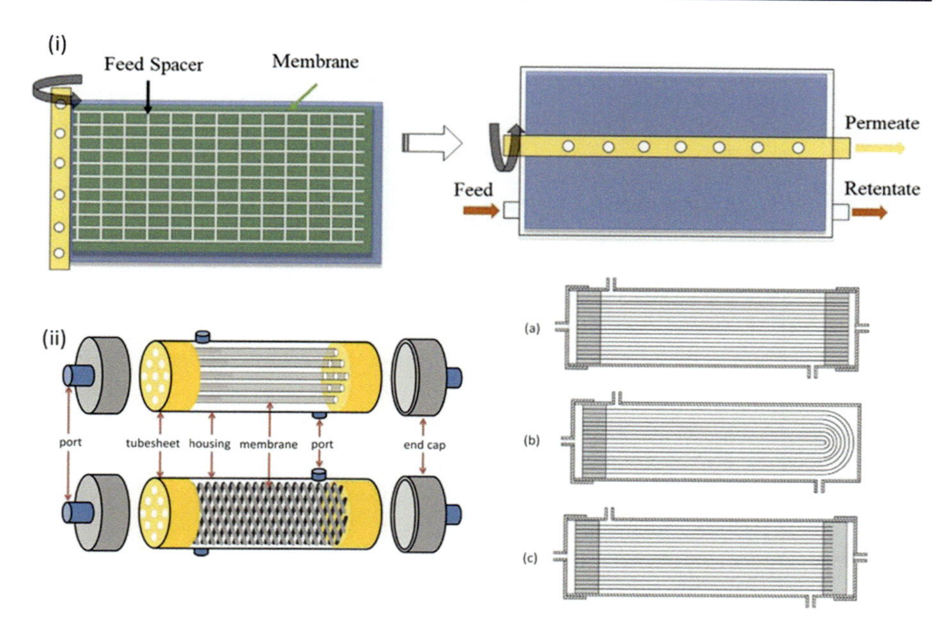

Figure 4.1 Graphic illustration of: (i) spiral wound membrane before and after spinning. (ii) hollow fiber membrane modules with a parallel membrane arrangement and a crisscross membrane arrangement and with different tubes arrangements; (a) two tubesheets at both ends, (b) a single tubesheet and a "U" shaped bundle, and (c) one tube sheet and one sealed end (Wan et al., 2017).
Source: (A) Adapted from Rivero, J. R., Nemetz, L. R., Da Conceicao, M. M., Lipscomb, G., & Hornbostel, K. (2023). Modeling gas separation in flat sheet membrane modules: Impact of flow channel size variation. Carbon Capture Science and Technology, *6* (2023). (B and C) From Wan, C. F., Yang, T., Lipscomb, G. G., Stookey, D. J., & Chung, T. S. (2017). Design and fabrication of hollow fiber membrane modules. *Journal of Membrane Science*, *538*, 96–107. https://doi.org/10.1016/j.memsci.2017.05.047.

configurations have been the preferred choice for fabricating gas separation membranes because they offer a larger membrane surface area within a specific module volume, resulting in increased productivity per membrane unit compared to spiral wound membranes (Goh et al., 2020). Hollow fiber configuration also has the advantage of being self-supporting, which simplifies manufacturing and replacement processes. Additionally, these membranes are known for their higher resistance to damage compared to other module types. Fig. 4.1 illustrates the two different modules of nanocomposite membranes suitable for CO_2 removal.

4.2.1 Structure of nanocomposite membranes

Nanocomposite membranes, or mixed matrix membranes, are generally classified depending on the fabrication techniques, membrane structure, and location of nanoparticles in the membranes (Fig. 4.2). In gas separation, two prominent types of

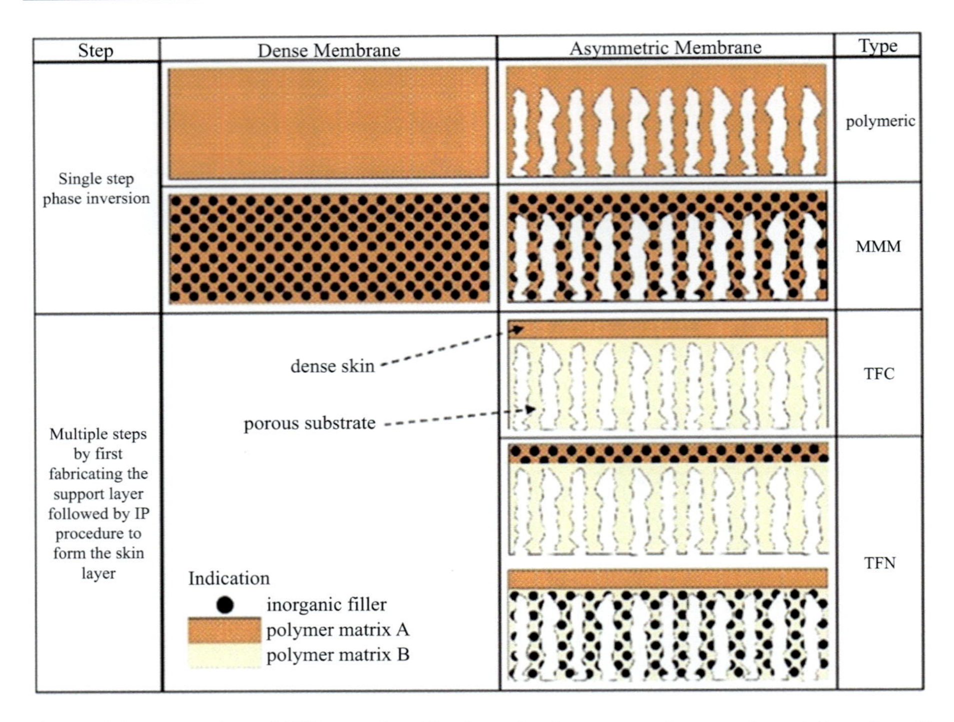

Figure 4.2 Illustration of different classification of polymer membranes adapted from Ismail and team.
Source: From Wong, K. C., Goh, P. S., & Ismail, A. F. (2016). Thin film nanocomposite: The next generation selective membrane for CO_2 removal. *Journal of Materials Chemistry A*, *4*(41), 15726–15748. https://doi.org/10.1039/c6ta05145f.

nanocomposite membranes are thin film composite (TFC) membranes and thin film nanocomposite (TFN) membranes. Understanding the differences between these two membrane types, their fabrication methods, and the role of nanoparticles is essential for optimizing their performance. TFC membranes are semipermeable membranes that consist of two distinct layers. The active separation layer, which is responsible for the separation process, is typically a polyamide thin film. This active layer is deposited onto a porous support layer, often made of a polymer like polysulfone or polyethersulfone. TFN membranes represent a more advanced approach to nanocomposite membrane design. These membranes incorporate nanoparticles or nanofillers directly into the active separation layer, providing unique advantages. The content of nanoparticles in TFN membranes can occupy a significant proportion, with some formulations reaching up to 50% of the total membrane weight (Chung et al., 2007). This increased nanoparticle content has a profound impact on the membrane's performance. The choice of nanofillers used in nanocomposite membrane fabrication plays a crucial role in achieving the desired separation properties. These nanofillers can be categorized as porous or nonporous, each with distinct characteristics. Porous nanofillers include materials like zeolites, carbon molecular sieves, activated carbon, MOFs, and carbon nanotubes (CNTs). Nonporous nanofillers, on the other hand, comprise materials such as silica, titanium oxide (TiO_2), and fullerene (C60). In addition to filler selection, another key consideration in nanocomposite membrane fabrication is the controlled embedment of nanofillers within the active layer, often referred to as the "skin" layer. To fully harness the potential of nanofillers for high selectivity and permeability, their distribution within this skin layer should be precisely controlled (Wong et al., 2016). TFN membranes, characterized by a substrate bottom layer formed through phase inversion and a selective top layer created via interfacial polymerization (IP), offer a solution to this challenge. The phase inversion process allows the distribution of nanofillers throughout the substrate layer, whereas the IP process controls their distribution within the skin layer (Seman et al., 2010). It is also possible to introduce multiple types of fillers during the IP process (Fig. 4.3).

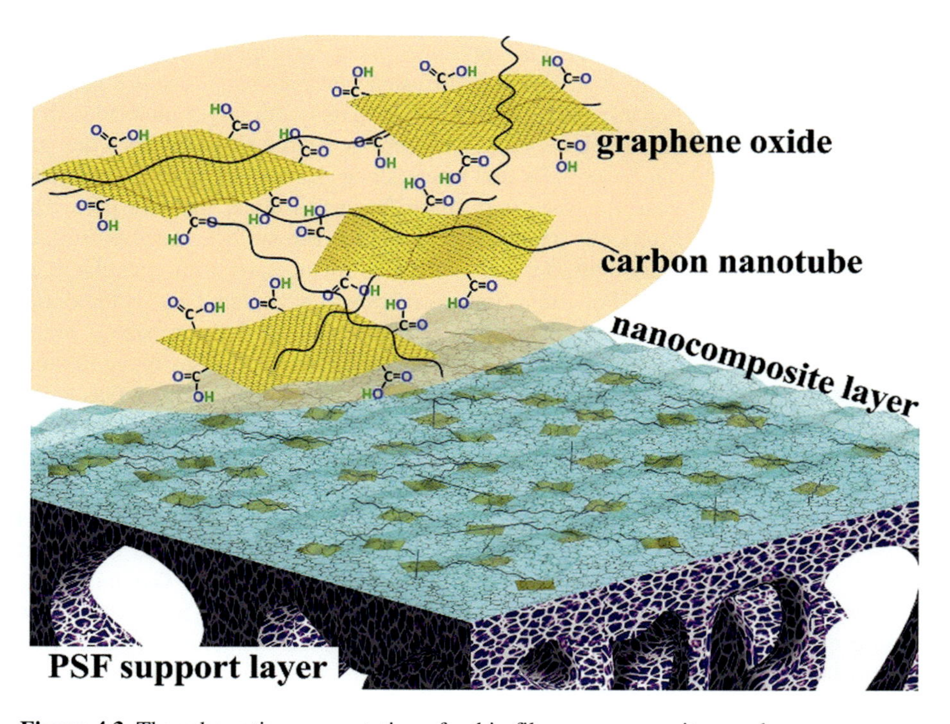

Figure 4.3 The schematic representation of a thin film nanocomposite membrane demonstrates the uniform dispersion of a dual-filler comprising graphene oxide and carbon nanotube within the polyamide layer (Wong et al., 2019).
Source: From Wong, K. C., Goh, P. S., Taniguchi, T., Ismail, A. F., & Zahri, K. (2019). The role of geometrically different carbon-based fillers on the formation and gas separation performance of nanocomposite membrane. *Carbon*, *149*, 33–44. https://doi.org/10.1016/j.carbon.2019.04.031.

4.2.2 Transport mechanism in nanocomposite membranes

In the process of gas separation using membranes, the movement of substances occurs when external pressure is applied to the components on the feed side. The membrane acts as a barrier and offers varying levels of resistance to the transport of different gas molecules present in a mixture of two or more gases. The extent of resistance is influenced by the chemical and physical characteristics of the gas molecules, including their polarity, interactions, and affinity toward the membrane materials, as well as their shape and molecular sizes. The membrane's effectiveness and efficiency in gas separation are evaluated based on two key factors: permeability, which refers to the ability of gases to pass through the membrane, and selectivity, which indicates the membrane's capability to preferentially allow certain gases to permeate while hindering others.

Transport mechanisms are different between porous and nonporous membranes. Gas transport in porous membranes typically occurs through three main mechanisms: Knudsen diffusion, Knudsen/surface flow, or Poiseuille flow. Knudsen diffusion applies to gases in the rarefied regime, where the pore size is comparable to or smaller than the mean free path of the gas molecules. Knudsen diffusion is governed by molecular collisions with the pore walls. Gas molecules move through the pores in a series of random collisions, resulting in a slower diffusion rate compared to other mechanisms. This mechanism becomes more significant for gases with larger molecular sizes or for membranes with smaller pore sizes. Knudsen, or surface flow, occurs when gas molecules collide with the pore walls and move along the surface of the membrane. This mechanism is relevant for gases with mean free paths larger than the pore size, allowing them to follow tortuous paths along the membrane surface. Knudsen/surface flow can contribute to gas transport, particularly in membranes with irregular pore structures or rough surfaces. Poiseuille flow, also known as viscous flow, is the dominant mechanism for gas transport in porous membranes when the pore size is larger than the mean free path of the gas molecules. In this mechanism, gases move through the larger pores of the membrane under the influence of a pressure gradient. Gas molecules flow through the pores in a laminar manner, following streamlines and experiencing viscous drag as they interact with the pore walls. Poiseuille flow is described by Darcy's law and is influenced by pore size, pore connectivity, and gas viscosity.

Gas transport through nonporous polymeric membranes relies on molecular-level solution-diffusion (Cong et al., 2007). It involves the dissolution of gas molecules into the polymer matrix, followed by diffusion through the polymer. Gas molecules first permeate the surface of the membrane, dissolve in the polymer matrix, and then diffuse through the polymer chains to the other side of the membrane. This mechanism is driven by concentration gradients and is influenced by factors such as the solubility of the gas in the polymer and the diffusivity of the gas within the polymer.

In a mixture of gas with two components, A and B, the selectivity coefficient is defined as the ability of a membrane to separate gases A and B. It is given as (Kim & Lee, 2015)

$$\alpha_{A/B} = \frac{P_A}{P_B} = \left(\frac{S_A}{S_B}\right)\left(\frac{D_A}{D_B}\right)$$

where P is the gas permeability, S is the gas solubility, and D is the gas diffusivity. The permeability unit is represented by Barrer. 1 Barrer is equivalent to 3.348×10^{-16} mol · m/(m^2 · s · Pa). The primary factor influencing diffusivity is the kinetic diameter of the penetrating gas molecule. A small kinetic diameter indicates a high diffusivity. For example, the kinetic diameters of H_2, CO_2, N_2, and CH_4 molecules are 2.89, 3.30, 3.64, and 3.82 Å (1 Å = 10^{-10} m), respectively (Zhao et al., 2014). Hence, it can be concluded that the diffusivity for $H_2 > N_2 > CO_2 > CH_4$. One interesting phenomenon to be observed when transporting a gas mixture is the competition among the "slow" and "fast" gas species. The "slow" gas can reduce the permeability of the "fast" gas; literally, the presence of the "fast gas" increases the permeability of the "slow" gas (Visser et al., 2005). Gas solubility is related to the affinity between gas molecules and the polymer matrix, while gas diffusivity is related to the free volume of polymers and the molecular size of the penetrating gases. The characteristics that govern permeability properties in dense polymeric membranes include factors like matrix attributes such as density, whether the membranes exhibit a rubbery or glassy state, and the presence of free space, among others (Favvas et al., 2017). Another pivotal factor contributing to discrepancies in gas permeability across distinct polymers arises from variations in the physicochemical interactions occurring between gas molecules and the polymer chains within different polymer types. Some nonporous membranes, particularly those made of certain polymers or hydrogels, can undergo swelling in the presence of certain gases or vapors (Liu et al., 2008). When a gas or vapor is in contact with the membrane, it can cause the polymer to absorb the gas, resulting in swelling. This absorption-induced swelling creates temporary pathways for the gas molecules to permeate through the membrane. Once the gas molecules have diffused through the swollen polymer, the membrane may return to its original state.

In crystalline polymers, gas absorption within the polymer crystallites is minimal. Conversely, amorphous polymers exhibit high permeability to small gas molecules, with little influence from polymer chain conformations (resulting in no swelling effects) and a constant D coefficient. Larger molecules, however, can trigger swelling in the polymer matrix, alter diffusion rates, and cause the D coefficient to rely on the permeate concentration. Elastomeric polymers, characterized by highly mobile polymer chains, experience faster swelling compared to diffusion, leading to D coefficient dependence solely on the permeate concentration. In contrast, polymers with stiff, rigid chains undergo slower swelling, causing diffusive molecules to rely on macromolecule relaxation rates (Favvas et al., 2017). Fig. 4.4 illustrates the possible gas transport mechanisms occurring in nanocomposite membranes. The relative importance of these gas transport mechanisms depends on various factors, such as the membrane pore size distribution, gas properties (such as molecular size and viscosity), and operating conditions (such as pressure and temperature). The interplay between these mechanisms determines the overall gas transport behavior of porous and nonporous membranes and influences their performance in applications such as gas separation, filtration, and purification.

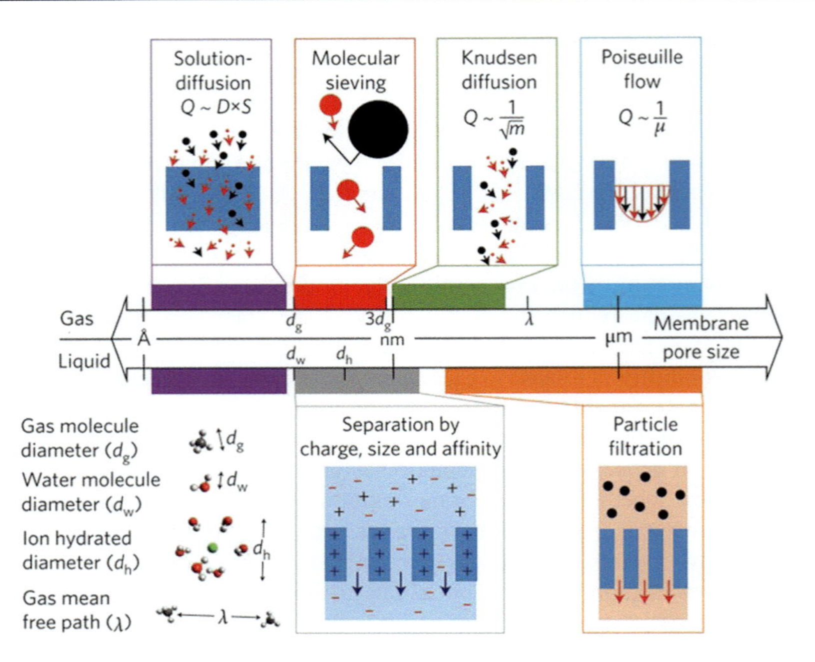

Figure 4.4 Illustration of transport mechanisms in nanocomposite membranes, relative scales of gas and water molecules, hydrated ions and gas mean free path are depicted on bottom left (Wang et al., 2017). *Q*: flux; *D*: diffusivity; *S*: sorption coefficient; *m*: molecular mass; μ: viscosity.
Source: Adapted from Wang, L., Boutilier, M. S. H., Kidambi, P. R., Jang, D., Hadjiconstantinou, N. G., & Karnik, R. (2017). Fundamental transport mechanisms, fabrication and potential applications of nanoporous atomically thin membranes. *Nature Nanotechnology*, *12*(6), 509–522. https://doi.org/10.1038/nnano.2017.72.

4.3 Nanofillers used in nanocomposite membranes fabrication

4.3.1 Classification of nanofillers

Polymer nanocomposites are materials that consist of nanosized particles dispersed within a polymer matrix or phase. The nanosized particles, also known as nanofillers, can be classified based on their shapes and dimensions and broadly categorized into four groups, namely 0D, 1D, 2D, and 3D. Fig. 4.5 illustrates the dimensions of different nanomaterials applied for gas separation.

0D refers to nanomaterials with all their dimensions below 100 nm. This category includes spherical nanoparticles, cubes, nanorods, polygons, hollow spheres, metallic and core-shell nanoparticles, quantum dots (QDs), dense nanoparticles (e.g., silica and fumed silica), metal oxides (e.g., TiO_2, MgO, and Al_2O_3), and polyhedral oligomeric silsesquioxanes. Upon incorporation into polymeric matrices, most 0D nanofillers tend to lead to an increase in CO_2 permeability. This increase

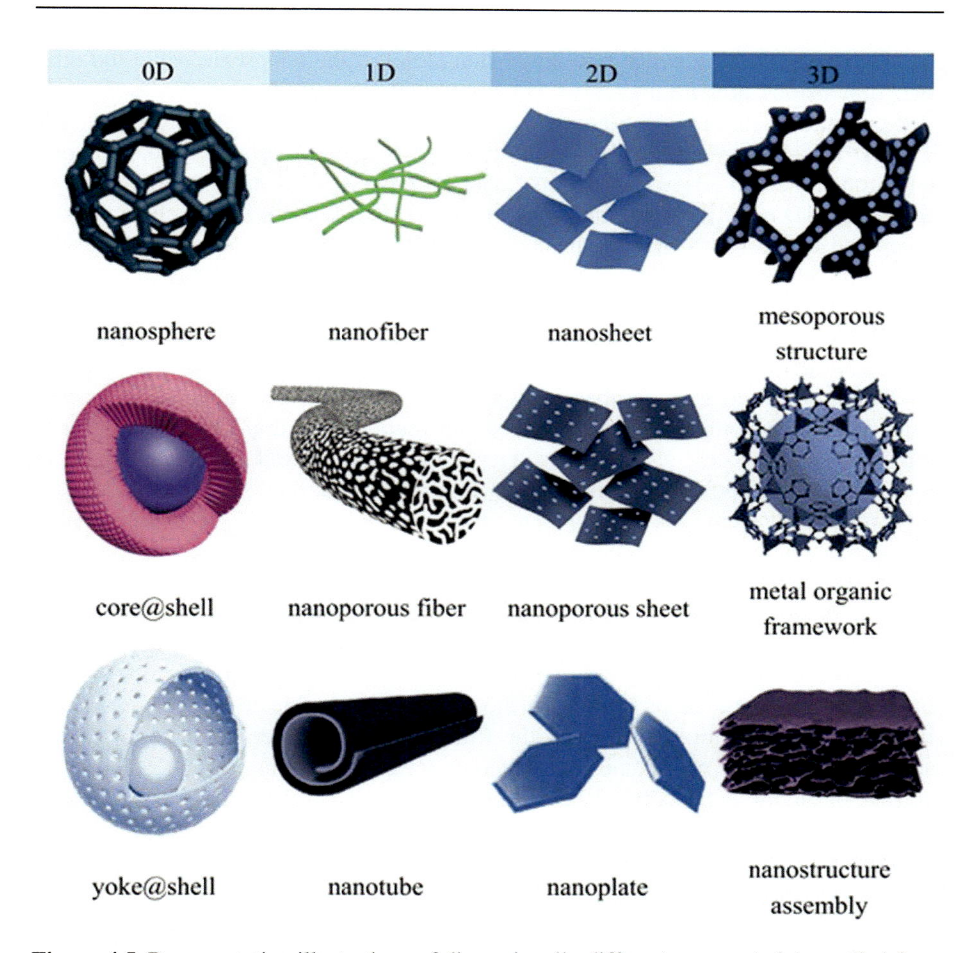

Figure 4.5 Representative illustrations of dimensionally different nanomaterials applied for gas separation and liquid separation nanocomposite membrane.
Source: Adapted from Goh, P. S., Wong, K. C., & Ismail, A. F. (2020). Nanocomposite membranes for liquid and gas separations from the perspective of nanostructure dimensions. *Membranes*, *10*(10), 1–29. https://doi.org/10.3390/membranes10100297.

can be attributed to two primary factors: heightened CO_2 sorption, dependent on the filler's nature or surface modification, and augmented CO_2 diffusion facilitated by the formation of voids at the interface between the polymer and nanoparticles. In contrast, metal oxide nanoparticles have shown favorable dispersibility in polymers and possess an intrinsic affinity toward CO_2. As a result, they have been extensively utilized in preparing nanocomposite membranes with a wide range of polymeric materials. TiO_2, in particular, has undergone massive investigation due to its CO_2-philic nature. On the other hand, other metal oxides with higher CO_2 adsorption capacity (e.g., ZnO, Al_2O_3, and MgO) have not received equal attention yet, but they hold potential for further development.

1D are materials in which one dimension is not at the nanoscale while the other two dimensions are. Examples of 1D nanomaterials include metallic, polymeric, and ceramic nanofibers or filaments, nanotubes, nanorods, nanowires, and CNTs. The 1D morphology inherently results in a high surface-to-volume ratio, providing numerous potential adsorption or functionalization sites with strong binding energy. Generally, 1D nanofillers exhibit a more disruptive effect on polymer chain packing compared to 0D fillers, leading to increased CO_2 permeability, albeit with varying effects on selectivity. However, in the case of CNTs, despite their high mass transport rate within their inner walls, effectively leveraging this property inside polymer matrices becomes challenging due to issues related to alignment and potential entry and end effects (Janakiram et al., 2018).

2D has one dimension on the nanoscale, while the other two dimensions are not. This category encompasses single-layered and multilayered thin films, crystalline or amorphous nanoplates, nanocoatings, graphene/graphene oxide (GO), molybdenum disulfide (MoS_2), polyaniline nanosheets, and other structures. 2D materials play a vital role in advancing the development of innovative nanocomposite membranes for gas separation. Their planar shape offers the most favorable arrangement for creating thin selective layers, allowing for full utilization of their properties even at lower filler loadings. However, beyond a certain threshold, increasing the filler loading tends to decrease CO_2 permeability due to the barrier effect of the nanoplatelets (Janakiram et al., 2018). On the positive side, the CO_2 selective properties are observed to increase in proportion to the filler concentration. Enhanced interaction improves the compatibility between the filler and the polymer, resulting in a more stable and effective membrane. 3D materials have dimensions beyond 100 nm and consist of multiple nanocrystals combined in various directions. Examples of 3D nanomaterials include mesoporous structures, MOF, foams, fibers, carbon nanobuds, nanotubes, fullerenes, pillars, polycrystals, honeycombs, and layered skeletons (Rizwan et al., 2021; Saleh, 2020). 3D nanofillers possess a higher surface area compared to their 0D or 1D counterparts. This increased surface area provides more adsorption sites for CO_2 molecules, enhancing their interaction with the filler and increasing the overall separation efficiency (Dashti et al., 2020). Some 3D nanofillers, such as MOFs, possess a porous structure with precisely defined pore sizes. These nanoscale pores act as molecular sieves, allowing selective passage of CO_2 molecules while blocking other gas molecules, leading to improved selectivity (Liu et al., 2012).

Nanofillers can also be categorized into porous and nonporous types. Porous fillers can disturb the polymer chain packing and modify the gas transport pathway by creating transport channels at the polymer-filler interface. Additionally, the intrinsic separation capability of porous fillers can further influence the transport properties of the resulting nanocomposite membranes.

4.3.2 Function of nanofillers

Nanofillers act as physical barriers, disrupting the straightforward path of gas molecules as they attempt to traverse the membrane. This tortuous pathway forces gas molecules to follow a more intricate and convoluted route, imposing a greater energy barrier for certain molecules, especially those with larger sizes or different properties. This inherent selectivity prevents some gases from easily permeating through the membrane, effectively screening and capturing them, as observed in the study by Regmi et al. (2022). Conversely, nanofillers might allow certain gases to navigate this intricate network with less hindrance due to their compatibility with the nanoparticle structure. These gases, often the target components for separation, find the pathway less obstructed and thus permeate more readily. This selective behavior is important for achieving higher separation efficiency in nanocomposite membranes. Also, some nanofillers, such as MOFs or zeolites, exhibit a strong attraction to particular gas molecules due to their surface chemistry and molecular structure. This attraction results in the adsorption of gas molecules onto the nanofiller's surface, effectively increasing the solubility selectivity of the membrane (reference is needed here). In the context of CO_2 separation, nanofillers with a high affinity for CO_2 can selectively adsorb CO_2 molecules from a gas mixture, allowing other gases to permeate through the membrane. Nanofillers can also contribute to an increase in the effective surface area of the membrane. This is achieved through the intricate and high-surface-area structures of many nanofillers, such as nanoparticles, CNTs, or zeolites. The greater surface area allows for more gas molecules to come into contact with the membrane simultaneously, promoting a higher rate of permeation. Furthermore, the incorporation of nanofillers can modify the morphology of the polymer matrix. By interacting with the polymer at a microscopic level, nanofillers can alter the structure of the matrix in ways that promote efficient gas transport. This modification might involve reducing the density of polymer chains in certain regions, opening up pathways for gas molecules to travel, or introducing favorable interactions between the polymer and gas molecules. These changes collectively result in improved gas transport rates.

In terms of durability, the incorporation of nanofillers can improve the mechanical strength and stability of the membrane (Sonpingkam & Pattavarakorn, 2014). They reinforce the polymer matrix, reducing the risk of membrane damage, deformation, or structural failure during operation. The various nanofillers available can tailor the surface properties of the membrane. The surface modification can enhance the membrane's compatibility with the targeted gas components, improve wettability, and reduce the occurrence of surface-related issues. Nanofillers are also able to introduce additional functional properties to the membrane. For example, nanoparticles with catalytic or photocatalytic properties can facilitate specific reactions or degradation processes during gas separation and convert into valuable compounds or the degradation of pollutants (Baniamer et al., 2021). In CO_2 capture and utilization processes, nanocomposite membranes equipped with catalytic nanofillers can promote the transformation of captured CO_2 into synthetic fuels or chemicals, contributing to carbon recycling and reducing emissions.

4.4 Ideal nanocomposite membranes for CO_2 removal

The essential criteria for a membrane to be suitable and practical for CO_2 gas separation can be summarized into three key requirements, which are as follows:

(1) achieving high permeability and selectivity; (2) demonstrating mechanical, thermal, and chemical stability; and (3) being economically competitive.

4.4.1 Permeability and selectivity

Permeability and selectivity of polymeric membranes are always a trade-off in gas separation. A relationship between selectivity and permeability of different gas pairs (CO_2, H_2, N_2, O_2, He, and CH_4) was illustrated through Robeson's upper bound logarithmic diagram (Robeson, 2008). Fig. 4.6 illustrates Robeson's upper bound curves, showing a general trade-off between permeability and selectivity for CO_2/CH_4. Researchers use the Robeson chart as a reference to guide the development of new membrane materials and structures, aiming to create membranes that approach or surpass the theoretical upper bound line for specific gas separation applications. It was developed by Dr. Richard Robeson and is a valuable tool for assessing and comparing the performance of different types of membranes in gas separation applications. Permeability refers to the rate at which a specific gas can pass through a membrane.

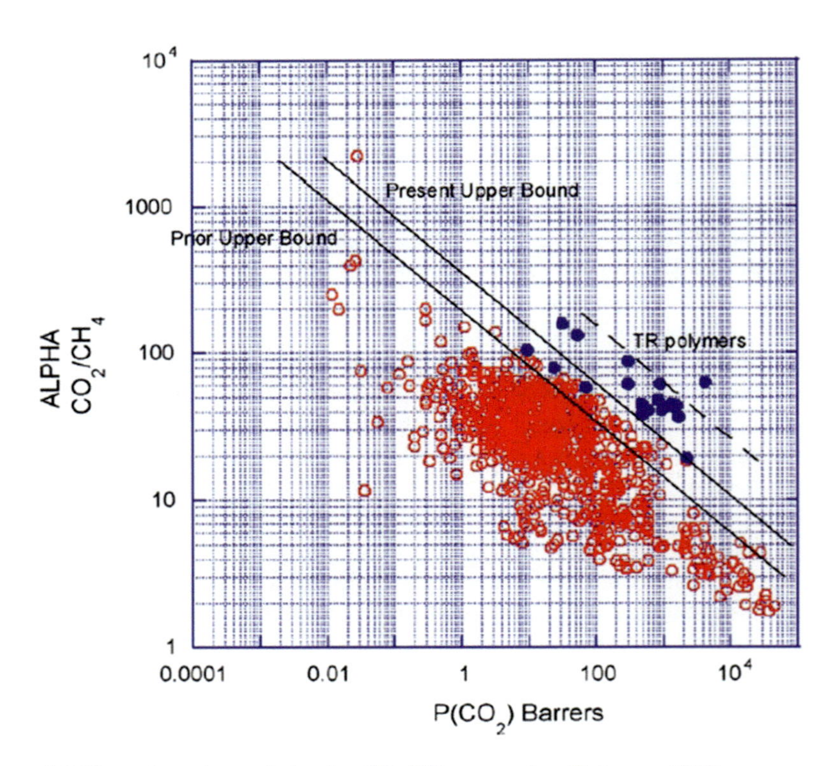

Figure 4.6 Upper bound correlation for CO_2/CH_4 separation (Robeson, 2008).
Source: Adapted from Robeson, L. M. (2008). The upper bound revisited. *Journal of Membrane Science*, *320*(1–2), 390–400. https://doi.org/10.1016/j.memsci.2008.04.030.

It is typically measured in Barrer (1 Barrer $= 3.35 \times 10^{-16}$ mol · m/m^2 · s · Pa), and higher permeability values indicate faster gas transport through the membrane. Selectivity represents the ability of a membrane to separate one gas from a mixture of gases. It is defined as the ratio of the permeabilities of the two gases of interest. A higher selectivity value indicates better separation performance. The Robeson chart is a plot where the x-axis represents selectivity (α) on a logarithmic scale and the y-axis represents permeability (P) on a logarithmic scale (Fig. 4.6). On this chart, different regions or areas are defined, and membranes that fall within these regions have certain characteristics. The upper bound line on the Robeson chart represents the theoretical limit of membrane performance. No membrane can exist above this line because it would violate the laws of physics. Membranes that surpass the upper bound line are considered highly efficient. Table 4.1 summarizes the recent work on CO_2 separation by nanocomposite membranes with different nanofillers, which has shown high permeability and/or selectivity.

In nanocomposite membrane development, the factors that affect the permeability and selectivity of CO_2 separation are a synergy between polymer and nanofiller. There are a few issues that must be addressed when discussing an optimum polymer—nanofiller pair. It includes the affinity of nanofillers towards CO_2 adsorption, the compatibility (dispersibility) of nanofillers with rubbery or glassy types of polymers, and the amount of nanofillers loading in the polymer matrix.

4.4.2 Selection of nanofillers and loading in nanocomposite membranes

Filler dispersion and the polymer—filler interaction are extremely important to ensure the homogeneity of the membrane (Lai & Tan, 2023). Particle agglomeration is always observed at high levels of nanofiller loading during nanocomposite membrane fabrication. Incompatibility between filler and polymer is the main cause of filler agglomeration, leading to the formation of nonselective voids and thus deterioration of the achievable gas separation performance (Lai & Tan, 2023). Vinoba and the team conducted an extensive review and introduced a new method of screening fillers, aiding in the development of new fillers and the fabrication of high CO_2 separation capacity nanocomposite membranes (Vinoba et al., 2017). One method to improve the dispersibility of nanofillers is by surface modification (functionalization) of nanofillers by incorporating functional groups such as $-NH_2$, $-SO_3H$, or $-COOH$ into the MOF structure (Xiang et al., 2017). Amine (NH_2), bromine (Br), and hydroxyl (OH) moieties are widely used in surface functionalization of fillers for CO_2 separation as they can interact with CO_2 efficiently (Lai & Tan, 2023). Modification of ZIF-8 (Shi et al., 2021; Suhaimi et al., 2019), UiO-66 (Anjum et al., 2015), and MIL-53 (Chen et al., 2013) with $-NH_2$ has been shown to exhibit excellent dispersibility in the polymer matrix as well as increase CO_2 permeability and CO_2/CH_4 selectivity. This was contributed by the dipole—quadrupole interaction between the NH_2-modified MOF and CO_2. Another method is to add polymers, for example, polydopamine, to the surface of

Table 4.1 Summary of high CO_2 gas permeability (> 1000 barrer) and selectivity of the most recent study.

Polymer	Nanofiller	Operating conditions	CO_2 permeability (barrer)	Selectivity CO_2/CH_4	Selectivity CO_2/N_2	References
Polymer of Intrinsic Microporosity (PIM-1)	UiO-66	4 bar/35°C	11261	–	28	Yu et al. (2023)
PIM-1	Pyrazole based MOF-303	2 bar/35°C	7528.2	27.6	–	Chen et al. (2023)
PIM-1	Zn-SiF6-py	1 bar/30°C	6268	–	26.3	Feng et al. (2023)
PIM-1	ZIF-62	2 bar/25°C	5914	67	–	Feng et al. (2023)
Polysulfone	GO grafted with PEG	1 bar/25°C	3169	15.8	37.4	Lee et al. (2023)
Polyvinylamine	Polymer-modified Schiff base network-1	1 bar/25°C	3092	–	184	Wang et al. (2023)
PVDF	Bentonite + amine-modified GO	1.5 bar/25°C	2679	326	307.3	Dong et al. (2023)
PDMS	Magnesium oxide nanosheet	2 bar/25°C	1929	3.4	12.7	Zainuddin et al. (2023)
6FDA-Durene	ZIF-8 hybrid with resorcinol & formaldehyde	10 bar/35°C	1332.1	28.6	–	Chen et al. (2023)
PDMS	SAPO-34 zeolite	20 bar/25°C	5753	4.92	31	Haider et al. (2021)
Pebax	COF	1 bar/25°C	1843	–	28.2	Ying et al. (2021)
Polyimide	ZIF-8 modified with isophthalic dihydrazide	3 bar/35°C	1683	10.4	-	Shi et al. (2021)
Pebax	GO nanosheets functionalized with PEG & PEI	1 bar/25°C	1330	45	120	Li et al. (2015)

the nanofillers. The polymer and filler's polydopamine interface improves their compatibility at the interface and boosts the polymer's ability to disperse in the nanocomposite membrane. Wang et al. reported the wrapping of ZIF-8 nanoparticles with a polydopamine (PD) layer with a controlled thickness at the nanoscale (Wang et al., 2016). The resulting ZIF-8@PD nanoparticles demonstrated high dispersity in various solvents. Wang et al. reported an approach to graft polyimide brushes on MOF surface (Wang et al., 2018). The grafted polyimide brushes largely improved the dispersity and compatibility of MOF particles. The introduction of ionic liquid into the pores of nanofillers can modify the surface properties of nanofillers by decorating the chemical onto the surface of the MOFs (Vu et al., 2019) or by the caging method (Ban et al., 2015). The electrostatic interaction, hydrogen bond, and π-π stacking are the intermolecular interactions established by the different charges and aromatic rings present in ionic liquids.

The selection of nanofillers with a high affinity for CO_2 adsorption is a crucial strategy in the design of nanocomposite membranes for CO_2 capture. Nanofillers, such as MOFs and CNTs, offer unique advantages due to their ability to adsorb and desorb CO_2, making them attractive choices for improving the separation efficiency of CO_2-containing gas mixtures. One of the key criteria in choosing an appropriate MOF for CO_2 adsorption is the compatibility of the MOF's pores with the kinetic diameter of CO_2 molecules (Ghanbari et al., 2020). The kinetic diameter of a gas molecule is a measure of its effective size, considering the molecular dimensions and the available pathways within the MOF structure. To enhance CO_2 adsorption, MOFs with pores of appropriate size and geometry are excellent candidates for facilitating CO_2 adsorption. Another significant criterion for selecting MOFs with superior CO_2 adsorption capabilities is the presence of polar functional groups within the MOF's structure. Functional groups such as $-OH$ (hydroxyl), $-N=N-$ (azobenzene), $-NH_2$ (amino), and $-N=C(R)$- (isocyanide) introduce polarity and induce quadrupole moments in the CO_2 molecules (Ghanbari et al., 2020). These quadrupole moments enable stronger interactions between the CO_2 molecules and the MOF, enhancing the adsorption capacity. The polar groups within the MOF create favorable binding sites for CO_2, leading to increased adsorption efficiency. Additionally, CNTs are another class of nanofillers known for their high CO_2 adsorption affinity. The hollow, cylindrical structure of CNTs provides a large internal surface area, enabling them to adsorb significant amounts of CO_2. The CNTs' extended, 1D channels offer ample space for CO_2 molecules to interact with the carbon atoms on the inner surface of the tubes. This structural advantage, coupled with the strong van der Waals interactions between CO_2 and the CNTs, results in enhanced CO_2 adsorption.

The impact of nanofiller loading in mixed matrix membranes on the permeability and selectivity of CO_2 gas separation is a complex and nonlinear relationship. This phenomenon has been extensively studied, and recent research has shed light on the nuances of this relationship. Wang et al. (2023) have discovered that the amount of nanofiller loading in nanocomposite membranes does not show a direct correlation with the effectiveness of CO_2 separation. In other words, simply increasing the quantity of nanofillers does not guarantee improved performance in terms of both permeability and selectivity. This highlights the importance of understanding the optimal nanofiller loading to achieve the desired gas separation outcomes. Shah Buddin and Ahmad (2021) further elucidated this complex relationship. They found that simultaneous improvements in permeability and selectivity in nanocomposite membranes are typically limited and tend to plateau at a specific threshold of nanofiller loading. Beyond this point, a declining trend in either permeability or selectivity, or even in both, can occur. This observation underscores the critical need for a careful balance between the quantity of nanofillers and the desired separation characteristics. A comprehensive review conducted by Vinoba et al. (2017) explored the impact of various fillers on nanocomposite membranes for CO_2 separation. Their findings indicate that, in general, permeability increases with an increase in the fraction of inorganic fillers. This suggests that the presence of nanofillers can indeed enhance the transport of CO_2 through the membrane. However, it is important to consider the overall performance, which involves a trade-off between permeability and selectivity. The order of fillers in terms of improving CO_2/CH_4 and CO_2/N_2 selectivity, as revealed by Vinoba and colleagues, is carbon > MOF > zeolites (Vinoba et al., 2017). This sequence can be attributed to the specific properties of these fillers. Carbon-based nanofillers, such as graphene, have interstitial channels and a large surface area, which allow them to preferentially adsorb CO_2 over lighter gases like CH_4 and N_2. MOFs and zeolites, while still effective, may not possess the same structural attributes as carbon-based fillers, making carbon nanofillers particularly advantageous for enhancing selectivity. The impact of nanofiller loading in mixed matrix membranes for CO_2 gas separation is not a linear relationship. It involves a delicate balance between permeability and selectivity. Simply increasing nanofiller loading may not lead to improved CO_2 separation performance. There is typically a threshold beyond which increasing the load may result in diminishing returns or even a decrease in performance. Understanding the unique properties of different nanofillers, such as carbon, MOFs, and zeolites, is crucial for tailoring mixed matrix membranes to achieve the desired CO_2 separation outcomes.

4.4.3 Structural robustness

Another critical aspect of an ideal nanocomposite membrane for application is its requirement for exceptional chemical and physical stability, especially during prolonged exposure to the harsh conditions of high temperatures and pressures in acid gas streams. Under such challenging circumstances, membrane failures often occur due to plasticization, swelling, aging, compaction, and surface damage.

Polymers are categorized into two main classes: glassy polymers, which possess glass transition temperatures (T_g) higher than room temperature, and rubbery polymers, with T_g values lower than room temperature. The T_g of a polymer plays a significant role in determining its selectivity performance. Below the T_g, polymer chain mobility is restricted, rendering the polymer rigid and durable (Roozitalab et al., 2023). Glassy polymers, exemplified by materials like polysulfone, polyethersulfone, polycarbonate, polyvinylidene fluoride, polyamide, polyetherimide (PEI), and polyimide, exhibit solid, glass-like structures and operate below their respective T_g. These glassy polymers are characterized by low permeability but high selectivity (Mannan et al., 2013). In contrast, rubbery polymers, including poly(dimethylsiloxane) (PDMS), ethylene propylene rubber, polychloroprene, nitrile butadiene rubber, polyurethane, polyisoprene, polyoctylmethyl siloxane, styrene butadiene rubber, and butyl rubber, are soft and flexible materials that function above their T_g values. Rubbery polymers are known for their high permeability but tend to exhibit low selectivity in gas separation processes (Mannan et al., 2013). The choice of polymer, influenced by its glassy or rubbery nature, is a critical factor in designing mixed matrix membranes for various applications, with glassy polymers typically favored for selectivity and rubbery polymers for permeability.

CO_2 is known to be the strongest plasticization agent (Kapantaidakis et al., 2003), especially under extreme conditions. Plasticization, in the context of polymers and membranes, refers to a process in which a polymer material becomes softer, more flexible, and less rigid when exposed to high temperatures or the presence of certain solvents or gases. When a polymer is exposed to plasticizing conditions, the polymer chains have increased mobility, leading to a reduction in intermolecular forces and a loosening of the polymer structure. As a result, the polymer becomes more permeable to gases and may lose its ability to retain its original mechanical and separation properties. This phenomenon can lead to a decrease in the selectivity of the membrane as well as reduced mechanical strength and stability, ultimately affecting its overall performance in gas separation applications. Polymers with a high T_g have better resistance to plasticization. The interaction between the polymer matrix and CO_2 molecules can cause swelling in the nanocomposite membranes (Suleman et al., 2016). When CO_2 molecules are in contact with the polymer matrix in the presence of certain fillers, they can be absorbed or adsorbed, leading to the expansion of the polymer structure. In certain cases, the nanofillers themselves may absorb CO_2 or undergo structural changes upon CO_2 exposure, causing them to swell and, in turn, inducing swelling in the entire membrane. Excessive swelling can lead to a loss of mechanical stability, reduced selectivity, and compromised membrane performance. Techniques to overcome plasticization in polymeric membranes include cross-linking (Du et al., 2011) and thermal treatment (Chen et al., 2013), polymer sulfonation (Khan et al., 2011), thermal rearrangement (Adewole et al., 2013), and polymer blending (Bos et al., 2001).

When a nanocomposite membrane is subjected to high pressures or mechanical stress, the polymeric matrix and the nanofillers can undergo compression, causing the void spaces within the membrane to shrink (Kadirkhan et al., 2022). It is a phenomenon where the structure of the membrane becomes denser or more compact, resulting in decreased pore size and altered transport properties. This compaction can lead to a reduction in pore size, limiting the passage of gas molecules through the membrane. As a result, the gas permeability and separation performance of the membrane may be adversely affected. Managing and minimizing compaction is important for maintaining the long-term performance and stability of nanocomposite membranes and ensuring their efficient operation in gas separation and other relevant applications. Researchers now actively develop new materials or improve the common ones that can surpass the upper bound of the Robeson diagram. Polyimide, polysulfone, and cellulose acetate are the most commonly used polymers for the processing of nanocomposite gas separation membranes (Favvas et al., 2017). These glassy polymers primarily dominate the upper bound region (Fig. 4.6), owing to their high solubility coefficient.

4.5 Related case studies on pilot-scale/full-scale applications

Although several membrane materials have been created at the lab scale as self-standing membranes, very few of these materials can be made into thin-film composite membranes that have workable CO_2 permeance, and even fewer of these materials have been used to create membrane modules that can be used for a pilot-scale test. An in-field CO_2 capture test was conducted using a prepilot scale hollow fiber membrane module with a polyvinyl alcohol (PVA)/amino acid salt (ProK) nanocomposite membrane carrying up to 40 wt.% amino acid salt (Fig. 4.7) (Dai et al., 2019). The membrane has an effective area of around 200 cm^2. The feed flue gas is generated from the rotary kiln with prefiltration of suspended particulate matter. A CO_2 permeate flux of 5×10^{-3} cm^3(STP) cm^{-2}/s and a CO_2 content of 50% in the permeate were recorded under optimal conditions, which is comparable to other facilitated transport membranes.

Figure 4.7 The location of the field test in the cement plant, (A) rotary kiln with five stages cyclone preheating tower and the gas emission stack; (B) sampling point of the flue gas from the emission stack and (C) permeation test rig located in the preheating tower.
Source: Adapted from Dai, Z., Fabio, S., Giuseppe Marino, N., Riccardo, C. & Deng, L. (2019). Field test of a pre-pilot scale hollow fiber facilitated transport membrane for CO_2 capture. *International Journal of Greenhouse Gas Control*, *86*, 191–200. https://doi.org/10.1016/j.ijggc.2019.04.027.

A membrane pilot-scale system was implemented and evaluated at the Norcem Cement factory in Norway with the primary goal of capturing CO_2 from a flue gas stream containing a high concentration of CO_2 (17 mol% on a wet basis) (Hägg et al., 2017). This system featured hollow fiber membrane modules using polyvinyl amine-based fixed-site carrier technology, with module sizes reaching up to 18 m^2. These membrane modules were initially obtained as commercially available components from Air Products, a company with a presence in both the United States and Norway, and then they were coated on-site at the Norwegian University of Science and Technology (NTNU). The results of the testing phase revealed that the membrane system achieved an impressive CO_2 purity level of 70 mol% in a single-stage operation. Furthermore, the membrane demonstrated remarkable stability, even when exposed to elevated concentrations of sulfur dioxide (SO_2) and nitrogen oxides (NO_x) for an extended period of 6 months. Importantly, this prolonged exposure to harsh conditions did not lead to any significant degradation in the membrane's performance.

Pohlmann et al. (2016) developed a pilot scale of the PolyActive™ TFC membrane for CO_2 separation unit with an effective area of 13 m^2 to separate flue gas streams containing mixed gases of 51–100 ppm SO_2, 75–91 ppm NOx, 14.0% H_2O, 14.6% CO_2, and 6.7% O_2 from a coal-fired power plant. They conducted two experiments. The first experiment applied a membrane-enveloped module with a feed flow rate of 50 m^3(STP) h^{-1} and 38 hours of operation. The second experiment was operated continuously for 31 days with a feed flow rate of 70 m^3(STP) h^{-1}. For both experiments, the permeate volumetric flow rate was between 2.5 and 3 m^3(STP) h^{-1} with 67%–73% CO_2 in permeate.

In a typical coal-fired power plant with a capacity of 600 megawatts (MWe), approximately 500 cubic meters per second (1540 million standard cubic feet per day) of flue gas, containing around 460 t/h (11,000 t/day) of CO_2, is generated. White et al. applied a pilot-membrane-sized system to separate a flue gas stream that contained about a ton of CO_2 day^{-1}, which represents just a small fraction (0.01%) of what would be needed for a commercial-scale plant (White et al., 2015). To make CO_2 capture from such large gas streams cost-effective, highly permeable membranes are essential. The pilot plant detailed in this paper employs a membrane specially developed by Membrane Technology and Research, Inc. (MTR) for this purpose, known as Polaris. This membrane exhibits remarkable CO_2 permeance values, ranging from 1000 to 2000 GPU at 50 psig and 23°C, with a CO_2/N_2 selectivity of 50–60. A primary objective of this testing program was to validate the ability to produce Polaris membranes at a commercial scale, with a width of 1 m and construct these membranes into full-scale modules capable of functioning effectively under field conditions. Even with these highly permeable membranes, the success of a full-scale carbon capture system depends on the membranes' ability to maintain stable operation during continuous exposure.

Sandru et al. (2013) designed a pilot-scale polymeric polyvinyl amine membrane configuration with stationary carrier separation layers at the Sines power plant of EDP in Portugal, aimed at CO_2 separation from flue gas generated by coal-fired power plants. The project encompassed various aspects, including membrane upscaling, material durability, and pilot testing at the power plant. Through gas permeation experiments and material analyses, it was confirmed that the membrane material and separation performance remained unaffected even when exposed to synthetic and real flue gas contaminants. During the 6.5-month continuous testing of a pilot-scale module equipped with a 1.5 m^2 NTNU membrane, consistent separation performances were observed. The permeate contained a maximum of 75% CO_2 and had a permeate flow rate of 525 l day^{-1}. CO_2 permeances of 0.2–0.6 m^3(STP)/(m^2hbar) were observed. These performances remained stable despite several challenges associated with power plant operations, such as high levels of NOx (600 mg/Nm^3) and 200 mg Nm^{-3} SO_2, as well as frequent power plant outages.

A membrane process consisting of four stages, including a dehumidification step, was designed, installed, and operated to showcase the viability of multistage membrane systems in the removal of CO_2 from flue gases (Choi et al., 2013). The separation layers of the hollow fiber membrane module were made of polyethersulfone. Initially, the modules were tested using a simulated CO_2/N_2 mixture. The membrane exhibited a CO_2 permeability of 40 GPU. Under these specific operating conditions, the membrane successfully captured 90% of the CO_2 with a purity level of 99%.

4.6 Future opportunities for improvement

For future work on nanocomposite membranes for CO_2 removal, several areas of research and development can be recommended to further enhance their performance, efficiency, and practical applications. The exploration and characterization of new types of nanofillers with unique properties that can enhance CO_2 selectivity and permeability must be continued. For future advancements in nanocomposite membranes for CO_2 removal, ongoing exploration and characterization of new types of nanofillers with unique properties that enhance CO_2 selectivity and permeability are essential. Investigate novel inorganic nanofillers, hybrid materials, and 2D nanomaterials with precise size sieving effects by engineering the design of the sheets and distance to identify those with superior gas separation properties. A focus must be placed on understanding and controlling the interactions between the nanofillers and polymeric matrix at the molecular level. The focus should be on investigating novel inorganic nanofillers, hybrid materials, and 2D nanomaterials with a precise size-sieving effect by designing the sheets and distances and understanding and controlling the interactions between the nanofillers and the polymeric matrix at the molecular level. Investigation of surface modifications and functionalization of nanofillers is also crucial to improve their compatibility with polymers and enhance interfacial interactions. Explore innovative fabrication methods that enable precise control of nanofiller homogenous dispersion and orientation within the polymeric matrix. Techniques such as 3D printing and layer-by-layer assembly may provide new possibilities for membrane design and performance enhancement. Nanocomposite membranes must be developed with improved mechanical strength and resistance to compaction under high pressure or stress conditions. This will help maintain the membrane's integrity and performance during long-term operation. Investigating membrane aging and durability must unravel how membrane aging affects the selectivity, permeability, and overall gas separation efficiency of the membranes. There is a need to develop more sophisticated theoretical prediction models to comprehensively simulate the permeation hindrance and transport pathways of gas molecules in 2D composite membranes for gas separation operations. Advanced characterization techniques to probe the structure and transport properties of nanocomposite membranes at different length scales can provide deeper insights into their gas separation mechanisms. In terms of scale-up and commercial viability, nanocomposite membranes developed must alleviate the need for high-cost modules, which are not appropriate for dealing with the emission of large volumes of gases. By addressing these research areas, nanocomposite membranes for CO_2 removal can be further optimized and advanced, contributing to more efficient and environmentally sustainable gas separation processes and facilitating the transition to a low-carbon future.

4.7 AI disclosure

During the preparation of this work, the author(s) used Quillbot to paraphrase sentences in a correct manner. After using this tool/service, the author(s) reviewed and edited the content as needed and take(s) full responsibility for the content of the publication.

References

Adewole, J. K., Ahmad, A. L., Ismail, S., & Leo, C. P. (2013). Current challenges in membrane separation of CO_2 from natural gas: A review. *International Journal of Greenhouse Gas Control*, *17*, 46–65. Available from https://doi.org/10.1016/j.ijggc.2013.04.012, http://www.elsevier.com/wps/find/journaldescription.cws_home/709061/description#description.

Ajasa, A. (2023). Unpublished content Carbon dioxide levels in atmosphere mark a near-record surge. *The Washington Post*. <https://www.washingtonpost.com/climate-environment/2023/06/05/carbon-dioxide-growing-climate-change/>.

Alqaheem, Y., Alomair, A., Vinoba, M., & Pérez, A. (2017). Polymeric gas-separation membranes for petroleum refining. *International Journal of Polymer Science*, *2017*. Available from https://doi.org/10.1155/2017/4250927, http://www.hindawi.com/journals/ijps/.

Anjum, M. W., Vermoortele, F., Khan, A. L., Bueken, B., De Vos, D. E., & Vankelecom, I. F. J. (2015). Modulated UiO-66-based mixed-matrix membranes for CO_2 separation. *ACS Applied Materials and Interfaces*, *7*(45), 25193–25201. Available from https://doi.org/10.1021/acsami.5b08964, http://pubs.acs.org/journal/aamick.

Ban, Y., Li, Z., Li, Y., Peng, Y., Jin, H., Jiao, W., Guo, A., Wang, P., Yang, Q., Zhong, C., & Yang, W. (2015). Confinement of ionic liquids in nanocages: Tailoring the molecular sieving properties of ZIF-8 for membrane-based CO_2 capture. *Angewandte Chemie -*

International Edition, *54*(51), 15483–15487. Available from https://doi.org/10.1002/anie.201505508, http://onlinelibrary.wiley.com/journal/10.1002/(ISSN)1521-3773.

Baniamer, M., Aroujalian, A., & Sharifnia, S. (2021). Photocatalytic membrane reactor for simultaneous separation and photoreduction of CO_2 to methanol. *International Journal of Energy Research*, *45*(2), 2353–2366. Available from https://doi.org/10.1002/er.5930, http://onlinelibrary.wiley.com/journal/10.1002/(ISSN)1099-114X.

Bos, A., Pünt, I., Strathmann, H., & Wessling, M. (2001). Suppression of gas separation membrane plasticization by homogeneous polymer blending. *AIChE Journal*, *47*(5), 1088–1093. Available from https://doi.org/10.1002/aic.690470515.

Chen, C. C., Miller, S. J., & Koros, W. J. (2013). Characterization of thermally cross-linkable hollow fiber membranes for natural gas separation. *Industrial and Engineering Chemistry Research*, *52*(3), 1015–1022. Available from https://doi.org/10.1021/ie2020729.

Chen, Ke, Ni, Linhan, Guo, Xin, Xiao, Chengming, Yang, Yue, Zhou, Yujun, Zhu, Zhigao, Qi, Junwen, & Li, Jiansheng (2023). Introducing pyrazole-based MOF to polymer of intrinsic microporosity for mixed matrix membranes with enhanced CO_2/CH_4 separation performance. *Journal of Membrane Science*, *688*, 122110. Available from https://doi.org/10.1016/j.memsci.2023.122110.

Chen, Ke, Ni, Linhan, Zhang, Hao, Xie, Jia, Yan, Xin, Chen, Saisai, Qi, Junwen, Wang, Chaohai, Sun, Xiuyun, & Li, Jiansheng (2021). Veiled metal organic frameworks nanofillers for mixed matrix membranes with enhanced CO_2/CH_4 separation performance. *Separation and Purification Technology*, *279*, 119707. Available from https://doi.org/10.1016/j.seppur.2021.119707.

Chen, Ruiqi, Chai, Milton, & Hou, Jingwei (2023). Metal-organic framework-based mixed matrix membranes for gas separation: Recent advances and opportunities. *Carbon Capture Science & Technology*, *8*, 100130. Available from https://doi.org/10.1016/j.ccst.2023.100130.

Chen, X. Y., Hoang, V. T., Rodrigue, D., & Kaliaguine, S. (2013). Optimization of continuous phase in amino-functionalized metal-organic framework (MIL-53) based co-polyimide mixed matrix membranes for CO_2/CH_4 separation. *RSC Advances*, *3*(46), 24266–24279. Available from https://doi.org/10.1039/c3ra43486a.

Choi, S. H., Kim, J. H., & Lee, Y. (2013). Pilot-scale multistage membrane process for the separation of CO_2 from LNG-fired flue gas. *Separation and Purification Technology*, *110*, 170–180. Available from https://doi.org/10.1016/j.seppur.2013.03.016.

Chung, T. S., Jiang, L. Y., Li, Y., & Kulprathipanja, S. (2007). Mixed matrix membranes (MMMs) comprising organic polymers with dispersed inorganic fillers for gas separation. *Progress in Polymer Science (Oxford)*, *32*(4), 483–507. Available from https://doi.org/10.1016/j.progpolymsci.2007.01.008.

Cong, H., Radosz, M., Towler, B. F., & Shen, Y. (2007). Polymer-inorganic nanocomposite membranes for gas separation. *Separation and Purification Technology*, *55*(3), 281–291. Available from https://doi.org/10.1016/j.seppur.2006.12.017.

Cuéllar-Franca, R. M., & Azapagic, A. (2015). Carbon capture, storage and utilisation technologies: A critical analysis and comparison of their life cycle environmental impacts. *Journal of CO_2 Utilization*, *9*, 82–102. Available from https://doi.org/10.1016/j.jcou.2014.12.001, http://www.journals.elsevier.com/journal-of-co2-utilization/.

Dai, Z., Fabio, S., Giuseppe Marino, N., Riccardo, C., & Deng, L. (2019). Field test of a pre-pilot scale hollow fiber facilitated transport membrane for CO_2 capture. *International Journal of Greenhouse Gas Control*, *86*, 191–200. Available from https://doi.org/10.1016/j.ijggc.2019.04.027, http://www.elsevier.com/wps/find/journaldescription.cws_home/709061/description#description.

Dashti, Amir, Bahrololoomi, Arash, Amirkhani, Farid, & Mohammadi, Amir H. (2020). Estimation of CO_2 adsorption in high capacity metal − organic frameworks: Applications to greenhouse gas control. *Journal of CO_2 Utilization*, *41*, 101256. Available from https://doi.org/10.1016/j.jcou.2020.101256.

Dong, S., Huang, W., Li, X., Wang, X., Yan, B., Zhang, Z., & Zhong, J. (2023). Synthesis of dual-functionalized mixed matrix membrane with amino-GO-Bent/PVDF for efficient separation of CO2/CH_4 and CO_2/N_2. *Journal of Applied Polymer Science*, *140*(5). Available from https://doi.org/10.1002/app.53405, http://onlinelibrary.wiley.com/journal/10.1002/(ISSN)1097-4628.

Du, N., Cin, M. M. D., Pinnau, I., Nicalek, A., Robertson, G. P., & Guiver, M. D. (2011). Azide-based cross-linking of polymers of intrinsic microporosity (PIMs) for condensable gas separation. *Macromolecular Rapid Communications*, *32*(8), 631−636. Available from https://doi.org/10.1002/marc.201000775.

Favvas, E. P., Katsaros, F. K., Papageorgiou, S. K., Sapalidis, A. A., & Mitropoulos, A. C. (2017). A review of the latest development of polyimide based membranes for CO_2 separations. *Reactive and Functional Polymers*, *120*, 104−130. Available from https://doi.org/10.1016/j.reactfunctpolym.2017.09.002.

Feng, Yang, Yan, Wei, Kang, Zixi, Zou, Xiaoqin, Fan, Weidong, Jiang, Yujie, Fan, Lili, Wang, Rongming, & Sun, Daofeng (2023). Thermal treatment optimization of porous MOF glass and polymer for improving gas permeability and selectivity of mixed matrix membranes. *Chemical Engineering Journal*, *465*, 142873. Available from https://doi.org/10.1016/j.cej.2023.142873.

Ghanbari, T., Abnisa, F., & Wan Daud, W. M. A. (2020). A review on production of metal organic frameworks (MOF) for CO_2 adsorption. *Science of the Total Environment*, *707*. Available from https://doi.org/10.1016/j.scitotenv.2019.135090, http://www.elsevier.com/locate/scitotenv.

Goh, P. S., & Ismail, A. F. (2020). *Introduction*, *12*. Available from https://doi.org/10.1016/b978-0-12-819406-5.00001-0.

Goh, P. S., Wong, K. C., & Ismail, A. F. (2020). Nanocomposite membranes for liquid and gas separations from the perspective of nanostructure dimensions. *Membranes*, *10*(10), 1−29. Available from https://doi.org/10.3390/membranes10100297, https://www.mdpi.com/2077-0375/10/10/297/pdf.

Haider, B., Dilshad, M. R., Akram, M. S., Islam, A., & Kaspereit, M. (2021). Novel polydimethylsiloxane membranes impregnated with SAPO-34 zeolite particles for gas separation. *Chemical Papers*, *75*(12), 6417−6431. Available from https://doi.org/10.1007/s11696-021-01790-w, http://www.springer.com/11696.

Hägg, M. B., Lindbråthen, A., He, X., Nodeland, S. G., & Cantero, T. (2017). Pilot demonstration-reporting on CO_2 capture from a cement plant using hollow fiber process. *Energy Procedia*, *114*, 6150−6165. Available from https://doi.org/10.1016/j.egypro.2017.03.1752, 18766102 Elsevier Ltd Norway, http://www.sciencedirect.com/science/journal/18766102.

IEA, CO_2 Emissions in 2022. 2023.

Janakiram, S., Ahmadi, M., Dai, Z., Ansaloni, L., & Deng, L. (2018). Performance of nanocomposite membranes containing 0D to 2D nanofillers for CO_2 separation: A review. *Membranes*, *8*(2). Available from https://doi.org/10.3390/membranes8020024, http://www.mdpi.com/2077-0375/8/2/24/pdf.

Kadirkhan, F., Sean, G. P., Ismail, A. F., Wan Mustapa, W. N. F., Halim, M. H. M., Kian, S. W., & Yean, Y. S. (2022). CO_2 plasticization resistance membrane for natural gas sweetening process: Defining optimum operating conditions for stable operation. *Polymers*, *14*(21). Available from https://doi.org/10.3390/polym14214537, http://www.mdpi.com/journal/polymers.

Kapantaidakis, G. C., Koops, G. H., Wessling, M., Kaldis, S. P., & Sakellaropulos, G. P. (2003). CO_2 plasticization of polyethersulfone/polyimide gas-separation membranes. *AIChE Journal*, *49*(7), 1702–1711. Available from https://doi.org/10.1002/aic.690490710.

Khan, A. L., Li, X., & Vankelecom, I. F. J. (2011). Mixed-gas CO_2/CH_4 and CO_2/N_2 separation with sulfonated PEEK membranes. *Journal of Membrane Science*, *372*(1–2), 87–96. Available from https://doi.org/10.1016/j.memsci.2011.01.056.

Kim, S., & Lee, Y. M. (2015). Rigid and microporous polymers for gas separation membranes. *Progress in Polymer Science*, *43*, 1–32. Available from https://doi.org/10.1016/j.progpolymsci.2014.10.005, http://www.sciencedirect.com/science/journal/00796700.

Lai, S. F., & Tan, P. C. (2023). Polyimide blend metal–organic framework-based mixed matrix membrane for gas separation: A review. *Asia-Pacific Journal of Chemical Engineering*. Available from https://doi.org/10.1002/apj.2970, http://onlinelibrary.wiley.com/journal/10.1002/(ISSN)1932-2143.

Lee, C. S., Moon, J., Park, J. T., & Kim, J. H. (2023). Engineering CO_2-philic pathway via grafting poly(ethylene glycol) on graphene oxide for mixed matrix membranes with high CO_2 permeance. *Chemical Engineering Journal*, *453*. Available from https://doi.org/10.1016/j.cej.2022.139818, http://www.elsevier.com/inca/publications/store/6/0/1/2/7/3/index.htt.

Li, Xueqin, Cheng, Youdong, Zhang, Haiyang, Wang, Shaofei, Jiang, Zhongyi, Guo, Ruili, & Wu, Hong (2015). Efficient CO_2 capture by functionalized graphene oxide nanosheets as fillers to fabricate multi-permselective mixed matrix membranes. *ACS Applied Materials and Interfaces*, *7*(9), 5528–5537. Available from https://doi.org/10.1021/acsami.5b00106.

Li, Xinxin, Jiao, Chengli, Zhang, Xiaoqian, Tian, Zhengbin, Xu, Xia, Liang, Fangyi, Wang, Guang-hui, & Jiang, Heqing (2022). A general strategy for fabricating polymer/nanofiller composite membranes with enhanced CO_2/N_2 separation performance. *Journal of Cleaner Production*, *350*, 131468. Available from https://doi.org/10.1016/j.jclepro.2022.131468.

Liu, J., Thallapally, P. K., Mc Grail, B. P., Brown, D. R., & Liu, J. (2012). Progress in adsorption-based CO_2 capture by metal–organic frameworks. *Chemical Society Reviews*, *41*(6), 2308–2322. Available from https://doi.org/10.1039/c1cs15221a.

Liu, L., Chakma, A., & Feng, X. (2008). Gas permeation through water-swollen hydrogel membranes. *Journal of Membrane Science*, *310*(1–2), 66–75. Available from https://doi.org/10.1016/j.memsci.2007.10.032.

Luo, Wenjia, Niu, Zhenhua, Mu, Peng, & Li, Jian (2022). Pebax and CMC@MXene-based mixed matrix membrane with high mechanical strength for the highly efficient capture of CO_2. *Macromolecules*, *55*(21), 9851–9859. Available from https://doi.org/10.1021/acs.macromol.2c01532.

Mannan, H. A., Mukhtar, H., Murugesan, T., Nasir, R., Mohshim, D. F., & Mushtaq, A. (2013). Recent applications of polymer blends in gas separation membranes. *Chemical Engineering and Technology*, *36*(11), 1838–1846. Available from https://doi.org/10.1002/ceat.201300342.

Pohlmann, J., Bram, M., Wilkner, K., & Brinkmann, T. (2016). Pilot scale separation of CO_2 from power plant flue gases by membrane technology. *International Journal of Greenhouse Gas Control*, *53*, 56–64. Available from https://doi.org/10.1016/j.ijggc.2016.07.033, http://www.elsevier.com/wps/find/journaldescription.cws_home/709061/description#description.

Regmi, C., Ashtiani, S., Průša, F., & Friess, K. (2022). Synergistic effect of hybridized TNT@GO fillers in CTA-based mixed matrix membranes for selective CO_2/CH_4 separation. *Separation and Purification Technology*, *282*. Available from https://doi.org/10.1016/j.seppur.2021.120128, http://www.journals.elsevier.com/separation-and-purification-technology/.

Rezakazemi, M., Ebadi Amooghin, A., Montazer-Rahmati, M. M., Ismail, A. F., & Matsuura, T. (2014). State-of-the-art membrane based CO_2 separation using mixed matrix membranes (MMMs): An overview on current status and future directions. *Progress in Polymer Science, 39*(5), 817–861. Available from https://doi.org/10.1016/j.progpolymsci.2014.01.003.

Rivero, J. R., Nemetz, L. R., Da Conceicao, M. M., Lipscomb, G., & Hornbostel, K. (2023). Modeling gas separation in flat sheet membrane modules: Impact of flow channel size variation. *Carbon Capture Science and Technology, 6*. Available from https://doi.org/10.1016/j.ccst.2022.100093, https://www.journals.elsevier.com/carbon-capture-science-and-technology.

Rizwan, M., Shoukat, A., Ayub, A., Razzaq, B., & Tahir, M. B. (2021). *Types and classification of nanomaterials. Nanomaterials: Synthesis, Characterization, Hazards and Safety* (pp. 31–54). Pakistan: Elsevier. Available from https://www.sciencedirect.com/book/9780128238233, 10.1016/B978-0-12-823823-3.00001-X.

Robeson, L. M. (2008). The upper bound revisited. *Journal of Membrane Science, 320*(1–2), 390–400. Available from https://doi.org/10.1016/j.memsci.2008.04.030.

Roozitalab, Atefeh, Hamidavi, Fatemeh, & Kargari, Ali (2023). A review of membrane material for biogas and natural gas upgrading. *Gas Science and Engineering, 114*, 204969. Available from https://doi.org/10.1016/j.jgsce.2023.204969.

Saleh, Tawfik A. (2020). Nanomaterials: Classification, properties, and environmental toxicities. *Environmental Technology & Innovation, 20*, 101067. Available from https://doi.org/10.1016/j.eti.2020.101067.

Sandru, M., Kim, T. J., Capala, W., Huijbers, M., & Hägg, M. B. (2013). Pilot scale testing of polymeric membranes for CO_2 capture from coal fired power plants. *Energy Procedia, 37*, 6473–6480. Available from https://doi.org/10.1016/j.egypro.2013.06.577, 18766102 Elsevier Ltd Norway, http://www.sciencedirect.com/science/journal/18766102.

Seman, M. N. A., Khayet, M., & Hilal, N. (2010). Nanofiltration thin-film composite polyester polyethersulfone-based membranes prepared by interfacial polymerization. *Journal of Membrane Science, 348*(1–2), 109–116. Available from https://doi.org/10.1016/j.memsci.2009.10.047.

Shah Buddin, M. M. H., & Ahmad, A. L. (2021). A review on metal-organic frameworks as filler in mixed matrix membrane: Recent strategies to surpass upper bound for CO_2 separation. *Journal of CO_2 Utilization, 51*, 101616. Available from https://doi.org/10.1016/j.jcou.2021.101616.

Shi, Yapeng, Wu, Shanshan, Wang, Zhenggong, Bi, Xiangyu, Huang, Menghui, Zhang, Yatao, & Jin, Jian (2021). Mixed matrix membranes with highly dispersed MOF nanoparticles for improved gas separation. *Separation and Purification Technology, 277*, 119449. Available from https://doi.org/10.1016/j.seppur.2021.119449.

Sonpingkam, S., & Pattavarakorn, D. (2014). Mechanical properties of sulfonated poly (ether ether ketone) nanocomposite membranes. *International Journal of Chemical Engineering and Applications, 5*(2), 181–185. Available from https://doi.org/10.7763/IJCEA.2014.V5.374.

Suhaimi, N. H., Yeong, Y. F., Ch'ng, C. W. M., & Jusoh, N. (2019). Tailoring CO_2/CH_4 separation performance of mixed matrix membranes by using ZIF-8 particles functionalized with different amine groups. *Polymers, 11*(12). Available from https://doi.org/10.3390/polym11122037, https://res.mdpi.com/d_attachment/polymers/polymers-11-02037/article_deploy/polymers-11-02037.pdf.

Suleman, M. S., Lau, K. K., & Yeong, Y. F. (2016). Plasticization and swelling in polymeric membranes in CO_2 removal from natural gas. *Chemical Engineering and Technology, 39*

(9), 1604−1616. Available from https://doi.org/10.1002/ceat.201500495, http://www3.interscience.wiley.com/journal/10008333/home.

Sun, W. S., Yin, M. J., Zhang, W. H., Li, S., Wang, N., & An, Q. F. (2023). Tailor-made microstructures lead to high-performance robust PEO membrane for CO_2 capture via green fabrication technique. *Green Energy and Environment*, *8*(5), 1389−1397. Available from https://doi.org/10.1016/j.gee.2022.01.016, http://www.keaipublishing.com/en/journals/green-energy-and-environment/.

Vinoba, M., Bhagiyalakshmi, M., Alqaheem, Y., Alomair, A. A., Pérez, A., & Rana, M. S. (2017). Recent progress of fillers in mixed matrix membranes for CO_2 separation: A review. *Separation and Purification Technology*, *188*, 431−450. Available from https://doi.org/10.1016/j.seppur.2017.07.051, http://www.journals.elsevier.com/separation-and-purification-technology/.

Visser, T., Koops, G. H., & Wessling, M. (2005). On the subtle balance between competitive sorption and plasticization effects in asymmetric hollow fiber gas separation membranes. *Journal of Membrane Science*, *252*(1−2), 265−277. Available from https://doi.org/10.1016/j.memsci.2004.12.015, http://www.elsevier.com/locate/memsci.

Voumik, L. C., Islam, M. A., Ray, S., Mohamed Yusop, N. Y., & Ridzuan, A. R. (2023). CO_2 emissions from renewable and non-renewable electricity generation sources in the G7 countries: Static and dynamic panel assessment. *Energies*, *16*(3). Available from https://doi.org/10.3390/en16031044, http://www.mdpi.com/journal/energies/.

Vu, M. T., Lin, R., Diao, H., Zhu, Z., Bhatia, S. K., & Smart, S. (2019). Effect of ionic liquids (ILs) on MOFs/polymer interfacial enhancement in mixed matrix membranes. *Journal of Membrane Science*, *587*. Available from https://doi.org/10.1016/j.memsci.2019.05.081, http://www.elsevier.com/locate/memsci.

Wan, C. F., Yang, T., Lipscomb, G. G., Stookey, D. J., & Chung, T. S. (2017). Design and fabrication of hollow fiber membrane modules. *Journal of Membrane Science*, *538*, 96−107. Available from https://doi.org/10.1016/j.memsci.2017.05.047, http://www.elsevier.com/locate/memsci.

Wang, H., He, S., Qin, X., Li, C., & Li, T. (2018). Interfacial engineering in metal-organic framework-based mixed matrix membranes using covalently grafted polyimide brushes. *Journal of the American Chemical Society*, *140*(49), 17203−17210. Available from https://doi.org/10.1021/jacs.8b10138, http://pubs.acs.org/journal/jacsat.

Wang, L., Boutilier, M. S. H., Kidambi, P. R., Jang, D., Hadjiconstantinou, N. G., & Karnik, R. (2017). Fundamental transport mechanisms, fabrication and potential applications of nanoporous atomically thin membranes. *Nature Nanotechnology*, *12*(6), 509−522. Available from https://doi.org/10.1038/nnano.2017.72, http://www.nature.com/nnano/index.html.

Wang, Weifan, Yuan, Ye, Shi, Fei, Li, Qinghua, Zhao, Song, Wang, Jixiao, Sheng, Menglong, & Wang, Zhi (2023). Enhancing dispersibility of nanofiller via polymer-modification for preparation of mixed matrix membrane with high CO_2 separation performance. *Journal of Membrane Science*, *683*, 121791. Available from https://doi.org/10.1016/j.memsci.2023.121791.

Wang, Z., Wang, D., Zhang, S., Hu, L., & Jin, J. (2016). Interfacial design of mixed matrix membranes for improved gas separation performance. *Advanced Materials*, *28*(17), 3399−3405. Available from https://doi.org/10.1002/adma.201504982, http://www3.interscience.wiley.com/journal/119030556/issue.

White, L. S., Wei, X., Pande, S., Wu, T., & Merkel, T. C. (2015). Extended flue gas trials with a membrane-based pilot plant at a one-ton-per-day carbon capture rate. *Journal of Membrane Science*, *496*, 48−57. Available from https://doi.org/10.1016/j.memsci.2015.08.003, http://www.elsevier.com/locate/memsci.

Wong, K. C., Goh, P. S., & Ismail, A. F. (2016). Thin film nanocomposite: The next generation selective membrane for CO_2 removal. *Journal of Materials Chemistry A*, *4*(41), 15726–15748. Available from https://doi.org/10.1039/c6ta05145f, http://pubs.rsc.org/en/journals/journalissues/ta.

Wong, K. C., Goh, P. S., Taniguchi, T., Ismail, A. F., & Zahri, K. (2019). The role of geometrically different carbon-based fillers on the formation and gas separation performance of nanocomposite membrane. *Carbon*, *149*, 33–44. Available from https://doi.org/10.1016/j.carbon.2019.04.031, http://www.journals.elsevier.com/carbon/.

Xiang, L., Sheng, L., Wang, C., Zhang, L., Pan, Y., & Li, Y. (2017). Amino-functionalized ZIF-7 nanocrystals: Improved intrinsic separation ability and interfacial compatibility in mixed-matrix membranes for CO_2/CH_4 separation. *Advanced Materials*, *29*(32). Available from https://doi.org/10.1002/adma.201606999, http://www3.interscience.wiley.com/journal/119030556/issue.

Xu, J., Wang, Z., Qiao, Z., Wu, H., Dong, S., Zhao, S., & Wang, J. (2019). Post-combustion CO_2 capture with membrane process: Practical membrane performance and appropriate pressure. *Journal of Membrane Science*, *581*, 195–213. Available from https://doi.org/10.1016/j.memsci.2019.03.052, http://www.elsevier.com/locate/memsci.

Ying, Yunpan, Yang, Ziqi, Shi, Dongchen, Peh, Shing Bo, Wang, Yuxiang, Yu, Xin, Yang, Hao, Chai, Kungang, & Zhao, Dan (2021). Ultrathin covalent organic framework film as membrane gutter layer for high-permeance CO_2 capture. *Journal of Membrane Science*, *632*, 119384. Available from https://doi.org/10.1016/j.memsci.2021.119384.

Yu, Junjian, Wang, Zhe, Yang, Cancan, Wang, Fei, Cheng, Yanyin, Wang, Song, Zhang, Yi, & Wang, Zhaoli (2023). Imperfect perfection: Selective induction of CO_2 during diffusion in mixed matrix membranes. *Journal of Environmental Chemical Engineering*, *11* (5), 110672. Available from https://doi.org/10.1016/j.jece.2023.110672.

Zainuddin, M. I. F., Ahmad, A. L., & Shah Buddin, M. M. H. (2023). Polydimethylsiloxane/magnesium oxide nanosheet mixed matrix membrane for CO_2 separation application. *Membranes*, *13*(3). Available from https://doi.org/10.3390/membranes13030337, http://www.mdpi.com/journal/membranes.

Zhao, D., Ren, J., Li, H., Li, X., & Deng, M. (2014). Gas separation properties of poly (amide-6-b-ethylene oxide)/amino modified multi-walled carbon nanotubes mixed matrix membranes. *Journal of Membrane Science*, *467*, 41–47. Available from https://doi.org/10.1016/j.memsci.2014.05.009, http://www.elsevier.com/locate/memsci.

Section 2

Energy applications

Roles and application of nanocomposite ion exchange resin for biodiesel production

Amizon Azizan[1], *Rahida Wati Sharudin*[1], *Rafidah Jalil*[2], *Shareena Fairuz Abdul Manaf*[1], *Norhasyimi Rahmat*[1], *Suhaila Mohd Sauid*[1], *Khairul Faezah Md Yunos*[3] *and Kamil Kayode Katibi*[3,4,5]

[1]School of Chemical Engineering, College of Engineering, Universiti Teknologi MARA, Shah Alam, Selangor, Malaysia, [2]Forest Products Division, Forest Research Institute Malaysia (FRIM), Kepong, Selangor, Malaysia, [3]Department of Process and Food Engineering, Faculty of Engineering, Universiti Putra Malaysia, Serdang, Selangor, Malaysia, [4]Department of Physics, Faculty of Science, Universiti Putra Malaysia, Serdang, Selangor, Malaysia, [5]Department of Agricultural and Biological Engineering, Faculty of Engineering and Technology, Kwara State University, Malete, Nigeria

5.1 Ion exchange with resin scenario catering nanocomposite perspectives

Ion exchange is a technique for separation that can be used in several nonwater processes, for instance, in biodiesel purification apart from the processing of chemicals, the production of food, the production of pharmaceuticals, and the treatment of water.

Fundamentally, ion exchange with the solid matrix is the act of switching ions back-and-forth between a solid (ion exchange material) and a liquid without permanently or irreversibly altering the solid's structure. For instance, specific resin can be embedded in the solid matrix as a means of production or purification. Here, a certain polymeric compound is used that can exchange one ion for another of the same sort in a solution. Normally, the resin or the polymer used for the ion exchange is embedded in the insoluble matrix. This technique can reduce the number of contaminants in the finished product and improve the overall standard of the item by exchanging unfavorable ions for favorable ones (Bhandari et al., 2016).

Ion exchange resins, which are created using components other than polymer—for instance, from natural polymers like cellulose and starch—served as the foundation for the first generation of these materials (Tanaka, 2015b). The first generation of these materials was produced using these natural components. The development of the ion exchange process reached a turning point with the introduction of synthetic resins in the late 1930s. This historic occurrence had a direct impact on how the ion exchange process operated. The performance and durability of these synthetic resins are far

Nanocomposites for Environmental, Energy, and Agricultural Applications. DOI: https://doi.org/10.1016/B978-0-443-13935-2.00005-X

superior to those of their natural equivalents such as cellulose or starch. Synthetic resins are better because of the basic building blocks of the synthetic resins, which exhibit a high level of resistance to the degrading effects of both chemical and mechanical stress, for instance, cross-linked polystyrene or polyacrylate.

There are a few examples of synthetic resin for ion exchange such as sulfonic acid ($R-S(=O)_2$-OH) which has strong cation exchange resin, other than the weak cation exchange resin such as carboxylic acid (−COOH) groups. Strong and weak anion exchange resins are also examples of resins for the ion exchange. Strong base anion exchange resins normally contain quaternary ammonium cation groups (highly reactive group) working together with weakly basic groups such as from chloride (Cl^-) and sulfate (SO_4^{2-}) ions, resulting in its removal from the intended solution like biodiesel solution.

Physically, conventional ion exchange resins consist of a cross-linked polymer matrix with an even distribution of ion-active sites throughout the structure. Typically, ion exchange materials are available as spheres or, on occasion, granules. The dimensions and shape of these forms conform to the specifications of a particular application. The vast majority is manufactured in a spherical (bead) shape, either as traditional resin with a poly-dispersed particle size distribution of approximately 0.3−1.2 mm (50−16 mesh) or as uniform particle-sized (UPS) resin with all beads falling within a specific particle size range. In their distended state (water-swollen), the specific gravity of ion exchange resins typically ranges between 1.1 and 1.5 (Kakihana et al., 1977). The bulk density of the product after it has been mounted in a column accounts for 35%−40% void volume typical of spherical products. The bulk densities of wet resinous materials range between 35 and 60 lb/ft^3, with densities between 560 and 960 g/L. These are the great physical criteria conventionally acting as heterogeneous matrices either as catalysts or means of separation from the solution with impurities. Recently, the heterogeneous matrix has been upgraded toward nanocomposite matrix having nanoscale physical structure.

Going deeper into the principle, ion exchange resins are an exchange of ions that have charges which are equivalent to one another. For instance, the positive ions sodium (Na^+) and calcium (Ca^{2+}) can be exchanged for hydrogen (H^+) or other forms of positive ions when using a cation exchange resin. Using anion exchange resins, negative ions such as chloride (Cl^-) or sulfate (SO_4^{2-}) will be exchanged for hydroxide (OH^-) or other types of negative ions. The reusability of the ion exchange material determines the utility of ion exchange. Principally, as an example, water softening follows the reaction: $2RNa^+ + Ca^{2+} \rightarrow R_2Ca^{2+} + 2Na^+$ as shown in Fig. 5.1. The exchanger R can take calcium from hard water and replace it with an equivalent amount of sodium when it is in the form of sodium ions (Tanaka, 2015a). The calcium-loaded resin can then be treated with a sodium chloride solution to regenerate it back to its sodium form and prepare it for another cycle of operation.

There are many types of anions and cations produced during the purification process. To start with, in terms of the application of ion exchange resin with biodiesel application, the ion exchange resin is used to remove cations like sodium ions and magnesium ions by using cation ion exchange such as styrene strong acid type of sulfonated coal resin. In biodiesel, if we remove anions like carbonate ions, silica oxides,

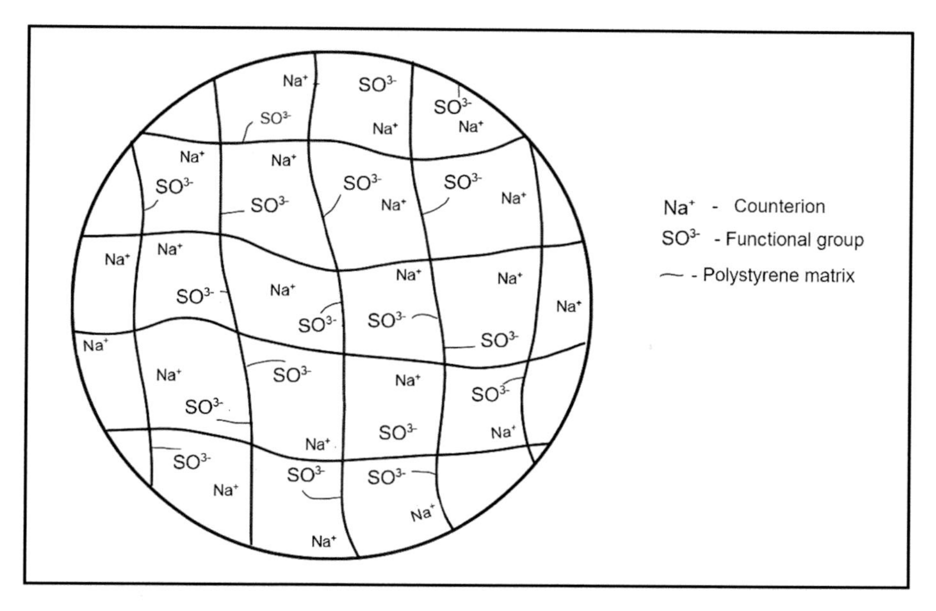

Figure 5.1 Schematic of cation exchange resin, illustrating the negatively charged matrix and the positive ions that can be swapped out.

formic acid, phenol, or fatty acids, a strong base anion exchange resin can be used. For instance, as mentioned earlier, the use of styrene quaternary ammonium salt as a strong base exchange resin could be adopted as an example of an ion exchange resin.

In general, in biodiesel purification, both anion and cation exchange resin can be beneficial at the same time to ensure the most purified biodiesel quality considering the many anions and cations involved in the purification. In purifying the glycerol, there are certain methods to ensure the adsorption occurs effectively. For instance, using the packing mechanism of the solid matrix in a vessel with beads containing the ion exchange resins either anion or cation types could be adopted for glycerol purification, leading toward easier handling for regeneration of the full capacity resins after long hours of ion exchange exposure of adsorption.

The adsorption of impurities on the ion exchange resin significantly depends on the few general chemical properties of resins. This includes the use of nanocomposite with the same concept of ion exchange resin in the improved solid matrix. These significant criteria may influence the yield of the adsorption outcome on any solid matrix of biodiesel purification. The three chemical properties are capacity, swelling, and kinetics of the ion exchange resin.

5.1.1 Capacity

There are a variety of ways to express one's ion exchange capacity. It simply means it is a relative comparison of how much number of ions can be exchanged per unit weight or volume of the used resins. However, the total capacity or total number of exchangeable sites in a material (so-called heterogeneous solid resin matrix) cannot be determined until it has undergone chemical regeneration procedures that transform it into a particular ionic form. After this phase, the ion is extracted chemically from a predetermined amount of the resin, and its concentration in solution is determined using standard analytical techniques. Total capacity can be expressed in terms of dry weight, wet weight, or wet volume.

Theoretically, the type of the polymer backbone and the environment in which the sample is placed can affect a resin's ability to absorb water or any nonwater solution, leading to its change in the wet weight and wet volume capacities. If the ion exchange occurs in a column, the term "operating capacity" is used to refer to the capacity of the ion exchange material can achieve when it is used in a column and put through a specific set of conditions (Riffat & Husnain, 2022). It depends on several variables, some of which are the resin's inherent (total) capacity, the degree of regeneration, the chemical compositions of the solution being treated, the flow rates through the column, temperature, particle size, and distribution.

In the sense of a nanocomposite ion exchange resin, the size or volume of the nanocomposite itself can affect the operating capacity of the ion exchange occurrence, despite knowing that the size and shape of a nanocomposite ion exchange resin can determine the optical, catalytic, and magnetic properties of the matrix (Domènech et al., 2012). This was explained by a study on a nanocomposite ion exchange material of a bifunctional polymer-metal whereby the location of the metal nanoparticles near the surface of the polymer affects the efficiency of the catalytic application. However, the size and shape of the nanoparticles in this bifunctional polymer-metal nanocomposite may determine the catalytic properties of the reaction or process (Domènech et al., 2012).

A number of variables—including the resin's chemical make-up (compositions), degree of cross-linking, and porosity—affect the resin's capacity to exchange ions. The most significant benefits of ion exchange resin technology are its superior selectivity, high capacity, and straightforward regeneration operations. The resin has the capacity to selectively remove undesired ions while keeping only the required ions in their place from a solution. Both the resin's surface area and its porosity, which can be modified to match or meet the requirements of a specific application, dictate the resin's capacity (Kakihana et al., 1977). In addition to being reusable, the resin can be "regenerated" by washing it in a solution with a higher concentration of the desired ions. This restores the resin's capacity to attract those ions.

In an upgraded scenario when a nanoscale perspective uses ion exchange resin in a nanocomposite, the surface area for the physical exchangeable sites on the matrix increases leads to higher effectiveness of the ions exchange with resins for biodiesel application. Other than that, the formed nanoparticles near the surface of the polymer content in an ion exchange resin may affect the accessibility and efficiency of any certain reaction during purification or production (Domènech et al., 2012).

5.1.2 Swelling

The hydration of the fixed ionic groups is primarily responsible for the water swelling that takes place in an ion exchanger. As the ion exchanger's capacity increases, up to the limits set by the polymer network, this swelling occurs more frequently. It simply means that the overall volume of the resin matrix increases when exposed to the intended solutions during production or purification. As a result, with a cation exchanger, there is a volume shift associated with the monovalent ion species in the following order: $Li^+ > Na^+ > K^+ > Cs^+ > Ag^+$. The volume of the resin moves as it is converted to ionic forms with variable degrees of hydration (Vinco et al., 2022). Na^+ is preferred over Ca^{2+} and Al^{3+} because of the increased osmotic pressure; less water is absorbed by solutions that have a higher concentration. This is because the cross-linking activity of polyvalent ions is harmful to hydration.

Drawbacks associated with employing conventional methods for ion exchange resin, such as utilizing resin beads with a certain physical configuration, encompass low mechanical strength and low thermal stability. Additionally, these methods may also cause swelling when the volume increases. Thus, there are many ways to overcome this by incorporating nanomaterials like nanotubes from carbon or graphene oxide, silica-based, or metal oxides-based polymeric resin matrix within. With this nanoscale in a nanocomposite, porosity, adsorption capacity, mechanical strength, and thermal stability can be enhanced. It was reported that biodiesel purification was improved using nanocomposite application (with smaller particle size), when the methanol, glycerol, and the free fatty acid contents in biodiesel were reduced significantly.

5.1.3 Kinetics

Kinetics is simply how fast the ions are exchanged during the process. The ion exchange process involves both diffusion within the resin particle and diffusion via the film of solution that is in close contact with the resins. Film diffusion (rate controlling) controls the rate at low concentrations, whereas particle diffusion (rate controlling) controls the rate at high concentrations. Regardless of whether the mechanism controlling the rate of diffusion is film diffusion or particle diffusion, the resin's particle size is likewise a deciding factor (Vinco et al., 2022). The kinetic performance of UPS resins is greatly enhanced because they do not contain the slower-moving larger beads seen in traditional polydisperse resins.

5.2 Construction methods, resin properties, and biodiesel application

5.2.1 Construction methods and resin properties

To create ion exchange resins, a cross-linked bead copolymer is prepared, followed by sulfonation (for strong acid cation resins) or chloromethylation (for anion resins), and amination. Two types of exchange resin are cations exchange resins and anion exchange resins.

5.2.1.1 Cations exchange resins

Weak acid cation exchange resins are mostly made of acrylic or methacrylic acid, cross-linked with a di-functional monomer (typically divinylbenzene, DVB). The ester of the acid in suspension may be polymerized first, and the resulting compound could then be hydrolyzed to produce the functional acid group. Weak acid resins are easily revived by strong acids due to their strong attraction to the hydrogen ion. Alkalinity and alkaline earth metals are both related to the acid-regenerated resin's high capacity for alkaline earth metals and poor capacity for alkali metals, respectively (Eisenacher et al., 2022).

Salt splitting is minimal when neutral salts are used. However, if the resin is not protonated (for instance, if it has been neutralized with sodium hydroxide), softening can still be achieved even in the presence of a high salt background. Strong acid resins are composed of styrene and DVB sulfonate copolymers. These materials are adaptable or versatile across the pH range due to their capacity to divide neutral salts or exchange cations.

An example from a cation exchange resin in biodiesel production was reported by using NKC-9, 001 X 7, and D61 for the esterification process to produce biodiesel generated from waste frying oils. The cation exchange resin uses the concept of heterogenous catalyst for the biodiesel yield. NKC-9 indicated the highest activity for the conversion of up to 90% from free fatty acids (FFA) using methanol in the cation ion exchange resin heterogeneously. This report proved that the cation exchange resin is advantageous to other methods over biodiesel production (Feng et al., 2010). However, the optimal reaction time using this cation exchange resin was 2 hours which gives ways for further improvement to shorten the yield time by using nanocomposite resins achieving of shorter time like less than 30 minutes. With the hope that with ion exchange capacity in nanocomposite resin, not only shorter time but also with the higher capacity of handling more ions during biodiesel production, can lead towards better yield.

5.2.1.2 Anion exchange materials

Weak base resins as the acid adsorbers lack exchangeable ionic sites. These resins may absorb strong acids and are rapidly regenerated with caustic. As a result, when combined with a strong base anion, they offer a very high working capacity and regeneration efficiency. An example of biodiesel production using anionic ion exchange resin was reported. It was found that lower cross-linking density, achieved by using smaller resins particle size can increase the reaction rate and result in a higher conversion rate of biodiesel of about 90% for triolein, a type of triglycerides. The presence of a heterogenous catalyst such as Diaion SA10A, SA20A, PA306, PA312, PA316, and PA318 was also reported to be the most efficient in biodiesel production. Nanocomposite of an anion exchange resin, or a combination of both the anion and cation in one setup, may be used to overcome the higher volume of oil needed (about 20%) for this anion exchange resin. This can lead to a reduction in the required amount of oil to less than 10% by weight, resulting in a higher conversion of FFAs (Shibasaki-Kitakawa et al., 2007).

5.3 Preliminary overview prospect of nanocomposite implementation in biodiesel production

Biodiesel is a mono-alkyl ester of long-chain fatty acid derived from triglycerides molecules during transesterification with the presence of a catalyst. Solid nanocomposite ion exchange resin (cationic or anionic based) (Shibasaki-Kitakawa et al., 2007) can be used to remove biodiesel byproducts/biowastes, that is, glycerol, FFA, water, and other contaminants, such as soap or metal ions, during biodiesel production process. Nanocomposite is formed by combining initially synthesized nanoparticles of the desired nanoparticle resins or other materials with specific functional groups on stable support or polymeric matrix through various techniques such as solution blending, melt mixing, in-situ polymerization, hydrothermal carbonization (Tran et al., 2016), and sulfonation (Tran et al., 2016) before mechanically shaping into its desired form through extrusion, compression molding or casting method.

Solid composite materials are used to adsorb biowastes by incorporating the nanoparticles or nanofillers into the matrix of the ion exchange resin/other materials' composite such as from silica, zeolites, activated carbon/clay/fibers (Atadashi, 2015; Tran et al., 2016), carbon materials like graphene or nanotubes or metal oxides as matrices to increase the adsorption capacity and efficiency. These are considered separation and purification mechanisms. Apart from that, the ion exchange resin may serve as the catalyst support (heterogeneous) in transesterification reactions via packings. For instance, the zinc-copper oxide (ZnCuO)/N-doped graphene nanocomposite catalyst for transesterification of waste cooking oil (Kuniyil et al., 2021) and iron oxide- manganese dioxide (Fe_3O_4/MnO_2) nanocomposite catalyst in a microwave-assisted transesterification process (Stegarescu et al., 2021) has been applied. This will be further elaborated in the next section.

Above all, the ion exchange resins application in biodiesel production can improve the selectivity of specific ions/contaminants, give higher adsorption capacity, promote higher overall adsorption efficiency, can enhance catalytic performance (Shibasaki-Kitakawa et al., 2007), be highly reusable, and can increase the stability of the biodiesel reaction.

There are a few challenges faced in converting lab-based research toward commercialization for industrial applications. Pacific Biodiesel and Axens have adopted adsorbents by applying ion exchange resins via adsorbent (Reis & Cardoso, 2016), using their understanding of the affinity of glycerine toward biodiesel purification proposed tools to be ventured. The few factors which are crucially affecting the smoothness of implementation cover the efficiency and cost-effectiveness of the adsorbent, relating to the upscaling methodology. Engineering parametric optimization for instance the contact time and the ion exchange dispersed resin with nanoparticles ratio, including any regeneration protocols to target certain waste dosage percentage, are very significant for upscaling methodology and ensuring high-quality biofuel production with enhanced viability and sustainability.

5.4 Biodiesel principle with ion exchange resins applications

5.4.1 Transesterification

Transesterification is one of the primary biodiesel production methods, which entails the reaction of triglycerides (found in vegetable oils or animal fats) with an alcohol in the presence of a catalyst to produce fatty acid alkyl esters (commonly known as biodiesel) (Salaheldeen et al., 2021). Methanol is the most common alcohol used for transesterification, however, ethanol and other alcohols may also be applied.

Several factors can influence the efficiency of the transesterification reaction such as molar ratio, reaction temperature and time, catalyst weight concentration, catalyst types, and feedstock quality (Sánchez et al., 2015). The alcohol-to-triglyceride molar ratio is an essential parameter in transesterification. To ensure complete conversion, the stoichiometric ratio is typically around 3:1 (alcohol: triglycerides). Excess alcohol contributes to the transition toward biodiesel production. However, using an excessive quantity of alcohol can be economically unfavorable and may interfere with the separation procedure (Rojas González et al., 2009). The temperature of a reaction has a substantial effect on its rate.

Transesterification processes are typically carried out at temperatures between 50°C and 65°C. A chemical reaction can be sped up by higher temperatures, but it can also cause unwanted side reactions (Mazubert et al., 2014). The duration of the reaction might vary from a few minutes to a few hours, depending on the catalyst, reactants, and reaction circumstances. The effectiveness of the reaction might be impacted by the quality of the feedstock, especially the triglycerides (vegetable oil or animal fats). According to researchers, soap generation and lower yield can be caused by a high concentration of FFAs in the substrate (Andrade et al., 2017). To avoid unwanted side effects, it is crucial to use feedstock with a low moisture content.

Effective transesterification requires a concentrated catalyst. For alkaline catalysis, such as utilizing sodium or potassium hydroxide, a typical catalyst concentration ranges from 0.5% to 2% by weight of the hydrocarbon feedstock (Díaz-Muñoz et al., 2022). Higher catalyst concentrations may cause detergent to form and make it more challenging to separate the biodiesel from the glycerol phase. The choice of catalyst impacts separation techniques, product quality, and reaction rate. Due to their quicker reaction rates, alkaline catalysts are more commonly utilized than acid catalysts. However, when the feedstock has a high concentration of FFAs, acid catalysts can be used for esterification (Lokman NolHakim et al., 2021).

Acid catalysts, such as sulfuric acid or hydrochloric acid, are effective in catalyzing esterification reactions, particularly when dealing with feedstocks with a high content of FFA. They can convert FFAs into esters, thereby facilitating the transesterification process (Kombe et al., 2013). Compared to base catalysts, acid catalysts can operate at lower temperatures and have reduced reaction times. However, as they can corrode equipment and pose safety risks, acid catalysts must be used with extreme care. In addition, acid-catalyzed reactions may result in enhanced soap formation due to the reaction between FFAs and alcohol, which can make the separation process more difficult. Because acid catalysts can convert FFAs into esters, they are particularly well-suited for feedstocks that include a high percentage of FFAs. To reduce FFA levels, the acid-catalyzed esterification phase can be performed prior to the transesterification procedure (Sondakh et al., 2018).

Base catalysts such as sodium hydroxide and potassium hydroxide are extensively employed in transesterification reactions due to their rapid reaction rates. Typically, base catalysts have excellent catalytic activity for transforming triglycerides into fatty acid alkyl esters, have greater yields, and can be carried out under relatively moderate reaction conditions (Essamlali et al., 2017). The elimination of FFAs is a prerequisite for the transesterification process when using base catalysts since these catalysts are unable to convert FFAs to esters in an efficient manner. Therefore, feedstocks with a high FFA content may require pretreatment procedures, such as acid esterification, to reduce FFA levels. Additionally, base-catalyzed reactions generate larger quantities of waste glycerol, which must be disposed of properly or processed further (Bazi et al., 2006).

Considering all the above challenges faced in biodiesel production, the use of heterogeneous catalysts, such as solid acid or base catalysts, has several distinct benefits. They have the potential to streamline the separation process by lowering the number of steps required for purification due to the ease with which the catalyst can be isolated from the reaction mixture. As opposed to homogeneous catalysts, heterogeneous catalysts typically demonstrate higher levels of stability as well as reusability. Nevertheless, the use of heterogeneous catalysts may result in lower reaction rates. In addition, locating an efficient heterogeneous catalyst for transesterification reactions can be challenging and the preparation can be more complex (Albayati & Doyle, 2015; Essamlali et al., 2017; Lee et al., 2009; Zeng et al., 2014).

Thus, ion exchange resin can be used as an esterification catalyst to improve glycerin purification during biodiesel production. For instance, the interesterification of palm oil with ethyl acetate to produce biodiesel and triacetin which improves biodiesel synthesis. The abovementioned acid and base catalyst use faced challenges in soap formation and possess sensitivity to water content, respectively, which results in difficult product separation. Ion exchange resin has advantages over the acid and base catalyst depending on the biodiesel production requirement.

Nanocomposite ion exchange resin was found to be more effective than the acid catalyst alone by improving the yield of biodiesel in a shorter reaction time for instance using waste cooking oil or soybean oil. This heterogeneous catalyst, via nanocomposite ion exchange resin having nanoparticles, proves to be a versatile catalyst for biodiesel production. The addition of nanoparticles to the resin improved its mechanical strength, thermal stability, and chemical resistance. It reduces the need to add more purification steps. The selectivity and adsorption capacity of the impurities are also improved since the addition of functionalities via metal nanoparticles acting as the active sites for the reaction of biodiesel can provide physical and chemical protection to overcome the challenge of mechanical strength and chemical fragility during production (Bano et al., 2020). Smaller particle size in the nanoscale increase the contact area and mass transfer for the catalytic activity between the glycerin-based materials during transesterification for biodiesel.

5.4.1.1 Microwave-assisted transesterification using synthesized Fe_3O_4/MnO_2 nanocomposite

This work reports the use of Fe_3O_4/MnO_2 nanocomposite acting as a catalyst during the transesterification assisted by microwave technology, which increased the reaction rate, producing a promising yield of biodiesel. The designed high specific surface area, optimal size, and porosity of the catalyst efficiently produced biodiesel (Stegarescu et al., 2021).

5.4.1.2 Transesterification using synthesized mesoporous bifunctional $MgO-SnO_2$ nanocatalyst in nanocomposite

This type of $MgO-SnO_2$ nanocomposite has a high surface area with a particle size of 15 nm acting as a catalyst during the transesterification of waste cooking oil to produce biodiesel. With the help of an 18:1 methanol to oil ratio at thermal conditions reaching 60°C, it results in a biodiesel yield of 88%. This type of nanocomposite was reused or recycled for up to four cycles with minimal activity loss. This nanocomposite design was considered highly efficient, easy to separate, low in cost, and eco-friendly (Velmurugan & Warrier, 2022).

5.4.2 Biochemical conversion

Biological processes and microorganisms are utilized to convert feedstocks, such as vegetable oils, animal lipids, and algae, into biodiesel during biochemical conversion. This strategy offers an array of favorable conditions, including the use of renewable feedstocks, an offset in greenhouse gas emissions, and the potential use of refuse materials. Typically, the procedure involves two steps: lipid extraction and lipid conversion (Ghasemi Naghdi et al., 2016; Prommuak et al., 2012). The feedstock is processed to extract lipids or oils during the lipid extraction phase. This can be accomplished via mechanical compression, solvent extraction, or more advanced techniques such as supercritical fluid extraction. The extracted lipids are then purified to eliminate impurities and contaminants, ensuring that the feedstock for subsequent conversion procedures is of high quality. Through various biochemical pathways, lipids are converted into biodiesel in the second phase of lipid conversion (Prommuak et al., 2012).

In enzymatic transesterification, lipase enzymes serve as catalysts to facilitate the reaction between lipids and alcohol, resulting in the production of biodiesel and glycerol as metabolites (Nigam et al., 2014; Ognjanovic et al., 2010; Onyinyechukwu et al., 2018). The advantages of enzymatic transesterification include mild reaction conditions, high specificity, and the ability to manage a diverse array of feedstocks (Camino Feltes et al., 2011).

Notably, transesterification can also be performed using enzymatic catalysts, which involve the use of enzymes as opposed to chemical catalysts. Enzymatic transesterification has several advantages, such as milder reaction conditions, greater specificity, and potentially reduced energy consumption. However, enzymatic transesterification processes are not yet extensively commercialized and are still in the research and development phase (Nigam et al., 2014; Ognjanovic et al., 2010; Ribeiro et al., 2011; Taher et al., 2011).

Immobilized enzymes such as *Candida rugosa* lipase together with ion exchange resins can be used as heterogenous catalysts to synthesize biodiesel. For instance, in a packed bed reactor. This combination is shown to improve biodiesel production. With the use of ion exchange resin such as Lewatit M-64, Amberlite IRA410CI, and Dialon PK208LH, biodiesel yield of 94.06%, 90%, and 73.88% have been achieved, respectively. In addition, these resins can be reactivated up to a few regeneration cycles with a high biodiesel yield of up to 80% using Lewatit resin without significant loss in biodiesel yield (Hidayatullah et al., 2023).

5.5 Purification of biodiesel

Purification of biodiesel can be achieved by wet washing, dry washing, precipitation, or membrane separation. The purification process is essential to separate and remove impurities such as free glycerol, soap, excess alcohol, and residual catalysts from crude biodiesel. Wet washing is carried out by using distilled water as the wet washing agent for the removal of water-soluble impurities from crude biodiesel. Commonly, this method includes washing trans-esterified products with acid aiming to neutralize the catalyst and degrade residual soap. Phosphoric acid is commonly used in low-concentration acid water washing and neutralizing homogeneous base catalysts prior to water washing (Hingu et al., 2010; Manique et al., 2012).

Prior to the wet washing step, the remaining alcohol is removed from the ester using evaporation or distillation. In the water-washing process, the water is utilized either at room temperature or hot because of the ability of the water-wetting process to tolerate a wide range of temperatures (20°C–90°C) (Sabudak & Yildiz, 2010). This method is the most conventional, very simple, and effective in removing residual glycerol and methanol from the biodiesel. Higher mass transfer rate of glycerol from the biodiesel into water was attained at higher biodiesel-to-water volume ratio. This leads to an increase in the volumetric mass transfer coefficient due to the larger mass transfer area when water is added (Atadashi et al., 2011). After the wet washing process, the water needs to be removed in the second cycle which contributed to the time-intensive process. The major drawback of wet washing is the results of cost related to wastewater generated due to the consumption of large water that should be sufficiently treated before reuse or disposal.

In the dry-washing process, impurities from crude biodiesel are removed using water-free washing agents such as acid resins and adsorbents. Dry-washing method is favorable over wet washing for its environmentally friendly approach and no risk of water on biodiesel. The most common types of affinity-based separation for dry washing are adsorption and ion exchange. Commercial adsorbents, such as Magnesol, Purolite, Amberlite, and silica ion, accumulate impurities (adsorbate) from the solution on the surface of a solid (adsorbent) through adsorption and ion exchange processes (Faccini et al., 2011). Those adsorbents consist of acidic and basic sites that could attract polar substances, including glycerol and alcohol (Atadashi, 2015). The major hindrances to large-scale application of these commercial adsorbents are the cost of the adsorbents and their reuse and discard (Alves et al., 2016). Thus, further research is needed to understand the economics and the reusability of this type of adsorbent before it can be applied in industry.

For economic and sustainable purposes, more studies have emerged on utilizing low-cost and different natural adsorbents from lignocellulosic material for biodiesel purification. Natural adsorbents like cellulose and starch from different sources like rice husk ash (Manique et al., 2012), eucalyptus pulp (Squissato et al., 2015), sugarcane bagasse (Squissato et al., 2015) and cassava (Gomes et al., 2021) are efficient in removing glycerine, methanol and water contents from crude biodiesel. Furthermore, Fadhil et al. (2012) extracted activated carbon from spent tea waste for biodiesel purification and discovered that activated carbon was more effective than washing with water and silica gel while eliminating contaminants. Those studies have shown that natural adsorbents as an alternative to synthetic adsorbents are advantageous and attractive from both economic viewpoints and exhibit good adsorption capacity (Bhatnagar & Sillanpää, 2010; Wan Ngah & Hanafiah, 2008) other than using ion exchange resin or its nanocomposites.

Apart from the wet washing process, membrane-based separation offers a widely effective purification process with no production of wastewater. Membrane-based separations are widely employed in water purification and protein separations. These membrane technologies are currently used in commercial separations of aqueous solutions, but the potential of the membrane technology of nonaqueous fluids is less exploited and still in the development stage. Membrane technology has plenty of advantages in terms of low energy-intensive and operational flexibility and is easy to operate. Both organic and inorganic membranes have been reported to permeate fatty acid methyl esters (FAMEs) (Bateni et al., 2017).

Because the complex products of methanol/ethanol, water, glycerol, and catalyst are present in the mixture, it becomes complicated for the membrane filtration system to identify and separate the biodiesel. Ceramic ultrafiltration membranes are the most effective in purifying biodiesel compared to other organic and inorganic membrane systems (Chozhavendhan et al., 2020) owing to their chemical and thermal stabilities for use with organic solvents. However, it becomes a challenge when free glycerol accumulates on the surface and precipitates over the membrane pores (Wang et al., 2009). Therefore, further research is needed to explore the potential of purifying biodiesel product mixture and scaling up membrane separation for commercial application (Rawat et al., 2013). For instance, investigations into the use of ion exchange resin in membranes with nanocomposite applications are necessary.

Considering all the scenarios mentioned above including challenges, this justifies why the utilization of nanocomposite materials for ion exchange has garnered considerable attention due to their ability to enhance the properties and performance of ion-exchange resins, increasing biodiesel yield. Incorporating nanomaterials into the resin matrix as reinforcements shows assurance in improving the ion-exchange capacity, selectivity, and overall efficiency of the materials for applications in water or nonwater purification (Alabi et al., 2018; Yadav, 2021). Another challenge for the proposed nanocomposite, which includes specific ion exchange resins, is the regeneration of the designed nanoparticles. This is also significant for cost-effectiveness in biodiesel production.

5.6 Prospect of regeneration to increase cost-effectiveness of biodiesel production

In biodiesel production, a few key components need to be regenerated. For instance, the catalyst used, the ion exchange resin, and the adsorbents in a conventional process increase the cost-effectiveness. However, the use of nanocomposite ion exchange resin can replace the conventional ion exchange resin to enhance the adsorption capacity, improve selectivity, enhance reusability, and ultimately improve the optimization of biodiesel production. These improvements contribute to reduced operational costs and offer sustainability overall.

Even with multiple regeneration of the nanocomposite ion exchange resin materials, the catalytic activities are higher as compared to the use of cation-exchange resins, allowing for repeated use of the nanocomposite resulting in a high conversion rate (Shibasaki-Kitakawa et al., 2007) of biodiesel via batch or continuous transesterification reactions.

5.7 Advantages and disadvantages of nanocomposite ion exchange resin

Nanocomposite ion exchange resins are a form of ion exchange material composed of nanomaterials or nanoparticles. Nanocomposite ion exchange resins are a potential research field for the advancement of new ion exchange materials with increased characteristics and efficiency. Nanocomposite ion exchange resins offer several advantages over traditional ion exchange resins, as shown in Fig. 5.2. For instance, its simple

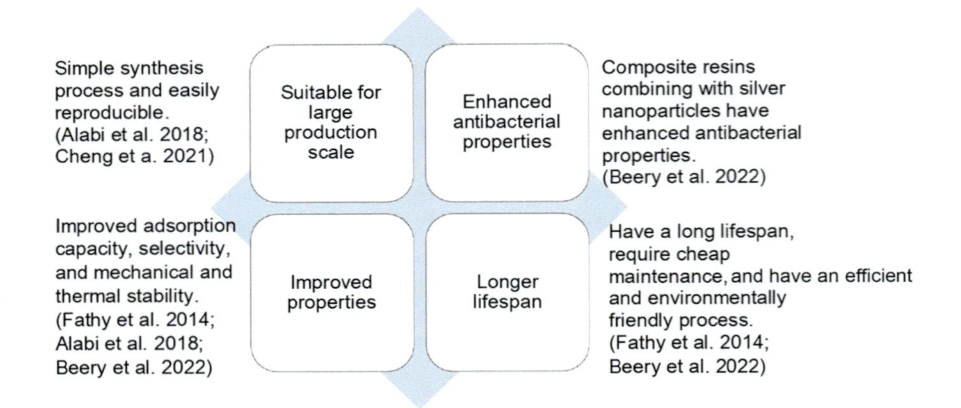

Figure 5.2 Advantages of nanocomposite ion exchange resin.

synthesis process and its easy reproducibility characteristic resulting in its suitability for large production-scale prospects (Alabi et al., 2018; Cheng et al., 2022). Its enhanced antibacterial properties, in which the composite resins are combined with silver nanoparticles, have enhanced the antibacterial properties of the surfaces (Alabi et al., 2018). Some other advantages of ion exchange resin nanocomposites include improved adsorption capacity, selectivity, mechanical and thermal stability. (Alabi et al., 2018; Beery et al., 2022; Fathy et al., 2014) Finally, the nanocomposite ion exchange resin has long lifespan with a reduced maintenance cost, offering an eco-friendly process in industrial applications (Beery et al., 2022; Fathy et al., 2014).

Overall, nanocomposite ion exchange resins perform better than conventional resins in terms of characteristics and efficiency, resulting in an attractive field of research for the exploration of advanced ion exchange materials. Furthermore, the placement of metal nanoparticles near the polymer surface improves the efficacy of their antimicrobial and catalytic applications (Domènech et al., 2012). Furthermore, hybrid organic-inorganic/inorganic ion exchangers, such as nanocomposite ion exchange resins, have popularly grown due to benefits over organic ion exchangers, such as improved stability (Sharma et al., 2018). It is also obtained that high arsenic sorption efficiency with final concentrations is below the World Health Organization (WHO)'s recommendation value (Urbano et al., 2012). According to Nabi et al. (2011), nanocomposite ion exchange resins have increased ion-exchange capacity and selectivity for hazardous heavy metals. Moreover, the use of nanocomposite ion exchange resins as an insoluble support material has several advantages over other matrix materials, including ease of handling, recycling, compatibility with water and other reaction solvents, and negligible metal leaching. Besides, these nanocomposite ion exchange resins are appropriate for column operations, resulting in increased ion exchange capacity and improved physicochemical qualities. It has a wide range of applications, including antibacterial activity, catalysis, ion-selective electrodes, bimolecular separations, chromatography, hydrometallurgy, and environmental scientific engineering (Abdelsalam et al., 2017; Gebrewold, 2017; Sharma et al., 2014). For biodiesel production, nanocomposite ion exchange resins can be used as one of the catalysts during the esterification of fatty acids to improve its properties in terms of esterification efficiency (Patiño et al., 2021).

However, developing nanocomposites ion exchange resins also has several disadvantages based on their challenges. One of the challenges is to prepare uniform dispersion of nanoparticles where it can affect the efficiency of their biocide and catalytic application (Domènech et al., 2012). It is also unstable under different conditions such as pH, pressure, and temperature conditions (Fathy et al., 2014). Besides, the cost of producing nanocomposites ion exchange resins can be higher than conventional ion exchange resins, making the overall production cost is high (Alexandratos, 2009). Although these resins have improved thermal stability, their stability is still poor and radiation stability is also limited. On the other hand, the mechanical strength and removal capacity of ordinary organic ion-exchange resins tend to decrease under high-temperature conditions (Khan et al., 2012; Kodispathi & Jacinth Mispa, 2021). According to Patiño et al. (2021), ion-exchange resins have not been used for sludge transesterification for biodiesel production. Because of that, the application of nanocomposites ion exchange resins as catalysts during biodiesel production from sludge materials has several disadvantages in terms of its stability and efficiency.

5.8 Conclusion

In conclusion, the use of nanocomposite ion exchange resin holds significant promise for enhancing the purification of biodiesel and advancing its industrial applications. By addressing the pros and cons, challenges, and solutions in this chapter, we have gained valuable insights by learning from previous experiences via basic knowledge into the diverse applications of ion exchange resin, ranging from chemicals, organics or inorganics, water treatment, biomolecule integration, to biodiesel purification itself. This understanding has paved the way for future improvements in biodiesel production by leveraging the traditional yet advanced fundamentals of ion exchange resin from a nanocomposite perspective. Nanocomposite ion exchange resin offers the potential for enhanced performance, including high adsorbent efficiency, high ion exchange capacity, and excellent chemical, mechanical, and thermal properties. In the hope of combining certain advancements, we can further optimize the purification process and improve the overall efficiency and sustainability of biodiesel production.

References

Abdelsalam, S., Kheirallaa, Z., Abo-Seif, F., & Asker, S. (2017). Abiotic factors and microbial communities fouling anion exchange resin causing performance deficiency in electric power plants. *Egyptian Journal of Microbiology*, *0*(0), 17–28. Available from https://doi.org/10.21608/ejm.2017.812.1017.

Alabi, A., AlHajaj, A., Cseri, L., Szekely, G., Budd, P., & Zou, L. (2018). Review of nanomaterials-assisted ion exchange membranes for electromembrane desalination. *Npj Clean Water*, *1*(1). Available from https://doi.org/10.1038/s41545-018-0009-7, Nature Research.

Albayati, T. M., & Doyle, A. M. (2015). Encapsulated heterogeneous base catalysts onto SBA-15 nanoporous material as highly active catalysts in the transesterification of sunflower oil to biodiesel. *Journal of Nanoparticle Research*, *17*(2). Available from https://doi.org/10.1007/s11051-015-2924-6.

Alexandratos, S. D. (2009). Ion-exchange resins: A retrospective from industrial and engineering chemistry research. *Industrial and Engineering Chemistry Research*, *48*(1), 388–398. Available from https://doi.org/10.1021/ie801242v.

Alves, M. J., Cavalcanti, Í. V., de Resende, M. M., Cardoso, V. L., & Reis, M. H. (2016). Biodiesel dry purification with sugarcane bagasse. *Industrial Crops and Products*, *89*, 119–127. Available from https://doi.org/10.1016/j.indcrop.2016.05.005.

Andrade, T. A., Errico, M., & Christensen, K. V. (2017). Transesterification of castor oil catalyzed by liquid enzymes: Optimization of reaction conditions. *Computer Aided Chemical Engineering*, 2863–2868. Available from https://doi.org/10.1016/b978-0-444-63965-3.50479-7, Elsevier.

Atadashi, I. M. (2015). Purification of crude biodiesel using dry washing and membrane technologies. *Alexandria Engineering Journal*, *54*(4), 1265–1272. Available from https://doi.org/10.1016/j.aej.2015.08.005, Elsevier B.V.

Atadashi, I. M., Aroua, M. K., & Aziz, A. A. (2011). Biodiesel separation and purification: A review. *Renewable Energy*, *36*(2), 437–443. Available from https://doi.org/10.1016/j.renene.2010.07.019.

Bano, S., Ganie, A. S., Sultana, S., Sabir, S., & Khan, M. Z. (2020). Fabrication and optimization of nanocatalyst for biodiesel production: An overview. *Frontiers in Energy Research*, *8*. Available from https://doi.org/10.3389/fenrg.2020.579014.

Bateni, H., Saraeian, A., & Able, C. (2017). A comprehensive review on biodiesel purification and upgrading. *Biofuel Research Journal*, *4*(3), 668–690. Available from https://doi.org/10.18331/BRJ2017.4.3.5.

Bazi, F., El Badaoui, H., Tamani, S., Sokori, S., Oubella, L., Hamza, M., Boulaajaj, S., & Sebti, S. (2006). Catalysis by phosphates: A simple and efficient procedure for transesterification reaction. *Journal of Molecular Catalysis A: Chemical*, *256*(1–2), 43–47. Available from https://doi.org/10.1016/j.molcata.2006.04.034.

Beery, D., Mottaleb, M. A., Meziani, M. J., Campbell, J., Miranda, I. P., & Bellamy, M. (2022). Efficient route for the preparation of composite resin incorporating silver nanoparticles with enhanced antibacterial properties. *Nanomaterials*, *12*(3). Available from https://doi.org/10.3390/nano12030471.

Bhandari, V. M., Sorokhaibam, L. G., & Ranade, V. V. (2016). Ion exchange resin catalyzed reactions-An overview. *Industrial Catalytic Processes for Fine and Specialty Chemicals*, 393–426. Available from https://doi.org/10.1016/B978-0-12-801457-8.00009-4, Elsevier Inc.

Bhatnagar, A., & Sillanpää, M. (2010). Utilization of agro-industrial and municipal waste materials as potential adsorbents for water treatment-A review. *Chemical Engineering Journal*, *157*(2–3), 277–296. Available from https://doi.org/10.1016/j.cej.2010.01.007.

Camino Feltes, M. M., De, D., Luiz, J., & de Oliveir, J. V. (2011). An overview of enzyme-catalyzed reactions and alternative feedstock for biodiesel production. *Alternative Fuel*. Available from https://doi.org/10.5772/24057, InTech.

Cheng, S., Qian, J., Zhang, X., Lu, Z., & Pan, B. (2022). Commercial gel-type ion exchange resin enables large-scale production of ultrasmall nanoparticles for highly efficient water decontamination. *Engineering*. Available from https://doi.org/10.1016/j.eng.2021.09.010.

Chozhavendhan, S., Singh, M. V. P., Fransila, B., Kumar, R. P., & Devi, G. K. (2020). A review on influencing parameters of biodiesel production and purification processes. *Current Research in Green and Sustainable Chemistry*, *2*, 1–6.

Díaz-Muñoz, L. L., Reynel-Ávila, H. E., Mendoza-Castillo, D. I., Bonilla-Petriciolet, A., & Jáuregui-Rincón, J. (2022). Preparation and characterization of alkaline and acidic heterogeneous carbon-based catalysts and their application in vegetable oil transesterification to obtain biodiesel. *International Journal of Chemical Engineering*, *2022*, 1–13. Available from https://doi.org/10.1155/2022/7056220.

Domènech, B., Bastos-Arrieta, J., Alonso, A., Macanás, J., Muñoz, M., & Muraviev, D. N. (2012). *Bifunctional polymer-metal nanocomposite ion exchange materials. Ion Exchange Technologies*. InTech. Available from http://doi.org/10.5772/51579.

Eisenacher, M., Venschott, M., Dylong, D., Hoelderich, W. F., Schütz, J., & Bonrath, W. (2022). Upgrading bio-based acetone to diacetone alcohol by aldol reaction using Amberlyst A26-OH as catalyst. *Reaction Kinetics, Mechanisms and Catalysis*, *135*(2), 971–986. Available from https://doi.org/10.1007/s11144-022-02168-z.

Essamlali, Y., Amadine, O., Larzek, M., Len, C., & Zahouily, M. (2017). Sodium modified hydroxyapatite: Highly efficient and stable solid-base catalyst for biodiesel production. *Energy Conversion and Management*, *149*, 355–367. Available from https://doi.org/10.1016/j.enconman.2017.07.028.

Faccini, C. S., Da Cunha, M. E., Moraes, M. S. A., Krause, L. C., Manique, M. C., Rodrigues, M. R. A., Benvenutti, E. V., & Caramão, E. B. (2011). Dry washing in biodiesel purification: A comparative study of adsorbents. *Journal of the Brazilian Chemical Society*, *22*(3), 558–563. Available from https://doi.org/10.1590/S0103-50532011000300021.

Fadhil, A. B., Dheyab, M. M., & Abdul-Qader, A. Q. Y. (2012). Purification of biodiesel using activated carbons produced from spent tea waste. *Journal of the Association of Arab Universities for Basic and Applied Sciences*, *11*(1), 45–49. Available from https://doi.org/10.1016/j.jaubas.2011.12.001.

Fathy, M., Moghny, A. T., Awad Allah, A. E., & Alblehy, A. E. (2014). Cation exchange resin nanocomposites based on multi-walled carbon nanotubes. *Applied Nanoscience (Switzerland)*, *4*(1), 103–112. Available from https://doi.org/10.1007/s13204-012-0178-5.

Feng, Y., He, B., Cao, Y., Li, J., Liu, M., Yan, F., & Liang, X. (2010). Biodiesel production using cation-exchange resin as heterogeneous catalyst. *Bioresource Technology*, *101*(5), 1518–1521. Available from https://doi.org/10.1016/j.biortech.2009.07.084.

Gebrewold, F. (2017). Advances in inorganic ion exchangers and their applications. A review article. *Chemistry and Materials Research*, *9*(3), 1–5. Available from http://www.iiste.org.

Ghasemi Naghdi, F., González González, L. M., Chan, W., & Schenk, P. M. (2016). Progress on lipid extraction from wet algal biomass for biodiesel production. *Microbial Biotechnology*, *9*(6), 718–726. Available from https://doi.org/10.1111/1751-7915.12360.

Gomes, M. G., Santos, D. Q., Morais, L. C. de, & Pasquini, D. (2021). Purification of biodiesel by dry washing and the use of starch and cellulose as natural adsorbents: Part II–study of purification times. *Biofuels*, *12*(5), 579–587. Available from https://doi.org/10.1080/17597269.2018.1510721.

Hidayatullah, I. M., Soelander, F., Suiyasa, P. V., Cognet, P., & Hermansyah, H. (2023). Ion exchange resin and entrapped candida rugosa lipase for biodiesel synthesis in the recirculating packed-bed reactor: A performance comparison of heterogeneous catalysts. *Energies*, *16*(4765).

Hingu, S. M., Gogate, P. R., & Rathod, V. K. (2010). Synthesis of biodiesel from waste cooking oil using sonochemical reactors. *Ultrasonics Sonochemistry*, *17*(5), 827–832. Available from https://doi.org/10.1016/j.ultsonch.2010.02.010.

Kakihana, H., Kotaka, M., Satoh, S., Nomura, M., & Okamoto, M. (1977). Fundamental studies on the ion-exchange separation of boron isotopes. *Bulletin of the Chemical Society of Japan*, *50*(1), 158–163.

Khan, A., Asiri, A. M., Rub, M. A., Azum, N., Khan, A. A. P., Khan, I., & Mondal, P. K. (2012). Review on composite cation exchanger as interdicipilinary materials in analytical chemistry. *Int. Journal of Electrochemistry Science*, *7*, 3854–3902. Available from http://www.electrochemsci.org.

Kodispathi, T., & Jacinth Mispa, K. (2021). Fabrication, characterization, ion-exchange studies and binary separation of polyaniline/Ti(IV) iodotungstate composite ion-exchanger for the treatment of water pollutants. *Environmental Nanotechnology, Monitoring and Management*, *16*100555. Available from https://doi.org/10.1016/j.enmm.2021.100555.

Kombe, G. G., Temu, A. K., Rajabu, H. M., Mrema, G. D., Kansedo, J., & Lee, K. T. (2013). Pre-treatment of high free fatty acids oils by chemical re-esterification for biodiesel production-A review. *Advances in Chemical Engineering and Science*, *03*(04), 242–247. Available from https://doi.org/10.4236/aces.2013.34031.

Kuniyil, M., Kumar, S. J. V., Adil, S. F., Assal, M. E., Shaik, M. R., Khan, M., Al-Warthan, A., & Siddiqui, M. R. H. (2021). Production of biodiesel from waste cooking oil using ZnCuO/N-doped graphene nanocomposite as an efficient heterogeneous catalyst. *Arabian Journal of Chemistry*, *14*(3)102982. Available from https://doi.org/10.1016/j.arabjc.2020.102982.

Lee, D.-W., Park, Y.-M., & Lee, K.-Y. (2009). Heterogeneous base catalysts for transesterification in biodiesel synthesis. *Catalysis Surveys from Asia*, *13*(2), 63–77. Available from https://doi.org/10.1007/s10563-009-9068-6.

Lokman NolHakim, M. A. H., Shohaimi, N. A. M., Mokhtar, W. N. A. W., Ibrahim, M. L., & Abdullah, R. F. (2021). Immobilization of potassium-based heterogeneous catalyst over alumina beads and powder support in the transesterification of waste cooking oil. *Catalysts*, *11*(8), 976. Available from https://doi.org/10.3390/catal11080976.

Manique, M. C., Faccini, C. S., Onorevoli, B., Benvenutti, E. V., & Caramão, E. B. (2012). Rice husk ash as an adsorbent for purifying biodiesel from waste frying oil. *Fuel*, *92*(1), 56–61. Available from https://doi.org/10.1016/j.fuel.2011.07.024.

Mazubert, A., Taylor, C., Aubin, J., & Poux, M. (2014). Key role of temperature monitoring in interpretation of microwave effect on transesterification and esterification reactions for biodiesel production. *Bioresource Technology*, *161*, 270–279. Available from https://doi.org/10.1016/j.biortech.2014.03.011.

Nabi, S. A., Shahadat, M., Bushra, R., Shalla, A. H., & Azam, A. (2011). Synthesis and characterization of nano-composite ion-exchanger; its adsorption behavior. *Colloids and Surfaces B: Biointerfaces*, *87*(1), 122–128. Available from https://doi.org/10.1016/j.colsurfb.2011.05.011.

Nigam, S., Mehrotra, S., Vani, B., & Mehrotra, R. (2014). Lipase immobilization techniques for biodiesel production: An overview. *International Journal of Renewable Energy and Biofuels*, 1–16. Available from https://doi.org/10.5171/2014.664708.

Ognjanovic, D. N., Petrovic, D. S., Bezbradica, I. D., & Knezevic-Jugovic, D. Z. (2010). Lipases as biocatalysts for biodiesel production. *Chemical Industry*, *64*(1), 1–8. Available from https://doi.org/10.2298/hemind100100lo.

Onyinyechukwu, J. C., Christiana, N. O., Chukwudi, O., & James, C. O. (2018). Lipase in biodiesel production. *African Journal of Biochemistry Research, 12*(8), 73–85. Available from https://doi.org/10.5897/ajbr2018.0999.

Patiño, Y., Faba, L., Díaz, E., & Ordóñez, S. (2021). Biodiesel production from wastewater sludge using exchange resins as heterogeneous acid catalyst: Catalyst selection and sludge pre-treatments. *Journal of Water Process Engineering, 44*. Available from https://doi.org/10.1016/j.jwpe.2021.102335.

Prommuak, C., Pavasant, P., Quitain, A. T., Goto, M., & Shotipruk, A. (2012). Microalgal lipid extraction and evaluation of single-step biodiesel production. *Engineering Journal, 16*(5), 157–166. Available from https://doi.org/10.4186/ej.2012.16.5.157.

Rawat, I., Ranjith Kumar, R., Mutanda, T., & Bux, F. (2013). Biodiesel from microalgae: A critical evaluation from laboratory to large scale production. *Applied Energy, 103*, 444–467. Available from https://doi.org/10.1016/j.apenergy.2012.10.004.

Reis, M. H. M., & Cardoso, V. L. (2016). Biodiesel production and purification using membrane technology. *Membrane Technologies for Biorefining*, 289–307. Available from https://doi.org/10.1016/B978-0-08-100451-7.00012-8, Elsevier Inc.

Ribeiro, B. D., de Castro, A. M., Coelho, M. A. Z., & Freire, D. M. G. (2011). Production and use of lipases in bioenergy: A review from the feedstocks to biodiesel production. *Enzyme Research, 2011*, 1–16. Available from https://doi.org/10.4061/2011/615803.

Riffat, R., & Husnain, T. (2022). *Fundamentals of wastewater treatment and engineering*. <https://doi.org/10.1201/9781003134374>.

Rojas González, A. F., Girón Gallego, E., & Torres Castañeda, H. G. (2009). Operations variables in the transesterification process of vegetable oil: A review - chemical catalysis. *Ingeniería e Investigación, 29*(3), 17–22. Available from https://doi.org/10.15446/ing.investig.v29n3.15177.

Sabudak, T., & Yildiz, M. (2010). Biodiesel production from waste frying oils and its quality control. *Waste Management, 30*(5), 799–803. Available from https://doi.org/10.1016/j.wasman.2010.01.007.

Salaheldeen, M., Mariod, A. A., Aroua, M. K., Rahman, S. M. A., Soudagar, M. E. M., & Fattah, I. M. R. (2021). Current state and perspectives on transesterification of triglycerides for biodiesel production. *Catalysts, 11*(9), 1121. Available from https://doi.org/10.3390/catal11091121.

Sánchez, B. S., Benitez, B., Querini, C. A., & Mendow, G. (2015). Transesterification of sunflower oil with ethanol using sodium ethoxide as catalyst. Effect of the reaction conditions. *Fuel Processing Technology, 131*, 29–35. Available from https://doi.org/10.1016/j.fuproc.2014.10.043.

Sharma, G., Pathania, D., & Naushad, M. (2014). Preparation, characterization and antimicrobial activity of biopolymer based nanocomposite ion exchanger pectin zirconium(IV) selenotungstophosphate: Application for removal of toxic metals. *Journal of Industrial and Engineering Chemistry, 20*(6), 4482–4490. Available from https://doi.org/10.1016/j.jiec.2014.02.020.

Sharma, S., Yadav, A., & Ahmad, W. (2018). *The classification, characterization, and application of ion exchange resins:A general survey*. <https://ssrn.com/abstract = 3299226>.

Shibasaki-Kitakawa, N., Honda, H., Kuribayashi, H., Toda, T., Fukumura, T., & Yonemoto, T. (2007). Biodiesel production using anionic ion-exchange resin as heterogeneous catalyst. *Bioresource Technology, 98*(2), 416–421. Available from https://doi.org/10.1016/j.biortech.2005.12.010.

Sondakh, R. C., Hambali, E., & Indrasti, N. S. (2018). Esterification bio-oil using acid catalyst and ethanol. *International Journal of Engineering and Management Research, 8*(5), 137–141. Available from https://doi.org/10.31033/ijemr.8.5.15.

Squissato, A. L., Fernandes, D. M., Sousa, R. M. F., Cunha, R. R., Serqueira, D. S., Richter, E. M., Pasquini, D., & Muñoz, R. A. A. (2015). Eucalyptus pulp as an adsorbent for biodiesel purification. *Cellulose*, *22*(2), 1263–1274. Available from https://doi.org/10.1007/s10570-015-0557-7.

Stegarescu, A., Soran, M. L., Lung, I., Opris, O., Gutoiu, S., Leostean, C., Lazar, M. D., Kacso, I., Silipas, T. D., Pana, O., & Porav, A. S. (2021). Nanocomposite based on Fe_3O_4/MnO_2 for biodiesel production improving. *Chemical Papers*, *75*(7), 3513–3520. Available from https://doi.org/10.1007/s11696-021-01590-2.

Taher, H., Al-Zuhair, S., Al-Marzouqi, A. H., Haik, Y., & Farid, M. M. (2011). A review of enzymatic transesterification of microalgal oil based biodiesel using supercritical technology. *Enzyme Research*, *2011*, 1–25. Available from https://doi.org/10.4061/2011/468292.

Tanaka, Y. (2015a). *Fundamental properties of ion exchange membranes. Ion exchange membranes* (pp. 29–65). Elsevier. Available from https://doi.org/10.1016/b978-0-444–63319-4.00002-x.

Tanaka, Y. (2015b). *Ion exchange membranes: Fundamentals and applications: Second edition*. Elsevier Inc. Available from https://doi.org/10.1016/C2013-0-12870-X.

Tran, T. T. V., Kaiprommarat, S., Kongparakul, S., Reubroycharoen, P., Guan, G., Nguyen, M. H., & Samart, C. (2016). Green biodiesel production from waste cooking oil using an environmentally benign acid catalyst. *Waste Management*, *52*, 367–374. Available from https://doi.org/10.1016/j.wasman.2016.03.053.

Urbano, B. F., Rivas, B. L., Martinez, F., & Alexandratos, S. D. (2012). Water-insoluble polymer-clay nanocomposite ion exchange resin based on N-methyl-d-glucamine ligand groups for arsenic removal. *Reactive and Functional Polymers*, *72*(9), 642–649. Available from https://doi.org/10.1016/j.reactfunctpolym.2012.06.008.

Velmurugan, A., & Warrier, A. R. (2022). Production of biodiesel from waste cooking oil using mesoporous $MgO-SnO_2$ nanocomposite. *Journal of Engineering and Applied Science*, *69*(1), 92. Available from https://doi.org/10.1186/s44147-022-00143-y.

Vinco, J. H., Botelho Junior, A. B., Duarte, H. A., Espinosa, D. C. R., & Tenório, J. A. S. (2022). Kinetic modeling of adsorption of vanadium and iron from acid solution through ion exchange resins. *Transactions of Nonferrous Metals Society of China (English Edition)*, *32*(7), 2438–2450. Available from https://doi.org/10.1016/S1003-6326(22)65916-8.

Wan Ngah, W. S., & Hanafiah, M. A. K. M. (2008). Removal of heavy metal ions from wastewater by chemically modified plant wastes as adsorbents: A review. *Bioresource Technology*, *99*(10), 3935–3948. Available from https://doi.org/10.1016/j.biortech.2007.06.011.

Wang, Y., Wang, X., Liu, Y., Ou, S., Tan, Y., & Tang, S. (2009). Refining of biodiesel by ceramic membrane separation. *Fuel Processing Technology*, *90*(3), 422–427. Available from https://doi.org/10.1016/j.fuproc.2008.11.004.

Yadav, M. D. (2021). *Advanced nanocomposite ion exchange materials for water purification. Handbook of nanomaterials for wastewater treatment: Fundamentals and scale up issues* (pp. 513–534). Elsevier. Available from https://doi.org/10.1016/B978-0-12–821496-1.00014-3.

Zeng, R., Sheng, H., Zhang, Y., Feng, Y., Chen, Z., Wang, J., Chen, M., Zhu, M., & Guo, Q. (2014). Heterobimetallic dinuclear lanthanide alkoxide complexes as acid-base difunctional catalysts for transesterification. *The Journal of Organic Chemistry*, *79*(19), 9246–9252. Available from https://doi.org/10.1021/jo5016536.

Section 3

Agricultural applications

Nanocomposite pesticides: a more efficient and ecologically friendly strategy to protect agricultural crops

Tortella Gonzalo[1,2], Javiera Parada[1], Olga Rubilar[1,2], Paola Durán[3], Paola Fincheira[1], Antonio Juárez-Maldonado[4], Adalberto Benavides-Mendoza[5] and Carlos Alberto Garza-Alonso[6]

[1]Centro de Excelencia en Investigación Biotecnológica Aplicada al Medio Ambiente, CIBAMA-BIOREN, Universidad de La Frontera, Temuco, Chile, [2]Departamento de Ingeniería Química, Facultad de Ingeniería y Ciencias, Universidad de La Frontera, Temuco, Chile, [3]Scientific and Technological Bioresource Nucleus, Universidad de La Frontera, Temuco, Chile, [4]Department of Botany, Autonomous Agrarian University Antonio Narro, Saltillo, Coahuila, Mexico, [5]Department of Horticulture, Universidad Autónoma Agraria Antonio Narro, Saltillo, Mexico, [6]Facultad de Agronomía, Universidad Autónoma de Nuevo León, Gral. Escobedo, Mexico

6.1 Introduction

The worldwide population has increased exponentially during the last decades. United Nations (2022) estimates that the worldwide population could grow to around 8.5 billion in 2030, 9.7 billion in 2050, and 10.4 billion in 2100. Therefore, it is expected that the demand for food will increase considerably. In this sense, fertilization and pesticide application have been traditional agricultural activities frequently realized to maintain sustainable food production. However, it has been well-reported in the last years that long-term fertilization causes a reduction in soil quality and crop yield declines, affecting the current agricultural development (Shinoto et al., 2020; Wang et al., 2020).

Organic fertilizers have been seen as an excellent alternative to reduce the chemical inorganic fertilizers application (Wu et al., 2020). Some studies have demonstrated that applying organic fertilizer for two years caused a reduction in the use of chemical fertilizer and enhanced soil properties and bacterial communities of the grape rhizosphere soil (Wu et al., 2020). A combination of organic and inorganic fertilization for a long-time (16 years) has shown that soil fertility and a healthy ecosystem were higher and maintained over time, favoring an increase of beneficial microorganisms and inhibiting pathogenic microorganisms (Gao et al., 2022).

Nanocomposites for Environmental, Energy, and Agricultural Applications. DOI: https://doi.org/10.1016/B978-0-443-13935-2.00006-1

However, inorganic fertilization alone increased the risk of plant infection with soil-borne diseases. Although organic fertilization has shown encouraging results, it is challenging to think that large crops can be maintained only with an organic application. It is mentioned, for example, that if we add up the nitrogen needs of all the crops, then more nitrogen is necessary than the amount available from animal manure (Wur, 2021). In addition, it has been estimated that the yield obtained from organic farming is 30% less than that obtained from traditional fertilization. Therefore, 50% more arable land must be required to produce the same quantity of food produced under organic farming (Kirchmann, 2019).

On the other hand, for many years, pesticides have been considered an essential tool in agriculture to control fungal or bacterial diseases in plants, favoring continuous and increased food production to meet global food requirements. Tudi et al. (2021) indicate that without pesticide application, the losses in agriculture could reach around 32%, 54%, and 78% in cereal, vegetable, and fruit production, respectively. However, an intensive application of pesticides has several disadvantages. One problem is that successive applications can cause resistance in the microorganisms (Hahn, 2014; Shahid & Khan, 2021). Therefore, higher concentrations are increasingly necessary for the control of pathogens. Moreover, the indiscriminate use of pesticides can cause adverse effects on human health and the water and soil environment. Due to intensive farming, these agrochemicals are continuously applied, and less than 40% of the pesticides directly affect target organisms (Fan, 2017). For example, Pimentel (1995) reported that less than 0.1% of the applied insecticides reach the target insect. Therefore, soil and microbial communities or other nontarget organisms are continuously exposed to these chemicals, resulting in damage at a structural and functional level (Scott-Fordsmand et al., 2022). Currently, pesticides are necessary for food production in agricultural lands. Their residual effect is extended to nontarget microorganisms, causing variations in the turnover rate of nutrients and microbial community structure, as well as alterations in functional metabolic activities such as β-glucosidase, phosphatase, and urease, among others (Abhishek et al., 2018; Nasreen et al., 2015; Walia et al., 2018).

As fertilizers and pesticides continue to be applied, it is necessary to use new tools that allow the most efficient use of these and to avoid losses and the contamination of the environment.

In this sense, nanotechnology has emerged as an excellent opportunity to solve various agricultural difficulties. Nanotechnology has helped us in postharvest management (Neme et al., 2021), plant disease diagnostics (Tomer et al., 2021), nanosensors to be applied in crop protection identifying diseases and residues of the pesticides (Kashyap et al., 2019; Panoth et al., 2022), and nanoformulations for applying fertilizers and pesticides to improve crops (Graily Moradi et al., 2019; Salama et al., 2021). In this last point, several nanocomposites have been studied to allow controlled delivery of agrochemicals, favoring low losses, high durability, and low application doses, making these the most suitable tools for agriculture. Although the efficient use of fertilizers is of great importance worldwide, the decrease in the volume of pesticides applied in the environment continues to be challenging because of the harmful effects on human health and the environment. Moreover, during the last years, this problem has increased notably due to the increase in pest occurrence caused by global climate change(Schneider et al., 2022).

The objective of this chapter is to comprehensively explore the advancements in nanotechnology about the development and application of nanocomposite pesticides. It aims to critically analyze the efficiency, effectiveness, and environmental impact of nanocomposite pesticides compared to traditional pesticide formulations. The chapter intends to elucidate how these nanocomposites enhance pest control, minimize ecological footprints, and contribute to sustainable agriculture. Additionally, it seeks to highlight the challenges and considerations for future research and development in this field. This book chapter delves into nanocomposite pesticides, presenting their development, mechanisms, and potential in advancing sustainable agriculture. It encompasses a concise overview of production techniques, efficacy analyses, environmental impacts, and regulatory considerations, while challenges and future research directions are also highlighted. The chapter aims to provide an integrated perspective on the transformative potential of nanocomposite pesticides in enhancing crop protection while mitigating ecological harm.

6.2 Impact of climate on pest occurrence in agriculture

Climate change has become one of the biggest challenges for humanity over the last decades, resulting in severe temperature and precipitation fluctuations and increased atmospheric carbon dioxide (FAO, 2017). The influence of these changes on pathogen virulence concerns the scientific community since the spread of different pests, such as insects, pathogens, or weeds, is on the rise. In fact, in China, a significant crop producer, the cropland area affected by pests has quadrupled during the last five decades, and it is expected to duplicate by the end of the 21st century (Wang et al., 2022). At field conditions, the pest or pathogen coexists with the environment and the host (i.e., the plant) as a tripartite interaction. Therefore, the temperature impacts on plant state as abiotic stress, as well as on the pathogen survival triggering virulence factors.

Among plant pathogens, particular interest exists in fungi, as this group is recognized as responsible for the most significant crop losses (Fones et al., 2020). Some fungi have genetically adapted to increased temperatures; as a consequence, novel and more aggressive strains have emerged and replaced the older ones (Nnadi & Carter, 2021). Other reported cases have shown that besides temperature, higher moisture, and carbon dioxide can turn some phytopathogen strains more aggressive (Elvira-Recuenco et al., 2021; Inman et al., 2008). In this sense, a study by Hanson et al. (2022) evidenced that a warmer and dryer climate may drastically impact the atmosphere microbiome, particularly on the airborne fungi amount and composition. This study also suggests that changes in temperature and water availability could affect the overwintering capability of fungi, potentially altering pathogen-crop dynamics pre- and postharvest. Indeed, an increase in the suitable habitat for plant pathogens has already been forecasted due to climate change, where the temperature has been the primary determinant of the disease impact of soil-borne fungal (Gilardi et al., 2018; Raza & Bebber, 2022). Climate change can also provoke soil acidification, which consequently alters soil fungal communities, facilitating their development over other inhibited organisms (Castaño et al., 2018).

Considering the imminent threat to worldwide agriculture production coupled with the constant increase of the global population, innovative pest control techniques are demanded to avoid the overuse of conventional agrochemicals (known to cause a nontarget impact on the environment), as the food supply must be maintained. Such is the case with nanotechnology, a discipline that has improved the efficiency of many pesticides by developing a great variety of products such as nanofertilizers, nanopesticides, and nanocomposites.

6.3 Impact of pesticides on the environment

Pesticides have been indispensable in protecting crops against pests, known for causing significant economic losses. The most common are herbicides, fungicides and bactericides, insecticides, nematicides, and rodenticides, which target weeds, pathogens, insects, and others. However, their massive use has led to a severe negative impact on the environment, such as pollution of aquatic and terrestrial resources or the presence of their residues in food. This raises great concern considering the reported risks to human health (FAO & WHO, 2019; Kim et al., 2017). The lack of control on pesticide use has also resulted in the induction of resistance genes on target pests, which have even been transferred from agriculture to strains of human health concern (Berger et al., 2017; Van Leeuwen et al., 2020).

In detail, among the different routes of pesticide exposure in the environment (Fig. 6.1), the application at field conditions is one of the most problematic, commonly performed by farmers using sprayer units, increasing the risks of human exposure. Indeed, different diseases and carcinogenic effects have been associated with the pesticide exposure of agricultural workers (Memon et al., 2019; Rodríguez & León, 2020). After the pesticide spraying on crops, some residues remain in the soil for long periods depending on the season or soil type, causing alterations in enzymatic activities and microbial communities (Karas et al., 2018; Satapute et al., 2019) or affecting the bees and other pollinators performance, survival and reproduction (Sánchez-Bayo et al., 2016). The contamination of surrounding water bodies by runoff is also facilitated, affecting the quality of drinking water for humans and livestock. Thus, pesticides can be transported through rivers and lakes and end up in seas, where aquatic animals have been found accumulating them on their bodies (Sumon et al., 2018). There is also an indirect route of human exposure from consuming food containing pesticide residues. In this sense, concentrations of pesticides have been found in different products, which has led to severe changes in regulations from different countries, such as South Korea, to reduce the maximum residue limits (MFDS, 2017). This has promoted the urgent need to develop new strategies to reduce the amount of pesticides released into the environment.

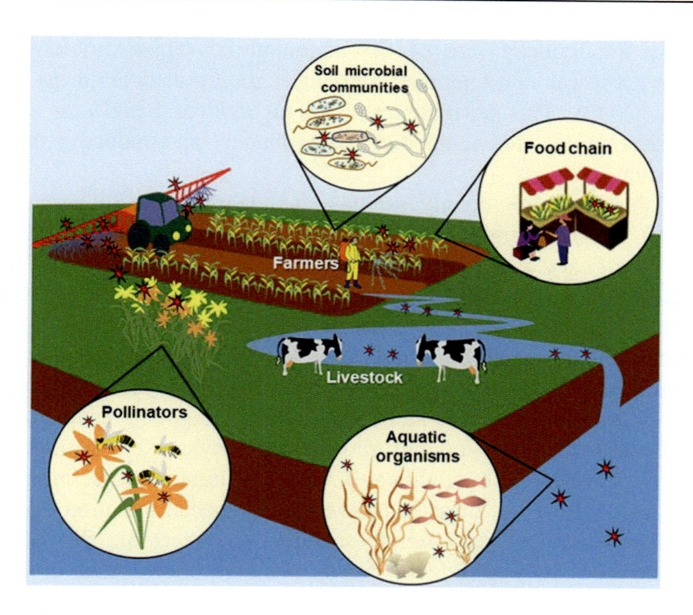

Figure 6.1 Main exposure routes to pesticides in the environment.

The uncontrolled use of pesticides lies in their chemical and physical properties. A considerable amount is commonly applied due to the low solubility of the active ingredients (AIs), which has led to nontarget effects on other organisms (Zaller & Brühl, 2019). This negative impact could also be maximized due to the use of large amounts of toxic solvents to dissolve them. For instance, volatile organic compounds, one of the most common air contaminants are emitted into the atmosphere using organic solvents (Shao & Zhang, 2017; Yang et al., 2023). What's more concerning is that approximately 90% of the dose applied may be lost through leaching, hydrolysis, or photodegradation, provoking terrestrial and aquatic contamination (Kah et al., 2018). Indeed, it has been mentioned that less than 0.1% of the amount applied may reach its target, which denotes the low efficiency of the application methods in field conditions (Kumar et al., 2014a, 2014b).

6.4 Nanocomposites for delivering agrochemicals

Implementing policy instruments to reduce pesticide use has gained ground during the last few years (Lee et al., 2019; Skevas, 2020). In this context, nanotechnology is highlighted as a promissory discipline to prolong the agrochemicals release (i.e., pesticides and fertilizers) by increasing their stability and efficiency, thus reducing their environmental impact (Grillo et al., 2021; Guha et al., 2020). Nanocomposites are multiphase materials containing one or more elements at the nanoscale that enhance the agrochemicals reactivity due to the higher surface area to volume facilitated by the reduced size of the nanomaterial (approximately 1–100 nm). Consequently, a lower agrochemical amount is required to maintain their activity (Kalia et al., 2020). This attribute provides an evident benefit in avoiding their immediate dissolution, enhancing their performance, and reducing their nontarget impact by leaching.

The available literature mainly refers to nanocomposites in terms of synthesis methods, their physicochemical behavior, and different characterization techniques. However, their efficacy as agricultural products has been evaluated to a lesser extent. In this regard, some studies have reported different nanocomposites based on various materials, mostly polymers, used as carrier systems for fertilizers and pesticides, including herbicides, fungicides, and insecticides (Table 6.1). Generally, the carrier type is chosen according to the hydrophilic or lipophilic nature of the compound being encapsulated. Thus, various metals (e.g., silica, magnetite, zinc, and aluminum), synthetic polymers (e.g., polyvinylpyrrolidone, polyethylene, ethylcellulose, and polydopamine) or biopolymers (e.g., alginates, chitosan, lignin, and cellulose) have been successfully used. However, these last have gained special attention due to their low cost, biodegradability, and eco-friendly nature (Kumar et al., 2014a; Orta et al., 2020); in some cases, they have facilitated a more controlled release of agrochemicals compared to conventional analogs (Campos et al., 2014).

Nowadays, the development of the so-called "smart" stimuli-responsive materials in nanocomposite formulations is gaining significant interest due to their capacity to regulate the release of active ingredients. This occurs depending on variations in specific conditions, such as temperature, pH, enzymes, or UV light (Camara et al., 2019). Consequently, agrochemicals can be more efficient by preventing their rapid degradation or release from nanocomposites.

6.5 Stimulation of plant defense system by nanoparticles for the control of pathogens

6.5.1 Mechanisms of action of nanopesticides

Nanocomposites represent an effective tool for the control of pathogenic organisms in plants. The way nanocomposites achieve this is divided into two groups: the direct effects derived from the nanocomposite-pathogen interaction and the indirect effects these compounds have on the defense system of plants (Fig. 6.2).

6.5.2 Direct mechanisms on target microorganisms

In general, the direct mechanisms of action of nanocomposites against pathogens can be encompassed in four significant effects: attraction of the nanocomposite to the cell wall of the pathogen, destabilization of the cell wall and membrane, toxicity due to an overproduction of ROS, and changes in the signaling pathways of pathogens (Susanti et al., 2022) (Fig. 6.2). This, in turn, depends on the physicochemical

Table 6.1 Nanocomposites based on fertilizers, herbicides, fungicides, and insecticides.

Type	Nanocomposite	Target organism	Main effects	References
Fertilizer	Chitosan-silicon	*Zea mays*	3.7-fold increased seedling vigor index and 45% higher test weight of plants compared to SiO_2.	Kumaraswamy et al. (2021)
	Carbon nanoparticles loaded with N and K	*Phaseolus vulgaris L.*	Plant height of 61 cm compared to 53 cm of control treatment, and enhanced flavonoid and phenols content in leaves and seeds.	Salama et al. (2021)
	Starch-g-poly(styrene-co-butylacrylate) with natural char nanoparticles (NCNPs) loaded with urea	*Solanum lycopersicum*	Improved plant growth and reduced N-release rate. Increased total N in aerial part of plants (21%) compared to pristine urea.	Salimi et al. (2021)
	Glauconite-Urea	*Avena sativa*	Slower plant growth compared to pure urea, reduced nitrate and ammonium release from the nanocomposite, and decreased nutrient leaching downstream to soil, saving 50% of the urea.	Rudmin et al. (2022)

(*Continued*)

Table 6.1 (Continued)

Type	Nanocomposite	Target organism	Main effects	References
Herbicide	Poly(ε-caprolactone) loaded with atrazine	– *Amaranthus viridis* – *Bidens pilosa*	50% inhibited photosystem II activity for both species, >50% decreased root and shoot growth for both species, and higher herbicidal activity compared with commercial atrazine.	Sousa et al. (2018)
	Pectin nanoparticles loaded with metsulfuron methyl	*Chenopodium album*	Reduced dry mass: from 48 g/m^2 in control up to 5 g/m^2 in plants treated with the nanocomposite. This contrasts with the 19 g/m^2 of the weed treated with commercial herbicide.	Kumar et al. (2017)
	Paraquat-loaded chitosan/ tripolyphosphate nanoparticles	*Spinacia oleracea*	Lipid peroxidation, photooxidizable P700 reaction center content, and NADPH/NADP$^+$ ratio levels significantly decreased in plants treated with nanoparticles compared to those with the nonencapsulated herbicide.	Pontes et al. (2021)
	Anionic clay (layered double hydroxide or LDH) and a cationic organoclay (Cloisite 10A, Clo10A) as nanocarriers for imazamox	*Brassica nigra*	Similar efficacy against *B. nigra* respect to commercial formulation. However, the imazamox amount in soil columns leachates decreased between 10% and 30% in the case of nanocomposites.	Khatem et al. (2019)

Fungicides	Poly(ethylene glycol) loaded with imidazole drugs (clotrimazole, econazole nitrate, and miconazole nitrate)	– *Rhizoctonia solani* – *Macrophomina phaseolina* – *Sclerotium rolfsii* – *Fusarium oxysporum*	Disease incidences from 11.11% to 27.38% in plants treated with nanoformulations compared to the 39.68% – 72.38% of the inoculated control	Tippannanavar et al. (2020)
	Mesoporous silica nanoparticles coated by pectin for the release of prochloraz (Pro@MSN-Pec)	*Magnaporthe oryzae*	Enhanced translocation in rice plants compared to commercial concentrate and a better fungicidal activity after 14 days: EC_{90} of 2.3 and 3.35 mg/L, respectively.	Abdelrahman et al. (2021)
	Iron-based metal–organic frameworks loaded with azoxystrobin	– *Fusarium graminearum* – *Phytophthora infestans*	Comparable fungicidal activity to two commercial formulations. Dual function since Fe enhanced the wheat growth (16.4% gain of height)	Shan et al. (2020)
	Nanomicelles based on poly (ethylene glycol) loaded with mancozeb	*Alternaria solani*	Reduced number of infected leaflets per tomato plant compared to the commercial formulation. Increased number of fruits per plant.	Majumder et al. (2020)

(*Continued*)

Table 6.1 (Continued)

Type	Nanocomposite	Target organism	Main effects	References
Insecticides	Chitosan-coated nanoliposomes as nanocarrier for a mixture of imidacloprid and lambda-cyhalothrin.	*Myzus persicae Sulzer*	Improved residual effect of the mixture in nanoliposomes compared to the use of each insecticide alone. The effect was dependent on the chitosan concentration.	Graily Moradi et al. (2019)
	Adhesive nanopesticide based on hollow mesoporous silica hybrid loaded with cyantraniliprole	– *Cnaphalocrocis medinalis* – *Chilo suppressalis*	The insecticidal efficacy was improved compared to commercial formulation and dependent on time after spraying. No side effects on the rice plants growth.	Gao et al. (2020)
	Esterase/glutathione (GSH) responsive photoactivated nano-pesticide loaded with photoactivated phloxine B	*Spodoptera frugiperda* (Sf9 cell line)	Exposure to esterase-6 and GSH stimulus triggered the destruction of the system and consequently the phloxine B release. Higher toxicity on Sf9 cell line compared to the *in vitro* light exposure itself.	Yin et al. (2020)
	Avermectin encapsulated in coordination polymers of copper and trimesic acid (AM@CuBTC)	*Bursaphelenchus xylophilus*	Higher mortality using AM@CuBTC (60.6%) compared to free avermectin (39.7%) after 8 h.	Liu et al. (2022)

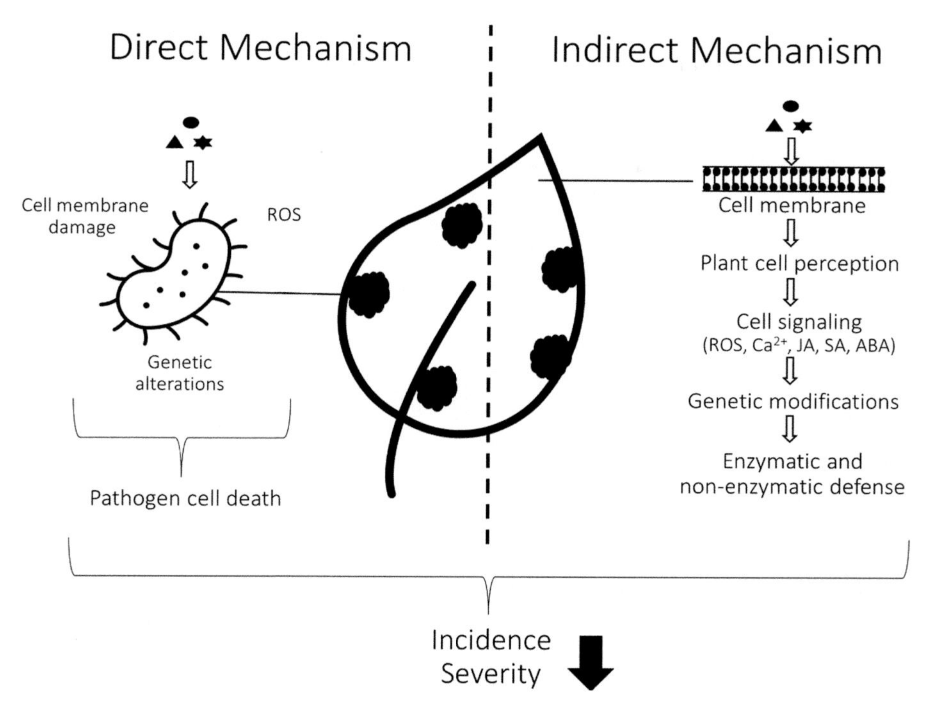

Figure 6.2 Direct and indirect mechanisms of action of nanocomposites against plant pathogens.

characteristics of the nanocomposites, mainly the size, shape, electric charge, and hydrophobicity, among others (Yin et al., 2023).

In the first instance, when nanocomposites are applied to plants, they can be attracted to the cell walls of pathogens by a difference in electrical charges, whereas nanocomposites with a net positive charge have more significant attraction due to the negative charge of the walls of the pathogen (Sánchez-López et al., 2020). Such interactions between the electrical charge of nanocomposites and their interaction with the external charges of pathogens result in membrane instability until cell lysis is produced (Adisa et al., 2019). Additionally, the small size of nanocomposites provides a large specific surface area, which increases contact points with microorganisms (Yin et al., 2023). After the first interaction of nanocomposites with pathogens and initial damage to the cell wall and membrane, smaller materials could quickly enter cells and cause damage to organelles such as ribosomes, mitochondria, and nucleus, causing the death of the microorganism (Hazarika et al., 2022). When nanocomposites enter the nucleus of cells, they produce genotoxicity due to chromosomal aberration and alterations in the gene expression of pathways involved in protein synthesis and overall growth (Zhao et al., 2022). A study by Ahmed et al. (2021) revealed that applications of ZnO NPs in *Burkholderia glumae* and *Burkholderia gladioli* produced membrane damage, cytoplasmic disorganization, and expulsion of genetic material, which was observed by transmission electron microscopy (TEM).

On the other hand, nanocomposites can also produce the so-called "mechanical folding," which consists of enveloping and immobilizing pathogens, inhibiting their growth (Khan et al., 2022). In addition to the above, nanocomposites can alter cells' proteins and nucleic acids through interactions of sulfated and nitrogenous compounds, resulting in organelle denaturation and cell death (Jasrotia et al., 2022). These interactions are mainly due to the release of ions and perforations in the cell wall that alter metabolic processes (Tortella et al., 2023). In addition, nanocomposites can inhibit the formation of bacterial biofilms, thereby reducing the ability of these organisms to protect themselves from external adverse factors (Li et al., 2023).

The application of nanocomposites has been reported to produce ROS, which in turn cause damage to pathogen structures (Ahmed et al., 2021). An overaccumulation of ROS results in DNA damage, lipid peroxidation, inhibition of enzymes, and cell apoptosis (Adisa et al., 2019). In addition, some reports indicate that nanocomposites cause damage to pathogens through inhibition of H^+-ATPase activity, inhibition of messenger RNA, blocking of nutrients, disruption of membranes, and inhibition of toxin production (Kutawa et al., 2021).

Another direct mechanism is the ability of nanocomposites to produce alterations in cell signaling pathways in pathogens, such as inhibition of phosphorylation of proteins essential to produce antioxidant enzymes and other defense compounds, resulting in a null response of pathogens to the accumulation of ROS (Susanti et al., 2022). Another possible effect of the above is the inhibition of enzymes responsible for peptidoglycan synthesis, which are paramount for the stability of the cell walls of multiple pathogens (Adisa et al., 2019). Additionally, nanocomposites can encapsulate AIs and function as a slow-release agent, which gradually becomes available as particles enter the target organism, thus preventing product degradation, improving efficiency, and ensuring that the AIs directly attack the organism (Shan et al., 2023). It has been reported that once they enter the plants, AIs continue to be released into the tissues for several days after application, even in plant organs far from the original zone (Bueno et al., 2021).

6.5.3 Indirect mechanisms of action

In addition to the direct mechanisms of action of nanocomposites against pathogens, these compounds also have diverse interactions with plant tissues, triggering a complex defense mechanism in plants (Fig. 6.2).

The indirect mechanism by which nanocomposites provide greater tolerance to pathogen attack can be divided into two major stages: the first due to the contact of nanocomposites with plant cells and the second explained by the content of the nanomaterial core (Juárez-Maldonado et al., 2019). When nanocomposites come into contact with plant cells, they produce a series of mechanisms that provide resistance to pathogen attack, mainly through greater efficiency in the cell signaling system, production of antimicrobial compounds, and increased activity of antioxidant enzymes (Goswami et al., 2022). At the molecular level, the response of plants to exposure by nanocomposites consists mainly of alterations in the cell signaling system, mainly in the levels of ROS, Ca^{2+}, and jasmonic acid (JA), producing changes in gene expression (Wahab et al., 2023). One of the genetic changes in plants was reported by Maryška et al. (2023), who, when applying Pd NPs to Brassica napus plants, discovered the overexpression of *pathogenesis-related gene 1* (*PR1*), which indicates the activation of the plant defense system. Likewise, other reports indicate that the use of CNPs in *Solanum lycopersicum* plants resulted in modifications in the transcription factors of the WRKY family. In addition, there was overexpression of genes involved in the synthesis of glucanase, defensin, and chitinase, all of which are related to the plant's defense against pathogens (Abd-Ellatif et al., 2022).

Alterations in gene expression result in increased production of antioxidant compounds such as phenols, flavonoids, glutathione, and vitamin C, as well as increased activity of antioxidant enzymes such as superoxide dismutase (SOD), catalase (CAT), ascorbate peroxidase (APX), and glutathione peroxidase (GPX) (Garza-Alonso et al., 2023). In addition to promoting cell signaling by JA, nanocomposites also promote the production of other signaling phytohormones such as salicylic acid (SA) and abscisic acid (ABA) (Li et al., 2022). The increase in the levels of these compounds provides greater tolerance to the attack of pathogens in plants. Shang et al. (2023) reported that when applying SeNPs to *Lactuca sativa* plants infected with *Fusarium oxysporum*, levels of tricarboxylic acids, phenolic compounds, and JA was increased, as well as an increase in plant biomass. In addition, an increase in the population of beneficial endophytic bacteria was observed, which provides additional resistance to attack by pathogens.

In the second stage, when nanocomposites enter plant cells, they are biotransformed into ions (e.g., Ag, Cu), which have activity against pathogens (Chaud et al., 2021). An example of this is Si, an element with various functions that provide resistance of the plant to pathogens, mainly through the deposit of this element in the epidermis of the leaves, forming a physical barrier that reduces the attack of pathogens (Wang et al., 2022). Something similar happens with chitosan NPs, since when this compound is released inside plant cells, it functions as an elicitor, providing tolerance to biotic stress through the modulation of signaling by ROS, NO, Ca^{2+}, and phytohormones (Tortella et al., 2023).

In addition to the above, nanocomposites also increase the photosynthetic activity of plants, promoting overall plant growth and resulting in greater tolerance to pathogen attack (Zhang et al., 2023). Likewise, nanocomposites promote better assimilation of nutrients by plants, providing an adequate nutritional balance, in addition to the fact that a large part of the essential elements is indispensable for the synthesis of biocompounds involved in the response of plants to biotic stress (Zhao et al., 2022).

6.6 Impact of nanocomposites on the management of phytopathogenic microorganisms

In this section, some reports on applying nanocomposites to control diseases caused by pathogens in plant species are presented, where some of the direct and indirect effects mentioned above are shown. The nanocomposites were grouped into metal nanoparticles, chitosan nanoparticles, and carbon nanocomposites.

6.6.1 Metallic nanoparticles

The application of different metal nanoparticles (Co_3O, CuO, Fe_3O_4, NiO, and ZnO) showed a reduction in the incidence and severity of *F. oxysporum* in *Phaseolus vulgaris* plants, in addition to promoting general plant growth and increasing chlorophyll levels in the leaves (El-Sayed et al., 2023). A similar trend was reported when performing applications of ZnO NPs in *Dacus carota* where the reduction of the incidence of *Rhizoctonia solani* was observed, in addition to favoring the vegetative growth of plants and increasing the concentration of photosynthetic pigments (Khan et al., 2022). In a study conducted under in vitro conditions, applications of ZnO NPs were made against the pathogens *Xanthomonas oryzae*, *Bipolaris oryzae*, and *Sphaerulina oryzina*, where in the first of them an inhibition zone of 25 mm was observed when a concentration of 20 μg/mL was used, while in *B. oryzar* and *S. oryzina* reductions of 73% and 96%, respectively were observed, in mycelium growth, in addition to a decrease in spore germination and hyphae disintegration (Cheema et al., 2022). Another report indicates that ZnO NPs produced an inhibition zone of up to 2.83 cm of the pathogens *B. glumae* and *B. gladioli* under in vitro conditions, in addition to a reduction in the growth of bacteria (Ahmed et al., 2021). The same effect was reported by Ali et al. (2021) when applying ZnO NPs against *Alternaria alternata*, *Sclerotium rolfsii*, and *Stemphylium solani*. On the other hand, applications of Ag NPs showed antibacterial efficacy against *Ralstonia solanacearum* and inhibited the growth of *F. oxysporum* mycelium (Khan et al., 2021).

González-Merino et al. (2021) reported that applications of 100–3000 mg/L of ZnO NPs inhibited mycelium growth and sporulation of *F. oxysporum* under in vitro conditions, in addition to reducing the incidence and severity of disease in *S. lycopersicum* plants. On the other hand, foliar applications of ZnO NPs reduced the incidence of *Pectobacterium betavasculorum* and *R. solani* in Beta vulgaris plants, in addition to promoting the activity of the antioxidant enzymes SOD, CAT, and PAL (Khan & Siddiqui, 2020).

6.6.2 Chitosan-based nanoparticles

Giri et al. (2023) applied CNPs on *S. lycopersicum* plants infected with the pathogens *Pseudomonas syringae*, *Erwinia chrysanthemi*, and *Xanthomonas campestris*, where they observed that the nanocomposite reduced the incidence of diseases, in addition to promoting overall plant growth and concentration of photosynthetic pigments. In another study carried out in the same plant species, the application of CNPs promoted the production of lignin in the tissues, in addition to favoring the activity of PAL, POX, PPO, and CAT when the plants were in the presence of *R. solanacearum* (Narasimhamurthy et al., 2022). Abd-Ellatif et al. (2022) reported that the application of CNPs in *S. lycopersicum* plants increased the activity of SOD, CAT, and APX, in addition to favoring the accumulation of antioxidant compounds such as phenols and flavonoids, as well as reducing the incidence and severity of *Fusarium solani*. In another investigation, foliar applications of CNPs were made in *Triticum aestivum* plants infected with the pathogen *Puccinia triticina*, where the results showed modifications in the expression of the *PR1*, *PR5*, and *PR10* genes, as well as an increase in the levels of H_2O2 and O_2^-, and a greater activity of POX and CAT, which reduced the severity of the disease by said pathogen measured through the decrease in size and number of pustules on leaves (Elsharkawy et al., 2022). Hoang et al. (2022) observed that the applications of CNPs + SA and CNPs + Al reduced the incidence and severity of *A. alternata* in plants of *Manihot esculenta*, in addition to favoring the vegetative growth of this plant species.

Regarding the stimulation of the defense mechanism of plants, Kongala and Mamidala (2023) reported that using CNPs in *Nicotinia tabacum* plants promoted the activity of the antioxidant enzymes CAT, POX, GR and APX. In other research, applications of CNPs + Cu were found to decrease the incidence and severity of *Pseudomonas savastanoi* and *Curtobacterium flaccumfaciens* in *Glycine max* plants (Tarakanov et al., 2023).

6.6.3 Carbon nanocomposites

In a study by Bytešníková et al. (2022), it was found that the use of a nanocomposite of graphene + Cu + Ag NPs reduced the severity of *Xanthomonas euvesicatoria* in *S. lycopersicum* and *Capsicum annuum* plants, in addition to modifying the expression of some genes related to plant defense (*PR1* and *PoP*).

On the other hand, the foliar application of carbon nanotubes (CNT) in *S. lycopersicum* plants showed a decrease in the incidence and severity caused by the pathogen *Alternaria solani*, in addition to promoting the accumulation of antioxidant compounds such as flavonoids, increasing the activity of the GPX enzyme, and increasing the photosynthetic rate of plants (González-García et al., 2021).

Applications of carbendazim-loaded graphene showed inhibition of growth, destruction of mycelium, and loss of cellular integrity of *Magnaporthe oryzae*; in addition, it reduced the incidence of disease caused by this pathogen in *Oryza sativa* (Hu et al., 2021). Damage to the pathogen was attributed to possible damage to the glutathione compounds of the cell membrane. Using graphene in *Brassica rapa* plants reduced the incidence and severity caused by *Fusarium graminearum* and *Sclerotinia sclerotiorum* (Peng et al., 2022). Table 6.2 presents some recent studies on the use of nanocomposites for pathogen control in plants.

6.7 Main advantages of nanoformulations for controlling agricultural pests in agriculture

Currently, nanotechnology offers different chances to deal with climate change (Subramanian et al., 2020). The unique physicochemical properties of nanoformulations rely on their reduced size of 1–100 nm, favoring an increased reactivity due to their higher surface area. This fact has led to nanotechnology being involved in

Table 6.2 Recent studies on the application of nanocomposites to control pathogens in plants.

Nanocomposite	Target pathogen	Hospedant plant	Main Effects	References
Si NPs	*Phytophtora infestans*	*Solanum tuberosum*	Inhibition of pathogen growth and ROS production.	Chen et al. (2023)
Bi_2S_3 NPs	*Bipolaris zeicola, Phoma herbarum, Epicoccum nigrum, Alternaria alternata, Fusarium brachygibbosum*	In vitro	Inhibition of mycelium growth.	Akanmu et al. (2023)
Cu/Cu_2O NPs	*Corynespora cassiicola, Neoscytalidium dimidiatum*	In vitro	Growth inhibition.	Duong et al. (2023)
Se NPs	*F. oxysporum Colletotrichum gloeosporioides*	In vitro	Decreased mycelium growth.	Lazcano-Ramírez et al. (2023)
Graphene	*M. oryzae*	*Oryza sativa*	Reduction of lesions by the pathogen, inhibition of spore germination.	Hu et al. (2023)
Graphitic Carbon Nitride (gC_3N_4)	*Bipolaris zeicola, Phoma herbarum, Epicoccum nigrum, A. alternata Fusarium brachygibbosum*	In vitro	Inhibition of mycelium growth.	Akanmu et al. (2023)
Graphene-Cu nanocomposite	*F. oxysporum*	*S. lycopersicum*	Increase in concentration of glutathione, flavonoids, and anthocyanins. Increase in activity of GPX, PAL, and CAT.	Cota-Ungson et al. (2023)
CNT and graphene	*F. oxysporum*	*S. lycopersicum*	Decrease in incidence and severity, increase in ascorbic acid concentration, antioxidant capacity and GPX activity.	González-García et al. (2022)

CNPs	*F. oxysporum*	*Cicer arietinum*	Production of pathogen resistance proteins, antioxidants, lignin production.	Sravani et al. (2023)
CNPs	*Xanthomonas campestris*	*S. lycopersicum*	Increase of phenolic compounds, greater activity of antioxidant enzymes, overexpression of genes related to the defense against the pathogen.	Giri et al. (2023)
Ag-CNPs	*F. oxysporum, Aspergillus niger*	*Trtiticum aestivum*	Inhibition of the growth of pathogens, and increase in chlorophyll concentration.	Mondéjar-López et al. (2023)
Ni-CNPs	*Pyricularia oryzae*	*O. sativa*	Reduction of symptoms caused by the disease and increased plant growth.	Parthasarathy et al. (2023)
CSNPs	*A. alternata*	*Manihot esculenta*	Inhibition of pathogen growth.	Hoang et al. (2023)
CSNPs	*Ralstonia solanacearum*	*Nicotinia tabacum*	Increase in the activity of CAT, SOD, and PPO enzymes. Overexpression of genes involved in resistance protein synthesis.	Wang et al. (2023)
CNPs	*Pseudomonas fluorescens*	*S. lycopersicum*	Increase in the activity of antioxidant enzymes such as SOD, PPO, POX, and PAL. Increase in the activity of chitinase and glucanase.	Abdelraouf et al. (2023)

many disciplines, such as medicine, electronics, and industry. In terms of agriculture, the main focus is to mitigate the extensive use of agrochemicals. Many nanoformulations for controlling different pests have been developed during the last years: nanopesticides, nanoemulsions, metal nanoparticles, and nanocomposites, among others. They all differ in the type of delivery of AIs and their chemical stability, form, or the matrix where they are encapsulated (An et al., 2022; Zhang et al., 2021). Their main advantage over conventional pesticides is that fewer AIs target pests more efficiently, reducing the risk of environmental pollution and human exposition. Most studies generally compare the effectiveness of pesticides (or commercial formulation) with the nanoformulation, corresponding to the same AIs encapsulated on a nanocarrier or in other support containing the same agent at the nanoscale. For instance, some nanoformulations based on herbicides, such as atrazine or 2,4-D, have been evaluated against some weeds in the soil (Cao et al., 2018; Oliveira et al., 2015; Sousa et al., 2018). In general, the nanoformulation's enhanced postemergence herbicidal activity has been found compared with the equivalent concentration of the commercial formulation. Indeed, Oliveira et al. (2015) observed that the herbicidal activity was maintained when the nanoformulation was 10-fold diluted, which did not occur with the commercial formulation.

Similarly, nanoformulations based on fungicides such as hexoconazole, prochloraz, azoxystrobin have been more efficient as antifungals than their commercial formulation. These studies were performed under in vitro conditions (Chauhan et al., 2017; Liang et al., 2018; Mustafa et al., 2018; Xu et al., 2018). However, there is scarce information about the effectiveness of nanoformulations under in vivo conditions, such as in soil (Tippannanavar et al., 2020).

On the other hand, it is known that at field conditions, commercial formulations are more prone to lose effectiveness due to their photolytic degradation, leaching, and runoff. Less than 0.1% of the applied pesticide may reach its target (Lourthuraj et al., 2022). This fact denotes the low efficiency of pesticides currently marketed and promotes the urgent need for new strategies to reduce their release into the environment. The release behavior of nanoformulations is a relevant aspect of overcoming the issues of low pesticide efficiency. In this regard, a slower and more controlled release of AIs in time has been achieved, compared to the rapid burst release from commercial formulations (Kumar et al., 2021). The protective role of encapsulating agents facilitates enhanced permeability and stability for AIs in nanoformulations. For instance, chitosan-tripolyphosphate nanocapsules based on hexaconazole exhibited a full-time release in soil of 14 days versus 5 days of the commercial formulation. This finding was consistent with the higher in vitro inhibition of the nanocapsules against *R. solani* (approximately 80%) over commercial formulation (approximately 50%) (Chauhan et al., 2017). Similar findings were obtained for 2,4-D loaded on biochar as a nanocarrier, which exhibited a slower soil release than commercial pesticides (Evy Alice Abigail, 2019). This release was much slower in soil than in water. Then, more studies in soil are still demanded to accurately determine the effectivity of nanoformulations at realistic field conditions.

6.8 Why use a temperature-responsive system loaded with pesticides to control pests in agriculture?

The development of intelligent formulations has raised great interest due to their capacity to enhance pesticide solubility and bioavailability. Thus, as lower amounts of residual pesticides are released into the environment, the risk of pollution is reduced. In this regard, stimuli-responsive nanoformulations are valuable and intelligent systems since they can only release pesticides as demanded by specific parameters such as pH, light, temperature, or specific enzymes (Abdelrahman et al., 2021; Chen et al., 2018; Xiang et al., 2018; Zhang et al., 2019). In particular, temperature-responsive carriers can release pesticides due to the physicochemical properties of the polymer acting as support.

Among thermosensitive polymers, chitosan has been used as shell material to encapsulate fertilizers, pesticides, and other compounds (Atalay et al., 2022; Yu et al., 2021). Also, this polymer is biodegradable and cost-effective, demonstrating its suitability for agriculture, particularly in soil or soil-plant systems where no phytotoxic impact has been reported (Malerba & Cerana, 2016). For instance, Zhang et al. (2022) developed a chitosan-based microsphere to release the insecticide dinotefuran. They observed a cumulative burst release from microspheres of 85% at 30°C, whereas a lower cumulative release was observed at 10°C and 20°C (47% and 65%, respectively). As explained by the authors, this occurs due to the extension promotion in the chitosan molecular chain and the subsequent formation of gaps at temperatures higher than 20°C, which allow the diffusion of the encapsulated agent outside the microsphere. Similar tendencies in response to temperature were obtained for the fungicide prochloraz and the plant growth regulator gibberellic acid, both released from chitosan nanoparticles (Liang et al., 2018; Pereira et al., 2017).

For the gibberellic acid release, chitosan/tripolyphosphate (CS/TPP) nanoparticles were synthesized by the ionic gelation method, where the crosslinking occurs by inter- and intramolecular linkage of the opposite charges of chitosan (positive) and TPP (negative). Approximately 80% of gibberellic acid was released from CS/TPP nanoparticles after 8 hours at 30°C, compared to 60% at 25°C. This can be attributed to the TPP ionic groups responsible for expanding the chitosan molecule (Hassani et al., 2015). These antecedents suggest that altering the hydrophilicity and viscosity of chitosan can improve the release of the encapsulated agent. Therefore, considering the consistency with the most suitable temperature for the *F. oxysporum* growth (i.e., 25°C–30°C), chitosan can support antifungal agents against this fungus. Triazole fungicides have been the most used for controlling Fusarium in agriculture since the 1970s decade (Ribas e Ribas et al., 2016). One of the most common of this group is tebuconazole (Teb), whose use is still controversial because it has been considered a pesticide of low toxicity in China (Jónsdóttir et al., 2016). Contrastingly, many studies demonstrate that residues can persist in the environment for long periods, even in crops (El Azhari et al., 2018), exhibiting additional risks for human health that have been widely discussed (Li et al., 2019; Li et al., 2020). Overall, it can be assumed that fungus and fungicide agents can negatively affect crops infected by *F. oxysporum*.

The low solubility of Teb in water is one of its main disadvantages. Studies have demonstrated that Teb in nanoformulations, such as oil-in-water nanoemulsions, can be more efficient than commercial formulations (marketed as emulsifiable concentrates). For instance, Selyutina et al. (2020) formulated a nanopesticide based on natural polysaccharides that enhanced fivefold the Teb penetration concerning the commercial formulation at the stage of presowing seed treatment. Díaz-Blancas et al. (2016) formulated a Teb-based nanoemulsion to decrease the Teb residual dose in the environment. However, the effectiveness of the nanoemulsion as a fungicide was not evaluated. Lucia et al. (2021) also developed a Teb nanoemulsion for the control of *Gloeophyllum sepiarium* and *Pycnoporus sanguineus*. This study highlighted the use of the essential oil eugenol as the oil phase for Teb encapsulation due to its known antifungal activity against those wood fungi and its low environmental impact. From these findings, it can be presumed that encapsulating Teb in a nanoemulsion using essential oils can be an effective combined strategy to reduce the necessary amount of Teb, considering that many essential oils (including eugenol) have been able to inhibit the growth of *F. oxysporum* (Abd-Elsalam & Khokhlov, 2015; Moutassem et al., 2019; Soliman et al., 2022). Scarce information about nanoformulations involving essential oils combined with Teb would represent a novelty for controlling *F. oxysporum*. Therefore, more profound studies should focus on enhancing the stability of these antifungal agents in a complex matrix such as soil. To sum up, it can be generalized that temperature-responsive systems related to controlling *F. oxysporum* in soil are still unexplored. This fact represents a critical gap that requires more profound attention since this fungus can remain in soils for extended periods. It is imperative that the fungicide can act just as demanded according to variations in temperature.

6.9 Feasibility of nanocomposite application for the control of pests in agriculture

As previously explained, increasing the efficiency of applying pesticides for pest control is essential, since this achieves a considerable reduction in contamination, ecological impact, and risks to humans. In this sense, the so-called nanopesticides have been developed for the most efficient control of pests in agriculture, which mainly function as nanocarriers and controlled release systems for pesticides, but also have a direct effect on insects, and even function as pest insect deterrents, or activators of the plant defense system. However, in some cases, nanocomposites have several functions, for example, in some nanocomposites designed as nanocarriers, they also induce susceptibility in insects, resulting in increased efficiency of applied insecticides (Zhou et al., 2021).

One of the significant advantages of controlled release systems can reduce the total consumption of pesticides since efficiency is increased by minimizing the use of the active ingredient. These types of systems range from diverse nanomaterials (Si NPs, CNT, and GO) to natural biopolymers such as chitosan, alginate, starch, and cellulose, among others (Guha et al., 2020). Particularly, plant-derived nanopesticides through controlled and sustained release of toxic substances are more effective in reducing pest population and plant infestation levels compared to the bulk form of pesticides (Gahukar & Das, 2020). In any case, the beneficial effect of nanoformulations in the control of insect pests has been demonstrated, especially due to the advantages of nanopesticides such as high adsorption, reduced volatilization, improved tissue permeability, and controlled release, compared to conventional pesticides (Kannan et al., 2023).

In the case of nanoinsecticides aimed at improving the defense system of the plant or increasing the susceptibility of the insect to insecticides, positive results have been observed for the control of pest insects (Ji et al., 2023; Qi et al., 2021). However, more studies are required to increase knowledge for a potential commercial application in the medium or long term.

6.9.1 Nanocarriers and controlled release systems

Nanocarriers or controlled release systems are designed to increase the efficiency of insecticides and focus on certain characteristics such as improving dispersibility and contact surface with insects, generating damage to insect tissues, or controlling insecticide release.

Gao et al. (2021) developed graphene oxide (GO)-based nanopesticides and existing pyrethroid pesticides (cyhalothrin, bifenthrin, and fenpropatrin), which were loaded onto GO as a carrier through a simple physisorption process. These nanocomposites were designed so that the release of the pesticides was sensitive to temperature, and their bioactivity against mites (*Tetranychus urticae* Koch) was evaluated indoors and in the field. Results demonstrated that GO pesticide nanocomposites exhibited much higher bioactivity than individual pesticides (up to 3.55-fold) and could adsorb to *T. urticae* cuticle and bean leaf surface with highly uniform dispersibility. The authors suggest that nanocomposites can enhance the effects of pesticides by improving dispersibility to increase contact between the pesticide and mites.

Wang et al. (2019) evaluated GO nanocomposites with three pesticides [Pyridaben (Pyr), chlorpyrifos (Chl), and beta-cyfluthrin (Cyf)] on the control of spider mites (*Tetranychus truncatus* and *T. urticae* Koch). The results showed that the efficiency of the pesticides increased by 1.5−1.78-fold. According to the authors, GO can function as a carrier, since pesticides are physically loaded onto its surface through $\pi-\pi$ interactions, this allows pesticides to be adsorbed on the cuticle surface of spider mites and, therefore, improves the effectiveness and efficiency of pesticides, and reduces pesticide loss. Zhang et al. (2022) evaluated a nanocomposite of Ti_3C_2 nanosheets (MXene) as a carrier for insecticide abamectin against larvae of *Spodoptera frugiperda*. The results showed that the *S. frugiperda* larvae on the leaves sprayed with the nanocomposite were almost all dead after 7 days of feeding. According to the authors, the nanocomposite exhibits excellent stability and photothermal capacity, enabling it to display strong lethality and sustained efficacy against the *S. frugiperda* pest by combined photothermal contact and poisoning kill.

Sharma et al. (2017) developed antidrift nanostickers of GO decorated with copper selenide (rGO-$Cu_{2-x}Se$), which has photothermal and photocatalytic properties. The nanocomposite was designed for controlled and targeted delivery of the pesticide chlorpyrifos, as it is held together until the native larval gut condition appears. The advantage of this nanocarrier is that it exhibits drift resistance in addition to targeted release, which increased larval (*Pieris rapae*) mortality by more than 35% compared to the pesticide alone. The authors suggest that the increase in pesticide efficiency may be due to the compound's ability to resist runoff, enhanced uptake, and targeted delivery of pesticides. Furthermore, the rGO-$Cu_{2-x}Se$ composite retained 40% (w/w) of pesticide and remained as a reservoir in the leaf without drift loss, demonstrating that the nanocomposite has targeted and efficient pesticide delivery.

Li et al. (2022) evaluated GO nanocomposites in combination with insecticides [chlorantraniliprole (Chl), beta cypermethrin (Bet), ethoxyhydrazide (Met), and spinetoram (Spi)] in *S. frugiperda* larvae; for this, the pesticide was adsorbed on the surface of graphene oxide by physical adsorption. The results showed that the toxicity of the Chl-GO, Bet-GO, Met-GO, and Spi-GO nanocomposites for the third instar larvae of *S. frugiperda* increased 1.56, 1.54, 2.53, and 1.74-fold, respectively. The authors propose that the synergistic mechanism of GO and pesticides may be related to the fact that GO can cause mechanical damage to the insect's body skeleton, causing the insect to rapidly lose water. The damaged body wall provides a new channel for the penetration of pesticides into insects, and the GO adsorbed by the pesticide can adhere to the insect's body wall and enhance the action of pesticides.

Zhou et al. (2021) developed an acaricide delivery nanosystem that uses GO nanosheets as nanocarriers to deliver four acaricides (avermectin, bifenazate, etoxazole, and spirodiclofen) to control *Tetranychus cinnabarinus*. The results showed that GO adsorbed and damaged the mite cuticle by binding to a cuticle protein (CPR) and inhibiting CPR gene expression, which increased the mite cuticle permeability. This significantly improved the efficacy of miticides up to 68% compared to miticide alone. Furthermore, silencing the CPR gene by RNAi resulted in dehydration, disturbed cuticle layer construction, and increased cuticle permeability and sensitivity of mites to GO-acaricidal nanocomposites. The authors suggest that the molecular mechanism of the synergistic effects of GO on acaricides against *T. cinnabarinus* was mediated by low expression of the *CPR* gene. Ahmed et al. (2019) developed a nanopesticide based on silver nanoparticles (Ag NPs) as a carrier of Lambda-cyhalothrin (L-CYN) to control the cotton leafworm (*Spodoptera littoralis*). Their results demonstrated that the AgNPS@L-CYN nanocomposite pesticide was 37 times more effective than L-CYN alone against second-instar larvae of lab and field cotton leafworms. The mechanism of action of these nanocomposites is to generate midgut tissue damage in the treated larvae, which was much higher compared to chemical insecticides.

In this regard, Qi et al. (2021) demonstrated that the application of nanocomposites based on nanometer-scale arginine-modified carbon dots that combined with sodium selenite to form selenium-carbon dots (Se-CDs) decreased *P450* gene expression and enzyme activity P450 in mites (*T. cinnabarinus*), which in turn increased the susceptibility of the mites to the insecticide fenpropatrin. The authors suggest that inhibiting *P450* gene expression and repressing *T. cinnabarinus* detoxification was the molecular mechanism by which the acaricidal activity of fenpropatrin was promoted. Therefore, nanomaterials can function as adjuvants for insecticides and improve their efficiency.

6.9.2 Nanoinsecticides

In some cases, the same nanomaterials function directly as insecticides. However, this function may be directly related to damage caused to insect tissue, insect deterrent properties, or even activation of the plant defense system. Jameel et al. (2020) developed a ZnO NP-based nanocomposite with thiamethoxam and evaluated its effect on fourth instar larvae of *Spodoptera litura* (Lepidoptera: Noctuidae). The larvae were fed with the ZnO NP nanocomposite with thiamethoxam (10–90 mg/L), and an increase in larval mortality of up to 27% was observed compared to thiamethoxam alone. In addition, a malformation was observed in pupae and adults, late emergence, and reduced fecundity and fertility. The authors propose that the observed effect is due to the ZnO-thiamethoxam nanocomposite, where electrostatic adsorption of ZnO-thiamethoxam groups on insect cells and cellular internalization of ZnO NPs played an important role in the synergistic insecticidal activities that facilitated the interaction of ZnO NPs, thiamethoxam and its combination with different components of *S. litura* cells, which generated damage in insects.

Karthick Raja Namasivayam et al. (2018) developed a chitosan nanocomposite with insecticidal metabolites derived from the fungal biopesticide agent *Nomuraea rileyi*. The results demonstrated that the insecticidal activity against the larval stages of *S. litura* was effective against all larval stages in terms of high mortality, drastic reduction of the biochemical composition of hemolymph and midgut macromolecules. As mentioned by both studies, the insecticidal effect of nanocomposites is largely due to the damage caused to insect tissues. Other authors have obtained excellent results in the insecticidal efficiency of nanocomposites, even better than that observed with commercial insecticides. Abd El-Rahman et al. (2020) developed nanocomposites of Ag and GO with magnesium chlorophyllin (Mg-Chl/Ag and Mg-Chl/GO), and evaluated them at three concentrations (1, 10, 100 mL/L) against larvae of *S. littoralis*. The results showed that the mortality rate of the larvae was positively correlated with increasing concentrations of the nanocomposites and the time after exposure. In the dark treatment, mortality reached 58.3% after 10 days of treatment at the highest concentration of Mg-Chl/GO, while it reached 55% at the same concentration of Mg-Chl/Ag. Under sunlight, the highest concentration of Mg-Chl/GO and Mg-Chl/Ag reached a mortality of 96.67% and 91.60%, respectively, after 15 days of treatment.

Araujo et al. (2023) synthesized nanocomposites formed by sulfur nanoparticles, coated with eucalyptus and rosemary essential oil (at concentrations of 0.25%, 0.5%, and 0.75%), and determined the insecticidal effect on the control of nymphs of paratrioza (*Bactericera cockerelli* Sulc) in potato crop, and compared the efficacy with the commercial insecticide thiamethoxam at 0.25%. The results showed that the eucalyptus nanocomposites (0.25%, 0.5%, and 0.75%) and the rosemary nanocomposites (0.5%) have an insecticidal efficacy of 100% for the control of insect nymphs 24 hours after their application. The insecticidal efficacy of rosemary nanocomposites at concentrations of 0.25% and 0.75% increased over time, reaching 100% at 24 and 72 hours, respectively. In addition, it was observed that the nanocomposites were more effective in controlling paratrioza nymphs than the commercial insecticide thiamethoxam (it only reached 83.3% insecticidal efficacy).

It has also been shown that nanocomposites have an insecticidal effect and possess insect deterrent characteristics or activation of plant defense systems, which provides novel methods for pest insect control. Bapat et al. (2020) evaluated the silica nanoparticle-mediated release of the protease inhibitor in tomato plants and its effect on the insect pest *Helicoverpa armigera*. For this, they used Si NPs of 20 and 100 nm that were functionalized with (3-aminopropyl) triethoxysilane to obtain Si20APT and Si100APT, respectively, that were nontoxic toward plants. The results demonstrated that the STI bound to the particle inhibited bovine trypsin by 80% and *H. armigera* intestinal proteinase (HGP) activity by 50%. While second-instar larvae of *H. armigera* that ingested STI-laden particles (incorporated in the artificial diet or in the leaves) showed significant growth retardation. The authors also developed choice assays, where insect larvae avoided leaf disks applied with Si20APT-STI, thus they propose that the nanocomposite exhibits insect deterrent properties. Ji et al. (2023) evaluated the effects of chitosan-coated mesoporous silica nanoparticles (CS-MSN), chitosan nanoparticles (CNP), and MSN on broad bean plants against aphids (*Acyrthosiphon pisum*). The results showed that foliar application of 100 mg/L of CS-MSN reduced aphid reproduction in terms of aphid nymph population more effectively than CNP, MSN, and the pesticide acetamiprid. CS-MSN reduced reproduction by 55.1% after 7 days of aphid infestation compared to the control. In addition, it was observed that the application of CS-MSN efficiently activated the defense systems of plants by significantly increasing the production of endogenous signals related to plant defense (Ca^{2+} and phytohormones), the production of defense metabolites (coumarin, quercetin, and luteolin), and reduce the oxidative stress of the leaves in response to aphid attacks by increasing the activities of antioxidant enzymes (SOD, POD, and CAT). Therefore, the authors suggest that CS-MSN may act as an alternative control tool to simultaneously induce host plant resistance against herbivores.

6.10 Conclusion and future remarks

Using nanocomposite pesticides in agricultural activities has marked a substantial advancement in pest control methods, offering various benefits over conventional pesticides. The nanoscale dimension of these composites allows an efficient delivery and controlled release of active ingredients, ensuring targeted action against agricultural pests while minimizing additional damage to nontarget organisms and the environment. This accurate delivery system enhances the efficacy of the pesticides, potentially minimizing the quantity needed and lowering the risk of pesticide resistance development in pest populations. Additionally, the encaptulation of active ingredients in nanoparticles can protect them from early degradation due to environmental factors, leading to continued effectiveness and stability. However, while nanocomposite pesticides promise a revolution in agricultural practices, their deployment is not without challenges. Significant concerns remain regarding nanomaterials' potential risks and long-term impacts on human health, nontarget organisms, and ecosystems. The uncertainty surrounding the fate of nanoparticles in the environment underscores the need for comprehensive risk assessments and the development of regulatory frameworks to govern the use of nanotechnology in agriculture. Looking ahead, the field of nanocomposite pesticides is ripe for innovation and growth, but it must be navigated with caution and responsibility. Continued research is essential to unveil the mechanisms of nanoparticle interaction with biological systems and the environment. Some gaps that need to be redirected include developing biodegradable and eco-friendly products that guarantee minimum damage to the environment while maintaining or enhancing pest control efficacy. Related to the development of precision agriculture, combining nanocomposite pesticides with technologies, such as sensors and drones, is necessary for more accurate pest detection and targeted application, further reducing the quantities required and environmental impact. Another aspect is the establishment of clear regulatory guidelines based on robust scientific evidence to ensure the safe and responsible use of nanocomposite pesticides and achieve public engagement by educating farmers and the public about the benefits and potential risks of nanocomposite pesticides to foster informed decision-making and acceptance. Finally, it is necessary for a global collaboration, encouraging international collaboration in research, standardization, and regulatory efforts to address global agricultural challenges and ensure food security. In conclusion, using nanocomposite pesticides in agricultural activities holds great promise for enhancing crop protection, reducing environmental damage, and contributing to sustainable agriculture. However, realizing this potential requires a balanced approach integrating innovation with stringent safety assessments, regulatory oversight, and public engagement. As the technology evolves, all stakeholders must collaborate to harness the benefits of nanocomposites while safeguarding human health and the environment.

Acknowledgment

ANID/FOVI 220003, ANID/FONDECYT/1230529, ANID/FONDECYT/11220070, ANID/ATE/220038 projects.

References

Abd El-Rahman, S. F., Ahmed, S. S., & Abdel Kader, M. H. (2020). Toxicological, biological and biochemical effects of two nanocomposites on cotton leaf worm, *Spodoptera littoralis* (Boisduval, 1833). *Polish Journal of Entomology*, *89*, 101–112.

Abd-Ellatif, S., Ibrahim, A. A., Safhi, F. A., Abdel Razik, E. S., Kabeil, S. S. A., Aloufi, S., Alyamani, A. A., Basuoni, M. M., ALshamrani, S. M., & Elshafie, H. S. (2022). Green synthesized of thymus vulgaris chitosan nanoparticles induce relative WRKY-genes expression in *Solanum lycopersicum* against *Fusarium solani*, the causal agent of root rot disease. *Plants*, *11*, 1–17.

Abdelrahman, T. M., Qin, X., Li, D., Senosy, I. A., Mmby, M., Wan, H., Li, J., & He, S. (2021). Pectinase-responsive carriers based on mesoporous silica nanoparticles for improving the translocation and fungicidal activity of prochloraz in rice plants. *Chemical Engineering Journal*, *404*126440.

Abdelraouf, A. M. N., Hussain, A. A., & Naguib, D. M. (2023). Nano-chitosan encapsulated *Pseudomonas fluorescens* greatly reduces Fusarium wilt infection in tomato. *Rhizosphere*, *25*100676.

Abd-Elsalam, K. A., & Khokhlov, A. R. (2015). Eugenol oil nanoemulsion: antifungal activity against *Fusarium oxysporum* f. sp. vasinfectum and phytotoxicity on cottonseeds. *Applied Nanoscience*, *5*, 255–265.

Abhishek, W., Kamaljeet, S., & Sudesh, K. (2018). Effect of chlorpyrifos and malathion on soil microbial population and enzyme activity. *Acta Scientific Microbiology*, *1*(4), 14–22.

Adisa, I. O., Pullagurala, V. L. R., Peralta-Videa, J. R., Dimkpa, C. O., Elmer, W. H., Gardea-Torresdey, J. L., & White, J. C. (2019). Recent advances in nano-enabled fertilizers and pesticides: A critical review of mechanisms of action. *Environmental Science: Nano*, *6*, 2002–2030.

Ahmed, K. S., Mikhail, W. Z. A., Sobhy, H. M., Radwan, E. M. M., El Din, T. S., & Youssef, A. M. (2019). Effect of lambda-cyhalothrin as nanopesticide on cotton leafworm, *Spodoptera littoralis* (Boisd.). *Egyptian Journal of Chemistry*, *62*, 1663–1675.

Ahmed, T., Wu, Z., Jiang, H., Luo, J., Noman, M., Shahid, M., Manzoor, I., Allemailem, K. S., Alrumaihi, F., & Li, B. (2021). Bioinspired green synthesis of zinc oxide nanoparticles from a native bacillus cereus strain rnt6: Characterization and antibacterial activity against rice panicle blight pathogens *Burkholderia glumae* and *B. gladioli*. *Nanomaterials*, *11*.

Akanmu, A. O., Ajiboye, T. O., Seleke, M., Mhlanga, S. D., Onwudiwe, D. C., & Babalola, O. O. (2023). The potency of graphitic carbon nitride (gC_3N_4) and bismuth sulphide nanoparticles (Bi_2S_3) in the management of foliar fungal pathogens of maize. *Applied Sciences (Switzerland)*, *13*.

Ali, J., Mazumder, J. A., Perwez, M., & Sardar, M. (2021). Antimicrobial effect of ZnO nanoparticles synthesized by different methods against food borne pathogens and phytopathogens. *Materials Today: Proceedings*, *36*, 609–615.

An, C., Sun, C., Li, N., Huang, B., Jiang, J., Shen, Y., Wang, Chong, Zhao, X., Cui, B., Wang, Chunxin, Li, X., Zhan, S., Gao, F., Zeng, Z., Cui, H., & Wang, Y. (2022). Nanomaterials and nanotechnology for the delivery of agrochemicals: strategies towards sustainable agriculture. *Journal of Nanobiotechnology*, *20*11.

Araujo, L.S., Tigrero, J.O., Delgado, V.A., Aguirre, V.A., & Villota, J.N. (2023). Sulfur nanocomposites with insecticidal effect for the control of *Bactericera cockerelli* (Sulc) (Hemiptera: Triozidae) 1–22.

Atalay, S., Sargin, I., & Arslan, G. (2022). Slow-release mineral fertilizer system with chitosan and oleic acid-coated struvite-K derived from pumpkin pulp. *Cellulose*, *29*, 2513–2523.

Bapat, G., Zinjarde, S., & Tamhane, V. (2020). Evaluation of silica nanoparticle mediated delivery of protease inhibitor in tomato plants and its effect on insect pest *Helicoverpa armigera*. *Colloids Surfaces B Biointerfaces*, *193*111079.

Berger, S., Chazli, Y., El., Babu, A. F., & Coste, A. T. (2017). Azole resistance in *Aspergillus fumigatus*: A consequence of antifungal use in agriculture? *Frontiers in Microbiology*, *8*, 1–6.

Bueno, V., Gao, X., Abdul Rahim, A., Wang, P., Bayen, S., & Ghoshal, S. (2021). Uptake and translocation of a silica nanocarrier and an encapsulated organic pesticide following foliar application in tomato plants. *Environmental Science and Technology*.

Byteśníková, Z., Pečenka, J., Tekielska, D., Kiss, T., Švec, P., Ridošková, A., Bezdička, P., Pekárková, J., Eichmeier, A., Pokluda, R., Adam, V., & Richtera, L. (2022). Reduced graphene oxide-based nanometal-composite containing copper and silver nanoparticles protect tomato and pepper against *Xanthomonas euvesicatoria* infection. *Chemical and Biological Technologies in Agriculture*, *9*, 1–16.

Camara, M. C., Campos, E. V. R., Monteiro, R. A., Do Espirito Santo Pereira, A., De Freitas Proença, P. L., & Fraceto, L. F. (2019). Development of stimuli-responsive nano-based pesticides: Emerging opportunities for agriculture. *Journal of Nanobiotechnology*, *17*, 1–19.

Campos, E. V. R., de Oliveira, J. L., Fraceto, L. F., & Singh, B. (2014). Polysaccharides as safer release systems for agrochemicals. *Agronomy for Sustainable Development*, *35*, 47–66.

Cao, L., Zhou, Z., Niu, S., Cao, C., Li, X., Shan, Y., & Huang, Q. (2018). Positive-charge functionalized mesoporous silica nanoparticles as nanocarriers for controlled 2,4-dichlorophenoxy acetic acid sodium salt release. *Journal of Agricultural and Food Chemistry*, *66*, 6594–6603.

Castaño, C., Lindahl, B. D., Alday, J. G., Hagenbo, A., Martínez de Aragón, J., Parladé, J., Pera, J., & Bonet, J. A. (2018). Soil microclimate changes affect soil fungal communities in a Mediterranean pine forest. *New Phytologist*, *220*, 1211–1221.

Chaud, M., Souto, E. B., Zielinska, A., Severino, P., Batain, F., Oliveira-Junior, J., & Alves, T. (2021). Nanopesticides in agriculture: Benefits and challenge in agricultural productivity, toxicological risks to human health and environment. *Toxics*, *9*.

Chauhan, N., Dilbaghi, N., Gopal, M., Kumar, R., Kim, K.-H., & Kumar, S. (2017). Development of chitosan nanocapsules for the controlled release of hexaconazole. *International Journal of Biological Macromolecules*, *97*, 616–624.

Cheema, A. I., Ahmed, T., Abbas, A., Noman, M., Zubair, M., & Shahid, M. (2022). Antimicrobial activity of the biologically synthesized zinc oxide nanoparticles against important rice pathogens. *Physiology and Molecular Biology of Plants*, *28*, 1955–1967.

Chen, C., Zhang, G., Dai, Z., Xiang, Y., Liu, B., Bian, P., Zheng, K., Wu, Z., & Cai, D. (2018). Fabrication of light-responsively controlled-release herbicide using a nanocomposite. *Chemical Engineering Journal*, *349*, 101–110.

Chen, S., Guo, X., Zhang, B., Nie, D., Rao, W., Zhang, D., Lü, J., Guan, X., Chen, Z., & Pan, X. (2023). Mesoporous silica nanoparticles induce intracellular peroxidation damage of *Phytophthora infestans*: A new type of green fungicide for late blight control. *Environmental Science and Technology*, *57*, 3980–3989.

Cota-Ungson, D., González-García, Y., Cadenas-Pliego, G., Alpuche-Solís, Á. G., Benavides-Mendoza, A., & Juárez-Maldonado, A. (2023). Graphene–Cu nanocomposites induce tolerance against *Fusarium oxysporum*, increase antioxidant activity, and decrease stress in tomato plants. *Plants*, *12*, 2270.

Díaz-Blancas, V., Medina, D., Padilla-Ortega, E., Bortolini-Zavala, R., Olvera-Romero, M., & Luna-Bárcenas, G. (2016). Nanoemulsion formulations of fungicide tebuconazole for agricultural applications. *Molecules (Basel, Switzerland)*, *21*, 1271.

Duong, N. L., Nguyen, V. M., Tran, T. A. N., Phan, T. D. T., Tran, T. B. Y., Do, B. L., Phung Anh, N., Nguyen, T. A. T., Ho, T. G. T., & Nguyen, T. (2023). Durian shell-mediated simple green synthesis of nanocopper against plant pathogenic fungi. *ACS Omega*, *8*, 10968–10979.

El Azhari, N., Dermou, E., Barnard, R. L., Storck, V., Tourna, M., Beguet, J., Karas, P. A., Lucini, L., Rouard, N., Botteri, L., Ferrari, F., Trevisan, M., Karpouzas, D. G., & Martin-Laurent, F. (2018). The dissipation and microbial ecotoxicity of tebuconazole and its transformation products in soil under standard laboratory and simulated winter conditions. *Science of The Total Environment*, *637–638*, 892–906.

El-Sayed, E. S. R., Mohamed, S. S., Mousa, S. A., El-Seoud, M. A. A., Elmehlawy, A. A., & Abdou, D. A. M. (2023). Bifunctional role of some biogenic nanoparticles in controlling wilt disease and promoting growth of common bean. *AMB Express*, *13*.

Elsharkawy, M. M., Omara, R. I., Mostafa, Y. S., Alamri, S. A., Hashem, M., Alrumman, S. A., & Ahmad, A. A. (2022). Mechanism of wheat leaf rust control using chitosan nanoparticles and salicylic acid. *Journal of Fungi*, *8*.

Elvira-Recuenco, M., Pando, V., Berbegal, M., Manzano Muñoz, A., Iturritxa, E., & Raposo, R. (2021). Influence of temperature and moisture duration on pathogenic life history traits of predominant haplotypes of *Fusarium circinatum* on Pinus spp. in Spain. *Phytopathology®*, *111*, 2002–2009.

Evy Alice Abigail, M. (2019). Biochar-based nanocarriers: fabrication, characterization, and application as 2,4-dichlorophenoxyacetic acid nanoformulation for sustained release. *3 Biotech*, *9*317.

Fan, L.Y. (2017). China founds pesticide office to combat pollution, overuse. <https://www.sixthtone.com/news/1000987/china-founds-pesticide-office-to-combat-pollution%2Coveruse>.

FAO (2017). The future of food and agriculture: Trends and challenges.

FAO, WHO (2019). Pesticide residues in food 2018 - Report 2018.

Fones, H. N., Bebber, D. P., Chaloner, T. M., Kay, W. T., Steinberg, G., & Gurr, S. J. (2020). Threats to global food security from emerging fungal and oomycete crop pathogens. *Nature Food*, *1*, 332–342.

Gahukar, R. T., & Das, R. K. (2020). Plant-derived nanopesticides for agricultural pest control: Challenges and prospects. *Nanotechnology Environmental Engineering*, *5*, 1–9.

Gao, R., Duan, Y., Zhang, J., Ren, Y., Li, H., Liu, X., Zhao, P., & Jing, Y. (2022). Effects of long-term application of organic manure and chemical fertilizer on soil properties and microbial communities in the agro-pastoral ecotone of North China. *Frontiers in Environmental Science*, *10*993973.

Gao, X., Shi, F., Peng, F., Shi, X., Cheng, C., Hou, W., Xie, H., Lin, X., & Wang, X. (2021). Formulation of nanopesticide with graphene oxide as the nanocarrier of pyrethroid pesticide and its application in spider mite control. *RSC Advances*, *11*, 36089–36097.

Gao, Y., Li, Donglin, Li, Dongyang, Xu, P., Mao, K., Zhang, Y., Qin, X., Tang, T., Wan, H., Li, J., Guo, M., & He, S. (2020). Efficacy of an adhesive nanopesticide on insect pests of rice in field trials. *Journal of Asia-Pacific Entomology*, *23*, 1222–1227.

Garza-Alonso, C. A., Cadenas-Pliego, G., Juárez-Maldonado, A., González-Fuentes, J. A., Tortella, G., & Benavides-Mendoza, A. (2023). Fe_2O_3 nanoparticles can replace Fe-EDTA fertilizer and boost the productivity and quality of Raphanus sativus in a soilless system. *Scientia Horticulturae*, *321*.

Gilardi, G., Garibaldi, A., & Gullino, M. (2018). Emerging pathogens as a consequence of globalization and climate change: leafy vegetables as a case study. *Phytopathologia Mediterranea*, 57.

Giri, V. P., Pandey, S., Srivastava, S., Shukla, P., Kumar, N., Kumari, M., Katiyar, R., Singh, S., & Mishra, A. (2023). Chitosan fabricated biogenic silver nanoparticles (Ch@BSNP) protectively modulate the defense mechanism of tomato during bacterial leaf spot (BLS) disease. *Plant Physiology and Biochemistry*, *197*107637.

González-García, Y., Cadenas-Pliego, G., Alpuche-Solís, Á. G., Cabrera, R. I., & Juárez-Maldonado, A. (2021). Carbon nanotubes decrease the negative impact of *Alternaria solani* in tomato crop. *Nanomaterials*, *11*, 1–15.

González-García, Y., Cadenas-Pliego, G., Alpuche-Solís, Á. G., Cabrera, R. I., & Juárez-Maldonado, A. (2022). Effect of carbon-based nanomaterials on Fusarium wilt in tomato. *Scientia Horticulturae*, *291*110586.

González-Merino, A. M., Hernández-Juárez, A., Betancourt-Galindo, R., Ochoa-Fuentes, Y. M., Valdez-Aguilar, L. A., & Limón-Corona, M. L. (2021). Antifungal activity of zinc oxide nanoparticles in *Fusarium oxysporum-Solanum lycopersicum* pathosystem under controlled conditions. *Journal of Phytopathology*, *169*, 533–544.

Goswami, P., Mathur, J., & Srivastava, N. (2022). Silica nanoparticles as novel sustainable approach for plant growth and crop protection. *Heliyon*, *8*e09908.

Graily Moradi, F., Hejazi, M. J., Hamishehkar, H., & Enayati, A. A. (2019). Co-encapsulation of imidacloprid and lambda-cyhalothrin using biocompatible nanocarriers: Characterization and application. *Ecotoxicology and Environmental Safety*, *175*, 155–163.

Grillo, R., Fraceto, L. F., Amorim, M. J. B., Scott-Fordsmand, J. J., Schoonjans, R., Chaudhry, Q., Cătălin Balaure, P., Gudovan, D., Gudovan, I., Nehra, M., Dilbaghi, N., Marrazza, G., Kaushik, A., Sonne, C., Kim, K. H., & Kumar, S. (2021). Ecotoxicological and regulatory aspects of environmental sustainability of nanopesticides. *Journal of Hazardous Materials*, *401*123369.

Guha, T., Gopal, G., Kundu, R., & Mukherjee, A. (2020). Nanocomposites for delivering agrochemicals: A comprehensive review. *Journal of Agricultural and Food Chemistry*, *68*, 3691–3702.

Hahn, M. (2014). The rising threat of fungicide resistance in plant pathogenic fungi: Botrytis as a case study. *Journal of Chemistry and Biology*, *7*(4), 133–141.

Hanson, M. C., Petch, G. M., Ottosen, T.-B., & Skjøth, C. A. (2022). Climate change impact on fungi in the atmospheric microbiome. *Science of The Total Environment*, *830*154491.

Hassani, S., Laouini, A., Fessi, H., & Charcosset, C. (2015). Preparation of chitosan–TPP nanoparticles using microengineered membranes – Effect of parameters and encapsulation of tacrine. *Colloids and Surfaces A: Physicochemical and Engineering Aspects*, *482*, 34–43.

Hazarika, A., Yadav, M., Yadav, D. K., & Yadav, H. S. (2022). An overview of the role of nanoparticles in sustainable agriculture. *Biocatalysis and Agricultural Biotechnology*, *43*102399.

Hoang, N. H., Le Thanh, T., Sangpueak, R., Thepbandit, W., Saengchan, C., Papathoti, N. K., Treekoon, J., Kamkaew, A., Phansak, P., & Buensanteai, K. (2023). The effect of chitosan nanoparticle formulations for control of leaf spot disease on cassava. *Phytoparasitica*, *51*, 621–636.

Hoang, N. H., Le Thanh, T., Thepbandit, W., Treekoon, J., Saengchan, C., Sangpueak, R., Papathoti, N. K., Kamkaew, A., & Buensanteai, N. (2022). Efficacy of chitosan nanoparticle loaded-salicylic acid and-silver on management of cassava leaf spot disease. *Polymers*, *14*, 1–22.

Hu, P., Zhu, L., Deng, W., Huang, W., Xu, H., & Jia, J. (2023). ConA-loaded PEGylated graphene oxide for targeted nanopesticide carriers against *Magnaporthe oryzae*. *ACS Applied Nano Materials*.

Hu, P., Zhu, L., Zheng, F., Lai, J., Xu, H., & Jia, J. (2021). Graphene oxide as a pesticide carrier for enhancing fungicide activity against *Magnaporthe oryzae*. *New Journal of Chemistry*, *45*, 2649–2658.

Inman, A. R., Kirkpatrick, S. C., Gordon, T. R., & Shaw, D. V. (2008). Limiting effects of low temperature on growth and spore germination in *Gibberella circinata*, the cause of pitch canker in pine species. *Plant Disease*, *92*, 542–545.

Jameel, M., Shoeb, M., Khan, M. T., Ullah, R., Mobin, M., Farooqi, M. K., & Adnan, S. M. (2020). Enhanced insecticidal activity of thiamethoxam by zinc oxide nanoparticles: A novel nanotechnology approach for pest control. *ACS Omega*, *5*, 1607–1615.

Jasrotia, P., Nagpal, M., Mishra, C. N., Sharma, A. K., Kumar, Satish, Kamble, U., Bhardwaj, A. K., Kashyap, P. L., Kumar, Sudheer, & Singh, G. P. (2022). Nanomaterials for postharvest management of insect pests: Current state and future perspectives. *Frontiers in Nanotechnology*, *3*, 1–19.

Ji, H., Wang, J., Xue, A., Chen, F., Guo, H., Xiao, Z., & Wang, Z. (2023). Chitosan–silica nanocomposites induced resistance in faba bean plants against aphids (*Acyrthosiphon pisum*). *Environmental Science Nano*.

Jónsdóttir, S. Ó., Reffstrup, T. K., Petersen, A., & Nielsen, E. (2016). Physicologically based toxicokinetic models of tebuconazole and application in human risk assessment. *Chemical Research in Toxicology*, *29*, 715–734.

Juárez-Maldonado, A., Ortega-Ortíz, H., Morales-Díaz, A. B., González-Morales, S., Morelos-Moreno, Á., Cabrera-De la Fuente, M., Sandoval-Rangel, A., Cadenas-Pliego, G., & Benavides-Mendoza, A. (2019). Nanoparticles and nanomaterials as plant biostimulants. *International Journal of Molecular Sciences*, *20*(1), 19.

Kah, M., Kookana, R. S., Gogos, A., & Bucheli, T. D. (2018). A critical evaluation of nanopesticides and nanofertilizers against their conventional analogues. *Nature Nanotechnology*, *13*, 677–684.

Kalia, A., Sharma, S. P., Kaur, Harleen, & Kaur, Harsimran (2020). *Novel nanocomposite-based controlled-release fertilizer and pesticide formulations: Prospects and challenges. Multifunctional hybrid nanomaterials for sustainable agri-food and ecosystems*. Elsevier Inc.

Kannan, M., Bojan, N., Swaminathan, J., Zicarelli, G., Hemalatha, D., Zhang, Y., Ramesh, M., & Faggio, C. (2023). Nanopesticides in agricultural pest management and their environmental risks: A review. *International of Journal of Environmental Science & Technology*.

Karas, P. A., Baguelin, C., Pertile, G., Papadopoulou, E. S., Nikolaki, S., Storck, V., Ferrari, F., Trevisan, M., Ferrarini, A., Fornasier, F., Vasileiadis, S., Tsiamis, G., Martin-Laurent, F., & Karpouzas, D. G. (2018). Assessment of the impact of three pesticides on microbial dynamics and functions in a lab-to-field experimental approach. *Science of the Total Environment*, *637–638*, 636–646.

Karthick Raja Namasivayam, S., Arvind Bharani, R. S., & Karunamoorthy, K. (2018). Insecticidal fungal metabolites fabricated chitosan nanocomposite (IM-CNC) preparation for the enhanced larvicidal activity - An effective strategy for green pesticide against economic important insect pests. *International Journal of Biological Macromolecules*, *120*, 921–944.

Kashyap, P. L., Kumar, S., Jasrotia, P., Singh, D. P., & Singh, G. P. (2019). Nanosensors for plant disease diagnosis: Current understanding and future perspectives. In R. Pudake, N. Chauhan, & C. Kole (Eds.), *Nanoscience for Sustainable Agriculture*. Cham: Springer.

Khan, A. U., Khan, M., Malik, N., Parveen, A., Sharma, P., Min, K., Gupta, M., & Alam, M. (2022). Screening of biosynthesized zinc oxide nanoparticles for their effect on *Daucus carota* pathogen and molecular docking. *Microscopy Research and Technique*, *85*, 3365–3373.

Khan, M., Khan, A. U., Bogdanchikova, N., & Garibo, D. (2021). Antibacterial and antifungal studies of biosynthesized silver nanoparticles against plant parasitic nematode meloidogyne incognita, plant pathogens *Ralstonia solanacearum* and *Fusarium oxysporum*. *Molecules (Basel, Switzerland)*, *26*.

Khan, M. R., & Siddiqui, Z. A. (2020). Role of zinc oxide nanoparticles in the management of disease complex of beetroot (*Beta vulgaris* L.) caused by *Pectobacterium betavasculorum, Meloidogyne incognita* and *Rhizoctonia solani. Hortic. Environ. Biotechnol.*, *62* (2), 225–241.

Khatem, R., Celis, R., & Hermosín, M. C. (2019). Cationic and anionic clay nanoformulations of imazamox for minimizing environmental risk. *Applied Clay Science*, *168*, 106–115.

Kim, K. H., Kabir, E., & Jahan, S. A. (2017). Exposure to pesticides and the associated human health effects. *Science of the Total Environment*, *575*, 525–535.

Kirchmann, H. (2019). Why organic farming is not the way forward. *Outlook on Agriculture*, *48*(1), 22–27.

Kongala, S. I., & Mamidala, P. (2023). Harpin-loaded chitosan nanoparticles induced defense responses in tobacco. *Carbohydrate Polymer Technologies and Applications*, *5*100293.

Kumar, A., Kanwar, R., & Mehta, S. K. (2021). Eucalyptus oil-based nanoemulsion: A potent green nanowagon for controlled delivery of emamectin benzoate. *ACS Agricultural Science and Technology*, *1*, 76–88.

Kumar, S., Bhanjana, G., Sharma, A., Dilbaghi, N., Sidhu, M. C., & Kim, K. H. (2017). Development of nanoformulation approaches for the control of weeds. *Science of the Total Environment*, *586*, 1272–1278.

Kumar, S., Bhanjana, G., Sharma, A., Sarita., Sidhu, M., & Dilbaghi, N. (2014a). Herbicide loaded carboxymethyl cellulose nanocapsules as potential carrier in agrinanotechnology. *Science of Advanced Materials*, *7*.

Kumar, S., Bhanjana, G., Sharma, A., Sidhu, M. C., & Dilbaghi, N. (2014b). Synthesis, characterization and on field evaluation of pesticide loaded sodium alginate nanoparticles. *Carbohydrate Polymers*, *101*, 1061–1067.

Kumaraswamy, R. V., Saharan, V., Kumari, S., Chandra Choudhary, R., Pal, A., Sharma, S. S., Rakshit, S., Raliya, R., & Biswas, P. (2021). Chitosan-silicon nanofertilizer to enhance plant growth and yield in maize (*Zea mays* L.). *Plant Physiology and Biochemistry*, *159*, 53–66.

Kutawa, A. B., Ahmad, K., Ali, A., Hussein, M. Z., Wahab, M. A. A., Adamu, A., Ismaila, A. A., Gunasena, M. T., Rahman, M. Z., & Hossain, M. I. (2021). Trends in nanotechnology and its potentialities to control plant pathogenic fungi: A review. *Biology*, *10*.

Lazcano-Ramírez, H. G., Garza-García, J. J. O., Hernández-Díaz, J. A., León-Morales, J. M., Macías-Sandoval, A. S., & García-Morales, S. (2023). Antifungal activity of selenium nanoparticles obtained by plant-mediated synthesis. *Antibiotics*, *12*.

Lee, R., den Uyl, R., & Runhaar, H. (2019). Assessment of policy instruments for pesticide use reduction in Europe; Learning from a systematic literature review. *Crop Protection*, *126*104929.

Li, S., Jiang, Y., Sun, Q., Coffin, S., Chen, L., Qiao, K., Gui, W., & Zhu, G. (2020). Tebuconazole induced oxidative stress related hepatotoxicity in adult and larval zebrafish (*Danio rerio*). *Chemosphere*, *241*125129.

Li, S., Sun, Q., Wu, Q., Gui, W., Zhu, G., & Schlenk, D. (2019). Endocrine disrupting effects of tebuconazole on different life stages of zebrafish (*Danio rerio*). *Environmental Pollution*, *249*, 1049–1059.

Li, X., Wang, Q., Wang, X., & Wang, Z. (2022). Synergistic effects of graphene oxide and pesticides on fall armyworm, *Spodoptera frugiperda*. *Nanomaterials*, *12*, 1–14.

Li, Y., Zhang, P., Li, M., Shakoor, N., Adeel, M., Zhou, P., Guo, M., Jiang, Y., Zhao, W., Lou, B. Z., & Rui, Y. (2023). Application and mechanisms of metal-based nanoparticles in the control of bacterial and fungal crop diseases. *Pest Management Science*, *79*, 21–36.

Liang, Y., Fan, C., Dong, H., Zhang, W., Tang, G., Yang, J., Jiang, N., & Cao, Y. (2018). Preparation of MSNs-chitosan@prochloraz nanoparticles for reducing toxicity and improving release properties of prochloraz. *ACS Sustainable Chemistry & Engineering*, *6*, 10211–10220.

Liu, Y., Zhang, Y., Xin, X., Xu, X., Wang, G., Gao, S., Qiao, L., Yin, S., Liu, H., Jia, C., Shen, W., Xu, L., Ji, Y., & Zhou, C. (2022). Design and preparation of avermectin nanopesticide for control and prevention of pine wilt disease. *Nanomaterials*, *12*.

Lourthuraj, A. A., Hatshan, M. R., & Hussein, D. S. (2022). Biocatalytic degradation of organophosphate pesticide from the wastewater and hydrolytic enzyme properties of consortium isolated from the pesticide contaminated water. *Environmental Research*, *205*112553.

Lucia, A., Murace, M., Sartor, G., Keil, G., Cámera, R., Rubio, R. G., & Guzmán, E. (2021). Oil in water nanoemulsions loaded with tebuconazole for populus wood protection against white- and brown-rot fungi. *Forests*, *12*, 1234.

Majumder, S., Kaushik, P., Rana, V. S., Sinha, P., & Shakil, N. A. (2020). Amphiphilic polymer based nanoformulations of mancozeb for management of early blight in tomato. *Journal of Environmental Science and Health - Part B Pesticides, Food Contaminants, and Agricultural Wastes*, *55*, 501–507.

Malerba, M, & Cerana, R (2016). Chitosan Effects on Plant Systems. *Int J Mol Sci*, *17*(7), 996.

Maryška, L., Jindřichová, B., Siegel, J., Záruba, K., & Burketová, L. (2023). Impact of palladium nanoparticles on plant and its fungal pathogen. A case study: *Brassica napus-Plenodomus lingam*. *AoB Plants*, *15*, 1–11.

Memon, Q. U. A., Wagan, S. A., Chunyu, D., Shuangxi, X., Jingdong, L., & Damalas, C. A. (2019). Health problems from pesticide exposure and personal protective measures among women cotton workers in southern Pakistan. *Science of the Total Environment*, *685*, 659–666.

MFDS. (2017). Pesticide MRLs in Food. Available from https://www.fao.org/faolex/results/details/en/c/LEX-FAOC190507/.

Mondéjar-López, M., López-Jimenez, A. J., Ahrazem, O., Gómez-Gómez, L., & Niza, E. (2023). Chitosan coated - Biogenic silver nanoparticles from wheat residues as green antifungal and nanoprimig in wheat seeds. *International Journal of Biological Macromolecules*, *225*, 964–973.

Moutassem, D., Belabid, L., Bellik, Y., Ziouche, S., & Baali, F. (2019). Efficacy of essential oils of various aromatic plants in the biocontrol of *Fusarium wilt* and inducing systemic resistance in chickpea seedlings. *Plant Protection Science*, *55*, 202–217.

Mustafa, I. F., Hussein, M. Z., Saifullah, B., Idris, A. S., Hilmi, N. H. Z., & Fakurazi, S. (2018). Synthesis of (hexaconazole-zinc/aluminum-layered double hydroxide nanocomposite) fungicide nanodelivery system for controlling ganoderma disease in oil palm. *Journal of Agricultural and Food Chemistry*, *66*, 806–813.

Narasimhamurthy, K., Udayashankar, A. C., De Britto, S., Lavanya, S. N., Abdelrahman, M., Soumya, K., Shetty, H. S., Srinivas, C., & Jogaiah, S. (2022). Chitosan and chitosan-derived nanoparticles modulate enhanced immune response in tomato against bacterial wilt disease. *International Journal of Biological Macromolecules*, *220*, 223–237.

Nasreen, C., Mohiddin, G. J., Srinivasulu, M., et al. (2015). Interaction effects of insecticides on microbial populations and dehydrogenase activity in groundnut (*Arachis hypogeae* l.) planted black clay soil. *International Journal of Current Microbiology & Applied Science*, *4*, 135–146.

Neme, K., Nafady, A., Uddin, S., & Tola, Y. B. (2021). Application of nanotechnology in agriculture, postharvest loss reduction and food processing: Food security implication and challenges. *Heliyon*, *7*(12)e08539.

Nnadi, N. E., & Carter, D. A. (2021). Climate change and the emergence of fungal pathogens. *PLoS Pathogens*, *17*, e1009503.

Oliveira, H. C., Stolf-Moreira, R., Martinez, C. B. R., Grillo, R., de Jesus, M. B., & Fraceto, L. F. (2015). Nanoencapsulation enhances the post-emergence herbicidal activity of atrazine against mustard plants. *PLoS One*, *10*, e0132971.

Orta, M. del M., Martín, J., Santos, J. L., Aparicio, I., Medina-Carrasco, S., & Alonso, E. (2020). Biopolymer-clay nanocomposites as novel and ecofriendly adsorbents for environmental remediation. *Applied Clay Science*, *198*105838.

Panoth, D., Manikkoth, S.T., Jahan, F., Thulasi, K.M., Paravannoor, A., & Vijayan, B. (2022). Nanosensors for pesticide detection in soil. In Adil Denizli, Tuan Anh Nguyen, Susai Rajendran, Ghulam Yasin, Ashok Kumar Nadda (eds.), *Micro and nano technologies, nanosensors for smart agriculture* (vol. 11, pp. 237–258). Elsevier. ISBN 9780128245545.

Parthasarathy, R., Jayabaskaran, C., Manikandan, A., & Anusuya, S. (2023). Synthesis of nickel-chitosan nanoparticles for controlling blast diseases in Asian rice. *Applied Biochemistry and Biotechnology*, *195*, 2134–2148.

Peng, F., Wang, X., Zhang, W., Shi, X., Cheng, C., Hou, W., Lin, X., Xiao, X., & Li, J. (2022). Nanopesticide formulation from pyraclostrobin and graphene oxide as a nanocarrier and application in controlling plant fungal pathogens. *Nanomaterials*, *12*, 1–14.

Pereira, A. E. S., Silva, P. M., Oliveira, J. L., Oliveira, H. C., & Fraceto, L. F. (2017). Chitosan nanoparticles as carrier systems for the plant growth hormone gibberellic acid. *Colloids and Surfaces B: Biointerfaces*, *150*, 141–152.

Pimentel, D. (1995). Amounts of pesticides reaching target pests: Environmental impacts and ethics. *Journal of Agricultural & Environmental Ethics*, *8*, 17–29.

Pontes, M. S., Antunes, D. R., Oliveira, I. P., Forini, M. M. L., Santos, J. S., Arruda, G. J., Caires, A. R. L., Santiago, E. F., & Grillo, R. (2021). Chitosan/tripolyphosphate nanoformulation carrying paraquat: Insights on its enhanced herbicidal activity. *Environmental Science: Nano*, *8*, 1336–1351.

Qi, C., Xu, Z., Qian, K., Shen, G., Rong, S., Zhang, C., Zhang, P., Ma, C., Zhang, Y., & He, L. (2021). Sodium selenite-carbon dots nanocomposites enhance acaricidal activity of fenpropathrin: Mechanism and application. *The Science of the Total Environment*, *777*145832.

Raza, M. M., & Bebber, D. P. (2022). Climate change and plant pathogens. *Current Opinion in Microbiology*, *70*102233.

Ribas e Ribas, A. D., Spolti, P., Del Ponte, E. M., Donato, K. Z., Schrekker, H., & Fuentefria, A. M. (2016). Is the emergence of fungal resistance to medical triazoles related to their use in the agroecosystems? A mini review. *Brazilian Journal of Microbiology*, *47*, 793–799.

Rodríguez, A. G. P., & León, J. A. A. (2020). Methods for determination of pesticides and fate of pesticides in the fields. In R. K. R., S. Thomas, T. Volova, & J. K. (Eds.), *Controlled release of pesticides for sustainable agriculture* (pp. 41–58). Cham: Springer International Publishing.

Rudmin, M., Banerjee, S., Makarov, B., Belousov, P., Kurovsky, A., Ibraeva, K., & Buyakov, A. (2022). Glauconite-urea nanocomposites as polyfunctional controlled-release fertilizers. *Journal of Soil Science and Plant Nutrition*, *22*, 4035–4046.

Salama, D. M., Abd El-Aziz, M. E., El-Naggar, M. E., Shaaban, E. A., & Abd El-Wahed, M. S. (2021). Synthesis of an eco-friendly nanocomposite fertilizer for common bean based on carbon nanoparticles from agricultural waste biochar. *Pedosphere*, *31*, 923–933.

Salimi, M., Motamedi, E., Safari, M., & Motesharezadeh, B. (2021). Synthesis of urea slow-release fertilizer using a novel starch-g-poly(styrene-co-butylacrylate) nanocomposite latex and its impact on a model crop production in greenhouse. *Journal of Cleaner Production*, *322*129082.

Sánchez-Bayo, F., Goulson, D., Pennacchio, F., Nazzi, F., Goka, K., & Desneux, N. (2016). Are bee diseases linked to pesticides? - A brief review. *Environment International*, *89–90*, 7–11.

Sánchez-López, E., Gomes, D., Esteruelas, G., Bonilla, L., Lopez-Machado, A. L., Galindo, R., Cano, A., Espina, M., Ettcheto, M., Camins, A., Silva, A. M., Durazzo, A., Santini, A., Garcia, M. L., & Souto, E. B. (2020). Metal-based nanoparticles as antimicrobial agents: An overview. *Nanomaterials*, *10*, 1–39.

Satapute, P., Kamble, M. V., Adhikari, S. S., & Jogaiah, S. (2019). Influence of triazole pesticides on tillage soil microbial populations and metabolic changes. *Science of the Total Environment*, *651*, 2334–2344.

Schneider, R., Koch, J., Troldborg, L., Henriksen, H. J., & Stisen, S. (2022). Machine-learning-based downscaling of modelled climate change impacts on groundwater table depth. *Hydrol. Earth Syst. Sci.*, *26*(22), 5859–5877.

Scott-Fordsmand, J. J., Fraceto, L. F., & Amorim, M. J. B. (2022). Nano-pesticides: the lunch-box principle—deadly goodies (semio-chemical functionalised nanoparticles that deliver pesticide only to target species). *Journal of Nanobiotechnology*, *20*, 13.

Selyutina, O. Y., Khalikov, S. S., & Polyakov, N. E. (2020). Arabinogalactan and glycyrrhizin based nanopesticides as novel delivery systems for plant protection. *Environmental Science and Pollution Research*, *27*, 5864–5872.

Shahid, M., & Khan, M. S. (2021). Tolerance of pesticides and antibiotics among beneficial soil microbes recovered from contaminated rhizosphere of edible crops. *Current Research Microbial Science*, *3*100091.

Shan, P., Lu, Y., Liu, H., Lu, W., Li, D., Yin, X., Lian, X., Li, Zhongyu, & Li, Zhihui (2023). Rational design of multi-stimuli-responsive polymeric nanoparticles as a 'Trojan horse' for targeted pesticide delivery. *Industrial Crops and Products*, *193*116182.

Shan, Y., Cao, L., Muhammad, B., Xu, B., Zhao, P., Cao, C., & Huang, Q. (2020). Iron-based porous metal–organic frameworks with crop nutritional function as carriers for controlled fungicide release. *Journal of Colloid and Interface Science*, *566*, 383–393.

Shang, H., Ma, C., Li, C., Cai, Z., Shen, Y., Han, L., Wang, C., Tran, J., Elmer, W. H., White, J. C., & Xing, B. (2023). *Aloe vera* extract gel-biosynthesized selenium nanoparticles enhance disease resistance in lettuce by modulating the metabolite profile and bacterial endophytes composition. *ACS Nano*.

Shao, H., & Zhang, Y. (2017). Non-target effects on soil microbial parameters of the synthetic pesticide carbendazim with the biopesticides cantharidin and norcantharidin. *Scientific Reports*, *7*5521.

Sharma, S., Singh, S., Ganguli, A. K., & Shanmugam, V. (2017). Anti-drift nano-stickers made of graphene oxide for targeted pesticide delivery and crop pest control. *Carbon*, *115*, 781–790.

Shinoto, Y., Otani, R., Matsunami, T., & Maruyama, S. (2020). Analysis of the shallow root system of maize grown by plowing upland fields converted from paddy fields: Effects of soil hardness and fertilization. *Plant Production Science*, *24*, 297–305. Available from https://doi.org/10.1080/1343943X.2020.1863823.

Skevas, T. (2020). Evaluating alternative policies to reduce pesticide groundwater pollution in Dutch arable farming. *Journal of Environmental Planning and Management, 63*, 733–750.

Soliman, S. A., Hafez, E. E., Al-Kolaibe, A. M. G., Abdel Razik, E.-S. S., Abd-Ellatif, S., Ibrahim, A. A., Kabeil, S. S. A., & Elshafie, H. S. (2022). Biochemical characterization, antifungal activity, and relative gene expression of two mentha essential oils controlling *Fusarium oxysporum*, the causal agent of lycopersicon esculentum root rot. *Plants, 11*, 189.

Sousa, G. F. M., Gomes, D. G., Campos, E. V. R., Oliveira, J. L., Fraceto, L. F., Stolf-Moreira, R., & Oliveira, H. C. (2018). Post-emergence herbicidal activity of nanoatrazine against susceptible weeds. *Frontiers in Environmental Science, 6*, 1–6.

Sravani, B., Dalvi, S., & Narute, T. K. (2023). Role of chitosan nanoparticles in combating Fusarium wilt (*Fusarium oxysporum* f. sp. ciceri) of chickpea under changing climatic conditions. *Journal of Phytopathology, 171*, 67–81.

Subramanian, K. S., Karthika, V., Praghadeesh, M., & Lakshmanan, A. (2020). Nanotechnology for mitigation of global warming impacts. In V. Venkatramanan, S. Shah, & R. Prasad (Eds.), *Global climate change: Resilient and smart agriculture* (pp. 315–336). Singapore: Springer Singapore.

Sumon, K. A., Rashid, H., Peeters, E. T. H. M., Bosma, R. H., & Van den Brink, P. J. (2018). Environmental monitoring and risk assessment of organophosphate pesticides in aquatic ecosystems of north-west Bangladesh. *Chemosphere, 206*, 92–100.

Susanti, D., Haris, M. S., Taher, M., & Khotib, J. (2022). Natural products-based metallic nanoparticles as antimicrobial agents. *Frontiers in Pharmacology, 13*, 1–14.

Tarakanov, R., Shagdarova, B., Lyalina, T., Zhuikova, Y., Il'ina, A., Dzhalilov, F., & Varlamov, V. (2023). Protective properties of copper-loaded chitosan nanoparticles against soybean pathogens *Pseudomonas savastanoi* pv. glycinea and *Curtobacterium flaccumfaciens* pv. flaccumfaciens. *Polymers, 15*.

Tippannanavar, M., Verma, A., Kumar, R., Gogoi, R., Kundu, A., & Patanjali, N. (2020). Preparation of nanofungicides based on imidazole drugs and their antifungal evaluation. *Journal of Agricultural and Food Chemistry, 68*, 4566–4578.

Tomer, A., Singh, R., & Dwivedi, S. A. (2021). Nanotechnology for detection and diagnosis of plant diseases. In H. Sarma, S. J. Joshi, R. Prasad, & J. Jampilek (Eds.), *Biobased nanotechnology for green applications. Nanotechnology in the life sciences*. Cham: Springer.

Tortella, G., Rubilar, O., Pieretti, J. C., Fincheira, P., de Melo Santana, B., Fernández-Baldo, M. A., Benavides-Mendoza, A., & Seabra, A. B. (2023). Nanoparticles as a promising strategy to mitigate biotic stress in agriculture. *Antibiotics, 12*.

Tudi, M., Daniel Ruan, H., Wang, L., Lyu, J., Sadler, R., Connell, D., Chu, C., & Phung, D. T. (2021). Agriculture development, pesticide application and its impact on the environment. *International Journal of Environmental Research and Public Health, 18*(3), 1112.

United Nations Department of Economic and Social Affairs, Population Division (2022). World Population Prospects 2022: Summary of Results. UN DESA/POP/2022/TR/NO. 3.

Van Leeuwen, T., Dermauw, W., Mavridis, K., & Vontas, J. (2020). Significance and interpretation of molecular diagnostics for insecticide resistance management of agricultural pests. *Current Opinion in Insect Science, 39*, 69–76.

Wahab, A., Batool, F., Muhammad, M., Zaman, W., Mikhlef, R. M., & Naeem, M. (2023). Current Knowledge, Research Progress, and Future Prospects of Phyto-Synthesized Nanoparticles Interactions with Food Crops under Induced Drought Stress. *Sustainability, 15*, 14792.

Walia, A., Sumal, K., & Kumari, S. (2018). Effect of chlorpyrifos and malathion on soil microbial population and enzyme activity. *Acta Scientific Microbiology, 1*, 14–22.

Wang, J., Li, R., Zhang, H., Wei, G., & Li, Z. (2020). Beneficial bacteria activate nutrients and promote wheat growth under conditions of reduced fertilizer application. *BMC Microbiology*, *20*, 38.

Wang, L., Ning, C., Pan, T., & Cai, K. (2022). Role of silica nanoparticles in abiotic and biotic stress tolerance in plants: A review. *International Journal of Molecular Sciences*, *23*.

Wang, X., Xie, H., Wang, Z., & He, K. (2019). Graphene oxide as a pesticide delivery vector for enhancing acaricidal activity against spider mites. *Colloids Surfaces B Biointerfaces*, *173*, 632–638.

Wang, Yao, Yang, L., Zhou, X., Wang, Ye, Liang, Y., Luo, B., Dai, Y., Wei, Z., Li, S., He, R., & Ding, W. (2023). Molecular mechanism of plant elicitor daphnetin-carboxymethyl chitosan nanoparticles against *Ralstonia solanacearum* by activating plant system resistance. *International Journal of Biological Macromolecules*, *241*124580.

Wu, L., Jiang, Y., Zhao, F., et al. (2020). Increased organic fertilizer application and reduced chemical fertilizer application affect the soil properties and bacterial communities of grape rhizosphere soil. *Scientific Reports*, *10*9568. Available from https://doi.org/10.1038/s41598-020-66648-9.

Wur, 2021. https://www.wur.nl/en/show-longread/farming-without-fertiliser-is-it-possible.htm. Acceded 14/07/2023.

Xiang, Y., Zhang, G., Chen, C., Liu, B., Cai, D., & Wu, Z. (2018). Fabrication of a pH-responsively controlled-release pesticide using an attapulgite-based hydrogel. *ACS Sustainable Chemistry and Engineering*, *6*, 1192–1201.

Xu, C., Cao, L., Zhao, P., Zhou, Z., Cao, C., Li, F., & Huang, Q. (2018). Emulsion-based synchronous pesticide encapsulation and surface modification of mesoporous silica nanoparticles with carboxymethyl chitosan for controlled azoxystrobin release. *Chemical Engineering Journal*, *348*, 244–254.

Yang, J., Wu, G., Jiang, C., Long, W., & Liu, W. (2023). Potential emissions of insecticide VOCs and their correlations between agricultural emissions and meteorological factors. *Agriculture (Switzerland)*, *13*, 1–15. Available from https://doi.org/10.3390/agriculture13010066.

Yin, J., Su, X., Yan, S., & Shen, J. (2023). Multifunctional nanoparticles and nanopesticides in agricultural application. *Nanomaterials*, *13*, 1–14.

Yin, Y., Yang, M., Xi, J., Cai, W., Yi, Y., He, G., Dai, Y., Zhou, T., & Jiang, M. (2020). A sodium alginate-based nano-pesticide delivery system for enhanced in vitro photostability and insecticidal efficacy of phloxine B. *Carbohydrate Polymers*, *247*116677. Available from https://doi.org/10.1016/j.carbpol.2020.116677.

Yu, J., Wang, D., Geetha, N., Khawar, K. M., Jogaiah, S., & Mujtaba, M. (2021). Current trends and challenges in the synthesis and applications of chitosan-based nanocomposites for plants: A review. *Carbohydrate Polymers*, *261*117904.

Zaller, J. G., & Brühl, C. A. (2019). Non-target effects of pesticides on organisms inhabiting agroecosystems. *Frontiers in Environmental Science*, *7*, 1–3.

Zhang, J., Kothalawala, S., & Yu, C. (2023). Engineered silica nanomaterials in pesticide delivery: Challenges and perspectives. *Environmental Pollution*, *320*121045.

Zhang, P., Guo, Z., Ullah, S., Melagraki, G., Afantitis, A., & Lynch, I. (2021). Nanotechnology and artificial intelligence to enable sustainable and precision agriculture. *Nature Plants*, *7*, 864–876.

Zhang, Q., Du, Y., Yu, M., Ren, L., Guo, Y., Li, Q., Yin, M., Li, X., & Chen, F. (2022). Controlled release of dinotefuran with temperature/pH-responsive chitosan-gelatin microspheres to reduce leaching risk during application. *Carbohydrate Polymers*, *277*118880.

Zhang, Y., Chen, W., Jing, M., Liu, S., Feng, J., Wu, H., Zhou, Y., Zhang, X., & Ma, Z. (2019). Self-assembled mixed micelle loaded with natural pyrethrins as an intelligent nano-insecticide with a novel temperature-responsive release mode. *Chemical Engineering Journal*, *361*, 1381–1391.

Zhao, W., Liu, Y., Zhang, P., Zhou, P., Wu, Z., Lou, B., Jiang, Y., Shakoor, N., Li, M., Li, Y., Lynch, I., Rui, Y., & Tan, Z. (2022). Engineered Zn-based nano-pesticides as an opportunity for treatment of phytopathogens in agriculture. *NanoImpact*, *28*100420.

Zhou, H., Liu, S., Wan, F., Jian, Y., Guo, F., Chen, J., Ning, Y., & Ding, W. (2021). Graphene oxide-acaricide nanocomposites advance acaricidal activity of acaricides against: *Tetranychus cinnabarinus* by directly inhibiting the transcription of a cuticle protein gene. *Environmental Science Nano*, *8*, 3122–3137.

Nanocomposite fertilizers: a tool for a better and efficient nutrition of plants

7

Yolanda González-García[1], Emilio Olivares-Sáenz[1], Marissa Pérez-Alvarez[2] and Gregorio Cadenas-Pliego[2]
[1]Faculty of Agronomy, Center for Protected Agriculture, Autonomous University of Nuevo Leon, General Escobedo, Mexico, [2]Applied Chemistry Research Center, Saltillo, Mexico

7.1 Introduction

Currently, agricultural development strategies focus on intensification through increased application of fertilizers and agrochemicals, better use of irrigation and automation, and the use of improved varieties in the field to meet the global food demand of a constantly growing population (Khatri-Chhetri et al., 2023).

Conventional chemical fertilizers are widely used throughout the world to achieve maximum performance in agricultural systems, however, their efficiency does not exceed 50%, which is why fertilizers are applied in excessive amounts to be able to supply crop nutrient demand (Upadhyay et al., 2023). This has generated a reduction in the profit margin of farmers, in addition to irreversibly modifying mineral cycles, affecting the structure and microflora of soil and plants and, ultimately, food chains in all ecosystems (Gade et al., 2023).

Fertilizers commonly deliver nutrients in chemical forms that plants cannot readily absorb and these fertilizers provide most of the macro and micronutrients that are poorly soluble, leading to very low utilization (Gade et al., 2023). For example, a significant proportion (50%–70%) of the content of essential macronutrients such as nitrogen (N), phosphorus (P), and potassium (K) is lost from the soil before being utilized by plants (Yadav et al., 2023a), while micronutrients such as zinc (Zn), iron (Fe), manganese (Mn), and copper (Cu) may not be easily assimilated by plants through existing conventional fertilization methods (Wang et al., 2022). As a result of the low efficiency of fertilizers, there is a need arises for their repetitive applications. Although the form of nutrient absorption generally occurs through the roots of the plants, the low availability of these elements has led to farmers to the application of fertilizers also by foliar route (Carrasco-Correa et al., 2023). This results in a reduction in the profit margin for farmers, in addition to irreversibly modifying the mineral cycles, affecting the structure and microflora of the soil and plants and, ultimately, affecting food chains in all ecosystems (Gade et al., 2023).

World agricultural development policies focus on improving agricultural productivity through technological change and upgrading existing technologies, considering

Nanocomposites for Environmental, Energy, and Agricultural Applications. DOI: https://doi.org/10.1016/B978-0-443-13935-2.00007-3

the limited availability of natural resources and the need to reduce pollution and greenhouse gas emissions generated by the agricultural sector (Khatri-Chhetri et al., 2023). In this sense, nanofertilizers (NFs) have emerged as a promising alternative to improve plant production, offering greater efficiency and reduced environmental impact.

NFs can be made from conventional fertilizers, bulk fertilizer feedstocks, or plant extracts (Gade et al., 2023). NFs can be nanoparticles (NPs) or nanomaterials (NMs) of the element individually or in combination with other constituents, or nanocomposites that include the nutrients either in nanosize or in their conventional form (Bhardwaj et al., 2022; Saffan et al., 2022). Compared to conventional fertilizers, NFs have superior chemical stability, higher surface tension, greater absorption and mobility capacity, greater pH tolerance, and high ionizing power (Zahra et al., 2022). In addition, they can be designed to have special characteristics such as the controlled and/or directed release of one or several elements (macro or micro) to plants in a high surface area, at specific concentrations and with an adequate size (Saraiva et al., 2023; Zahra et al., 2022). These characteristics can significantly improve the efficiency of nutrient uptake, photosynthesis, photosynthate accumulation, and nutrient translocation (Gade et al., 2023). In addition to this, they can improve the quality of the soil by reducing its chemical load and the frequency of applications, which leads to a higher yield and quality of the crops (Feregrino-Perez et al., 2018).

The present review provides a comprehensive view of recent developments in the types of NF, the way they are absorbed by plants, their impacts on agricultural crops, and their advantages and disadvantages compared to traditional fertilizers.

7.2 Problems in the nutrition of agricultural crops

Agriculture is fundamental for the human being and the constitution of society. This activity is responsible for the livelihood of around 40% of the world population and, excluding icy areas, a third of the earth's surface is dedicated to agriculture, which demonstrates the impact and representativeness of this practice worldwide (Silva et al., 2023).

Worldwide, agricultural production and food security are in danger due to overpopulation, industrialization, unplanned urbanization, climate change, and different types of soil degradation (Jat Baloch et al., 2023). The demands to increase food production to supply the needs of a growing population are leading to the depletion of soil nutrients; for this reason, the application of synthetic mineral fertilizers capable of providing crops with the nutrients they require for their growth and adequate production has been promoted (Piash et al., 2021).

Chemical fertilizers are the basis for increasing agricultural production since the application of these can increase the yield in open-sky agriculture up to approximately 30% (Zhao et al., 2023). However, one of the most important problems in the food production chain is that farmers tend to add excess inorganic fertilizers during plant production based on the assumption that this action can improve yield.

Moreover, the use of chemical fertilizers alone increases crop yields during the first cycles, but if an adequate balance is not made, it generates a negative impact on long-term sustainability (Alzain et al., 2023). This is because the fertilizer manufacturing process is expensive, consumes too much energy, and also, the availability of minerals for its production is limited. As if that were not enough, the environmental cost of applying these chemical products has been substantial, since a series of problems have been generated, such as contamination of surface and groundwater, eutrophication, soil degradation due to acidification, salinization and decrease in microbiota, formation of photochemical smog, and emissions of greenhouse gases such as nitrous oxide and ammonia (Piash et al., 2021; Wang et al., 2023b). Some of the problems in the use of conventional fertilizers are described in detail below.

7.2.1 Low efficiency of conventional fertilizers

According to the Food and Agriculture Organization (FAO) data, it was estimated that in 2019 more than 220 million tons of commercial fertilizers and liming materials were applied worldwide on agricultural fields (Rashid et al., 2023). Of the amount of fertilizers applied to the soil, only a small percentage is uptake and used by plants, while the rest is eventually dragged into water bodies through leaching and surface runoff, or in smaller proportions is lost by volatilization (Rop et al., 2018). Key macronutrients, including N, P, and K, applied to soil have been reported to be lost by 40%–70%, 80%–90%, and 50%–90%, respectively, causing considerable waste of resources and excessive pollution (Zulfiqar et al., 2019).

Specifically, nitrogenous fertilizers, currently the most widely used in the world as nitrogen is the main nutrient that produces organic substances for the plant, are extremely prone to lose in their various forms (Wang et al., 2023b). For example, ammonium nitrate (NH_4NO_3) contains nitrogen in both nitrous and ammoniacal forms, therefore it is susceptible to both volatilization and soil drainage losses (Tsintskaladze et al., 2017). Drained nitrogen can greatly contaminate surface and groundwater, in addition to generating nutrient imbalance, acidification, and salinization of the soil; while the volatilized can contaminate the atmosphere due to the emission of ammonia, nitrogen oxides, and other gases, endangering the environment and human health (Tsintskaladze et al., 2017; Wang et al., 2023b).

Phosphorus is another macro element that is highly prone to losses, even though there are several phosphate fertilizers, among the most commercially used in the world are single superphosphate (SSP), triple superphosphate (TSP), monoammonium phosphate (MAP), and diammonium phosphate (DAP) (Guelfi et al., 2022). Due to immobilization of P when transforming to scarcely available forms, only 10%–20% of P is uptake by the plant after fertilizer application, while the remaining P (80%–90%) is easily adsorbed or precipitated in the soil. For example, most orthophosphate ions adsorb on the surface of clay-size minerals dominated by aluminum and iron oxides, or precipitate as calcium phosphates in calcareous soils (Fertahi et al., 2022). Due to their inefficiency, phosphate fertilizers are commonly overused in agricultural fertilization, and currently almost 21 million tons of phosphorus are applied to production systems through fertilization every year, increasing residuality and contamination (Xiao et al., 2023).

7.2.2 Impact of fertilizers on soil chemistry

Excessive use of chemical fertilizers often leads to acidification and degradation of soil quality, eutrophication of surface waters, and an increasing incidence of crop pests and diseases (Li et al., 2023b).

Soil acidity is one of the greatest limitations of plant production, and although it occurs naturally when it is generated by anthropogenic activity such as fertilization, it is considered one of the main forms of chemical degradation of the soil (Yescas-Coronado et al., 2022). The acidity of the soil is determined by the pH, which is the negative logarithm, in base 10, of the H^+ activity, log (H^+) that determines if the soil is acidic (pH below 6.5, higher concentration of H^+) or alkaline (pH above 7.5) (Yan et al., 2018). Soil acidification is an important aspect of soil quality and can cause numerous negative effects on the ecosystem. In addition to generating a nutritional imbalance for plants, it can decrease the number of plant roots and crop yields, and reduce biological diversity of soil (Wang et al., 2023a).

Soil pH affects the availability of macro and microelements, only when it is in the range of 5.5–7, it is adequate for the availability of nutrients for plant roots (Fertahi et al., 2022). However, with the application of fertilizers, mainly those containing chlorine (Cl^-), such as ammonium chloride (NH_4Cl) and potassium chloride (KCl), a large amount of Cl^- enters the ecosystem of agricultural soils as accompanying ions and participates in the nutrient cycle of the soil and plants. Although this anion is easily leached and will not accumulate in areas of excessive rainfall, long-term application will significantly increase soil Cl^- content, and being a strong acidic anion can directly cause soil acidification which will further displace a large amount of basic ions and will further aggravate soil acidification (Wang et al., 2023c). In the case of nitrogen and sulfur fertilizers, a large amount of nitric acid and sulfuric acid are produced in the soil causing acidification, and as a further result, the soil also loses calcium and magnesium ions (Li et al., 2023a).

Secondary salinization of soil and water is another serious problem derived from the excessive use of chemical fertilizers that negatively affects the quality of arable land, induces soil compaction over time, destroys the structure, and significantly impedes crop growth (Zhang et al., 2023). Currently, it is estimated that approximately 20% of the world's croplands are affected by salinity problems to varying degrees, generating a decrease in the land use capacity for agricultural production, and therefore a decrease in food production (Singh et al., 2022). Saline soil and water are the reservoirs of various salts, mainly from fertilizers containing sodium (Na^+), chlorides (Cl^-), calcium (Ca^{2+}), magnesium (Mg^{2+}), sulfates ($SO_4{}^{2-}$), and carbonates ($HCO_3{}^-$) (Yildiz et al., 2021).

Under saline conditions, the relationships between the soil and the plant become immensely complex since soil salinity reduces the water retention capacity, the organic matter content, weakens the structure, alters the stability of the aggregates, causes aeration deficient, increases the pH, increases the percentage of exchangeable sodium and its adsorption rate, as well as the reduction of the cation exchange capacity (Kong et al., 2021; Nehela et al., 2021). This generates big problems in plant productivity by inducing ionic, osmotic, oxidative stress, and nutritional and hormonal imbalances that together significantly decrease plant performance (Etesami et al., 2021; Kozłowska et al., 2021).

7.2.3 Pollution from excess application of fertilizers

Fertilizers play an essential role in improving agricultural productivity, but the excessive application of chemical fertilizers has caused a series of environmental pollution and eutrophication problems (Zheng et al., 2022).

One of the main problems is the contamination of groundwater, which is produced by the leaching of fertilizers (Huddell et al., 2023). For example, when the input of nitrate (NO_3^-) in the soil exceeds the nitrate consumption capacity of plants, it causes the loss of nitrogen to the environment, one part is volatilized and the rest accumulates in the soil and the water (both surface and groundwater). In water, when the NO_3^- concentration exceeds the threshold value of 3 mg L^{-1}, it is considered to be affected by human factors, mainly excessive soil fertilization and mining (Kimbi et al., 2022).

Eutrophication is another consequence of the excessive application of fertilizers, mainly nitrates and phosphates. This has led to the enrichment of water bodies through runoff, resulting in a dense proliferation of aquatic plants such as Nymphaeaceae and the proliferation of various types of algae (Akinnawo, 2023). These affect higher trophic levels by decreasing the cover of perennial macrophytes and the increasing turbidity, which impairs the ability of predators to capture prey. Additionally, it can deplete oxygen levels in deep waters, causing hypoxia to benthic organisms, and destroying their habitats (Salo & Salovius-Laurén, 2022).

Contamination by heavy metals (HMs) is another problem that derives from the excessive application of fertilizers since these are found in the raw materials used to produce inorganic fertilizers, such as the phosphate rock (phosphorite) used to produce superphosphate, SSP, DSP, TDP, rock phosphate, MAP, and DAP (Rashid et al., 2023). It should be noted that this rock was added to the list of critical raw materials by the European Commission since it contains amounts of HMs (Albert & Bloem, 2023). It is known that most fertilizers containing phosphorus (superphosphates), and lesser amounts of those containing copper (copper sulfate), iron (iron sulfates and chelates) and zinc (zinc sulfate), can contain contaminants such as Cd, Co, Pb, and Cr, which apart from being toxic to plants, have multiple negative effects on human, animal, and environmental health (Rashid et al., 2023).

Because the vast majority of raw materials for fertilizer production are extracted from mining waste, they can contain a variety of contaminants. Recently, the concentration of radioactive nucleotides in chemical fertilizers from the West Asian area was evaluated, and it was shown that potassium sulfate, potassium nitrophosphate, and ammonium superphosphate contained a high concentration of radium (^{226}Rr), thorium (^{232}Th), and potassium (^{40}K), which exceed the risk values for developing cancer. Derived from this, it is suggested that this type of raw materials should be restricted or limited for the production of fertilizers in order to minimize the dangerous effects on humans and the environment (Younis et al., 2023).

7.2.4 Impacts on the soil microfauna

Microorganisms play an important role in maintaining soil productivity, since they participate in biochemical processes such as the decomposition of organic matter and the recycling of nutrients (Wu et al., 2020). Unfortunately, derived from excess chemical fertilization, a series of problems originate in the soil, such as the decrease in organic matter, which affects fertility, salinization, acidification, and compaction accelerated, propagation of resistant phytopathogenic microorganisms, accumulation of fertilizer residues that include HMs, and ultimately induces depletion of the numerical and species composition of soil microorganisms (Esikova et al., 2021; Rahman & Singh, 2019; Wu et al., 2020). Particularly, the microbial community is the most important ecological indicator of soil quality. It reflects the state of the biocenosis and its response to various influences such as contamination by toxic substances (Rahman & Singh, 2019).

Over-application of fertilizers influences soil microbial diversity through direct effects as high application rates lead to temporarily very high osmotic potentials and potentially toxic ion concentrations for microorganisms (Geisseler & Scow, 2014). In addition, they can significantly reduce the pH of the soil, which is closely related to the decrease in microbial diversity since most microorganisms are sensitive to acidic pH (Yan et al., 2018).

The impacts of different types of fertilizers on the soil microbial community have been examined in several studies. Zhang et al. (2022) reported an inhibition of the bacterial community of sandy soil by applying ammonia, urea, and phosphoric acid twice a year for a period of 4 years. Yan et al. (2018) observed a significant decrease in microorganisms in temperate forest soil when applying ammonium nitrate for a period of 4 years. Both investigations agree that the observed effect is due to the decrease in soil pH due to the nitrification of ammonium NH_4^+ that produced H^+ ions.

In addition to this, soil microorganisms depend on organic C for energy and cell synthesis. Hence, the excessive application of fertilizers in the soil, by inhibiting the availability of organic C, has a negative impact on the reproduction of microorganisms (Fertahi et al., 2022). Also, as mentioned above, some fertilizers are an important source of HMs for the soil. The accumulation of HMs is known to have a strong negative influence on the microbial habitat, since most natural soil microbial communities, more than any other soil organisms, are sensitive or hypersensitive to HM exposure (Rahman & Singh, 2019).

7.2.5 Conventional fertilizer costs

Chemical fertilizers are the basis for increasing agricultural production since their application is directly linked to crop yields (Zhao et al., 2023). However, the world prices of fertilizers increased drastically at the end of 2021 and the first half of 2022, seriously impacting food systems, poverty, and global food security. This increase in costs resulted in approximately 20–28 million more people being pushed into poverty and hunger worldwide (Arndt et al., 2023).

The increase in world fertilizer prices has been largely driven by the consequences of the war in Ukraine and the sanctions imposed on Russia, in addition to bans on the export of fertilizers, and interruptions in the production and supply chain as a result of the COVID-19 pandemic (Arndt et al., 2023; Komarek et al., 2017). Moreover, some nitrogenous and potassium fertilizers are often not affordable or accessible to farmers in developing countries, as their prices have doubled since the turn of the century due to their production being dominated by Northern Hemisphere-based countries, such as Belarus, Canada, China, Germany, and Russia, making many developing countries almost entirely dependent on imports (Swoboda et al., 2022). Taking into account the high costs of fertilizers, in addition to the problems mentioned by the use of chemical fertilizers, the importance of developing alternative options becomes more evident.

7.3 Nanotechnology and nanomaterials

Nanotechnology is the science responsible for the synthesis, design, characterization, and use of particles, materials, assemblies, tools, and systems by directing the variation of morphology and size at the nanometric level between 1 and 100 nm (Neme et al., 2021). NMs can be inorganic (metals and metal oxides), organic (natural products), and combined in nature (He et al., 2019). Nanotechnology is a multidisciplinary science that is studied and applied in all areas of knowledge ranging from mechanics to medical applications, as well as agriculture (Saleem et al., 2021).

Nanotechnology is considered to have great potential in the different areas used, since materials behave more efficiently at the nanoscale level, mainly due to their high surface/volume ratio and their excellent chemical, physical, and biological properties (Vishwanath & Negi, 2021). One of the main differences between NMs and bulk materials is that in the case of NMs, a large number of atoms are exposed to the surface, which is not possible in the case of bulk materials; furthermore, NMs have better properties in terms of configurations, types, densities, and reactivity, with respect to adsorption and oxidation/reduction reactions (Yadav et al., 2023b).

One of the potential fields in which nanotechnology is being applied is agriculture since it provides sustainable crop management (Saritha et al., 2022). Nanotechnology increases agricultural productivity through various mechanisms, such as the release of pesticides through products based on NMs and NPs, which has been successfully applied in a controlled and specifically customized manner, resulting in a management system for pests and diseases clean, simple and efficient (Periakaruppan et al., 2023). Crop biostimulation is another mechanism to increase agricultural productivity, since the interaction of NMs with plants modifies the activity of different integral proteins and generates changes in the redox balance, energy metabolism, and ion transport, stimulating the activation of the secondary metabolism of plants and inducing resistance to various stressors (Juárez-Maldonado et al., 2021). This can also lead to increased postharvest life due to antimicrobial potential, enzyme mobilization and activation, and decreased degradation that increases the shelf life of crops (Neme et al., 2021). Additionally, the promotion of growth through the use of NF, since these have the capacity for controlled release and directed delivery, and can solve the challenges associated with conventional fertilizers, such as leaching, volatilization, and the inefficient use of nutrients (Ndaba et al., 2022).

7.3.1 Routes of application and transport of nanomaterials in plants

The entry of NP and/or NM into plants occurs mainly through the roots or leaves, however, the forms of entry through direct injection, seed incubation, and administration through bioballistics are other forms of application of NPs that are under study (Saritha et al., 2022).

Through the leaf, the NM entry pathways to the plant are stomatal and cuticular. Uptake by stomata may be limited by size, density, and opening cycles, so cuticular uptake is potentially a more efficient pathway for NM entry due to the larger cuticle area on leaves, compared to the area of the stomata (Lv et al., 2019). After crossing the cuticle, the NMs pass to the tissues of the epidermis and mesophyll, where they interact with these tissues and their structures for their subsequent translocation from the shoot to the root, which is achieved through the vascular systems of the plant (Hubbard et al., 2020).

Through the root, the NMs can penetrate the cell walls, the cytoplasmic membrane, and the Caspari band to enter the vascular system and translocate to the other organs of the plant (He et al., 2021; Jordan et al., 2018). After entering the plant, penetrating the cell wall and plasma membranes of the epidermis, and entering the interior of the vascular tissues, NMs enter the plant along with the uptake of water and other solutes, then move to the stems. and later to the leaves through the process of transpiration (Tripathi et al., 2017).

The transport of NMs in the plant can be via the apoplast, that is, through the cell wall, where the transport of larger particles (approximately 200 nm) is favored. By the symplast pathway, through cell to cell, mediated by plasmodesmata, this pathway favors the transport of smaller particles (less than 50 nm) and through the xylem and phloem (Lv et al., 2019; Schwab et al., 2016) (Fig. 7.1).

7.3.2 Uses of nanomaterials

NMs have been applied in agriculture to increase crop productivity through different uses such as nanosensors (NSs) which are real-time detection devices that are environmentally friendly and less invasive for plants compared to sensors conventional (Zain et al., 2023). NS can be designed to monitor water content, nutrients, and crop health in addition to measuring environmental conditions such as

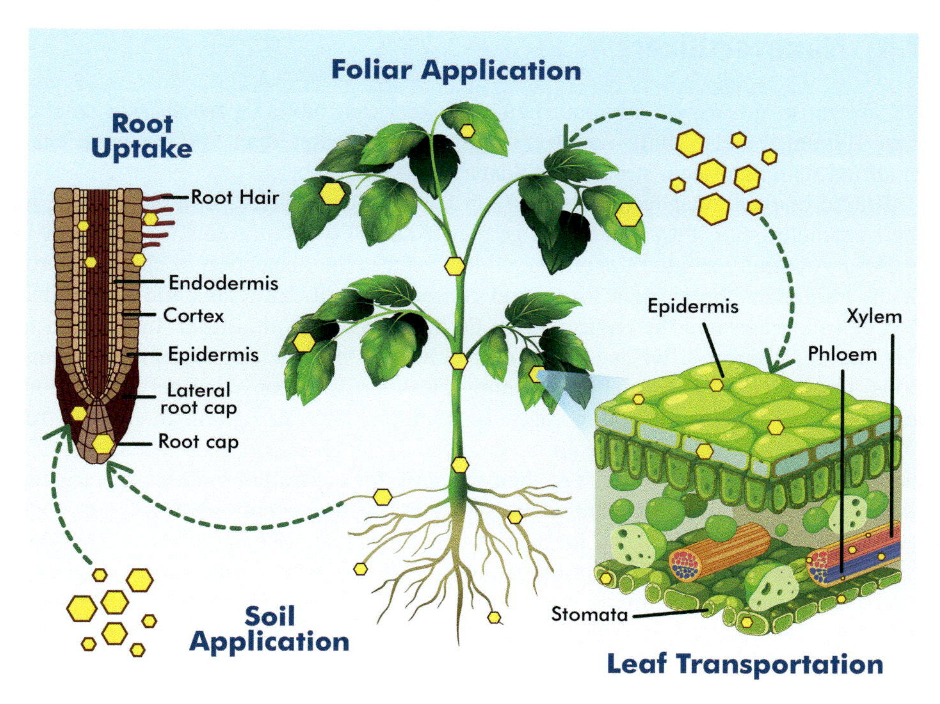

Figure 7.1 Routes of application of nanomaterials, and movement through the plant.

humidity, temperature, and soil nutrient levels, providing key information for crop management through the detection of electrochemical signals that translate chemical reactions into voltages and optical signals that measure changes in fluorescence emission for the quantification of analytes (Voke et al., 2021).

Nanopesticides are another excellent alternative offered by nanotechnology for agricultural applications due to the excellent capacity for targeted release and delivery of active substances with antimicrobial characteristics that optimize the control of plant pests and pathogens (Nisha Raj et al., 2021). Nanopesticides are classified into two types: type I—metal-based nanopesticides that act as active ingredients; and type II—active ingredients that are captured within nanocarriers (polymers, lipids, and clays) (Baliyarsingh & Pradhan, 2023). Both are excellent alternatives to traditional pesticides since most of these do not work properly and are lost through volatilization and leaching, causing them to not reach the target site and be released into the environment (Nisha Raj et al., 2021).

Unlike bulk materials, it has been shown that NMs can be used in plant systems to mitigate various types of stress abiotic such as soil salinity and sodicity, HMs, extreme temperatures, and high irradiance (Dilnawaz et al., 2023). This is due to their ability to act through different mechanisms such as the regulation of ion transport and absorption, increased antioxidant activity, improved water absorption and retention in plants, increased photosynthetic efficiency, among others (Junedi et al., 2023).

7.4 Nanofertilizers

NF are those macro- or micronutrient fertilizers that have a particle size smaller than 100 nm, or else, bulk fertilizers with a size greater than 100 nm have been modified with nanoscale structures (Ndaba et al., 2022).

Unlike conventional fertilizers, NF can increase crop yields, in addition, due to their characteristics, they generate a greater availability of nutrients for plants, it is necessary to apply smaller quantities, which reduces the dispersion of chemical products, minimizes the nutrient losses and consequently, it can reduce the contamination of soil, water and the environment (da Silva Júnior et al., 2022). In addition to this, the application of NF is usually much cheaper than the application of conventional fertilizers, since less labor is required, less fertilizer is needed per application, and there are higher absorption rates. In addition, NF can remain in the soil for long periods without being precipitated or fixed, which makes their use more efficient (Yadav et al., 2023a). Due to their capacity for controlled release and targeted delivery, they can solve the challenges associated with conventional fertilizers, such as leaching, volatilization, and inefficient use of nutrients (Ndaba et al., 2022). This is why NF have the ability to promote sustainable agriculture and improve agricultural yield because they improve efficiency in the use of nutrients both in the open field and in protected agriculture (Zahra et al., 2022).

7.4.1 Uptake of nanofertilizers by plants

NF enter the plant tissue mainly through the roots or tops and their movement is through the xylem and phloem (Seleiman et al., 2021). When NF are applied to the soil or through the nutrient solution, they come into contact with the root system of the plants, and like water and ions, their access to the root interior occurs mainly by simple diffusion (Ndaba et al., 2022). For uptake and translocation to occur, NF must cross some root physiological barriers, including the root surface cuticle, epidermis, cortex, endodermis, and Caspari band to reach the xylem vessels and finally transported to the organs of the plant (Lv et al., 2019).

Most NFs can be absorbed by plant roots, however, the root zone of plants generally has a negative charge, so NFs, like NP and NM with a positive surface charge, are more likely to be absorbed and accumulate on the root surface (Juárez-Maldonado et al., 2021).

Interacting with the root epidermis, the main route of absorption is the apoplast pathway in which NFs first penetrate the cell wall pores and then diffuse into the space between the cell wall and the plasma membrane or cross the intercellular space without crossing the cell membrane (Lv et al., 2019). However, the diameter of the cell wall pores, which usually ranges between 5 and 20 nm, restricts the passage of those NFs that have a diameter greater than 20 nm. However, there is evidence that some materials generate a slight degradation of the cell membrane to increase the size of the pores and some induce the generation of new pores in the cell wall, which improves absorption (Seleiman et al., 2021). Another absorption route is the symplast pathway that involves the transport of water, and solutes between the cytoplasm of adjacent cells that are linked by pores of sieve plates and plasmodesmata, which facilitate the entry of NF with a diameter of between 20 and 50 nm. Through cell-to-cell movements (Ndaba et al., 2022). In some cases, NF undergoes endocytosis, so they enter the cell through the folding of the outer layer of the plasma membrane to form a kind of sac and are transported to different compartments of the cell. They can also bind to some proteins that act as transporters for internalization and uptake within the plant cell (Ndaba et al., 2022; Wong et al., 2016).

When the application of NF is carried out by the foliar route, the first contact of these is with the cuticle of the leaves, which has two entry points, that is, the lipophilic or cuticular route, and the hydrophilic or stomatal route (Seleiman et al., 2021). By the lipophilic pathway, NF and nonpolar solutes can enter the leaves through diffusion, while the hydrophilic pathway is for NF and polar solutes, so that NF or its small-sized aggregates can easily enter through the diffusion cuticular pathway, whereas larger and polar NFs can enter through the hydrophilic or stomatal pathway (Lv et al., 2019; Seleiman et al., 2021). Once inside the plant, NF are transported through the vascular system, however, since plant vascular systems are unidirectional and noncirculatory, the NF applied by the foliar route only has the option of transport through the phloem for the absorption and translocation of these toward the different plant organs (Avellan et al., 2019).

7.4.2 Classes of nanofertilizers

The NF is the product of a new technology, which is why their classification system is complex, so there are several classes depending on specific characteristics. In accordance with Tarafder et al. (2020) and Zahra et al. (2022), NF can be classified based on their preparation method into three groups: (1) Nanoscale fertilizers, which refers to macro and microelements smaller than 100 nm in size; (2) nanoscale additive fertilizer, which are bulk fertilizers added with one or several NMs; and (3) nanoscale coating fertilizer, bulk or nanoscale nutrients that are covered by NMs or are introduced into the nanoscale pores of a host material. But other classifications can be based on their mechanism of action, nutrient composition, or based on the consistency of the materials used in their preparation. Derived from this, different classes of NF are mentioned as: controlled release NF, NF for targeted delivery, NF that stimulates plant growth, NF that controls water and nutrient loss, inorganic and organic NF, hybrid NF, NF nutrient-loaded, coated NF, nanocarrier-based NF, and consistency-based NF (Yadav et al., 2023a).

However, the current diversity of NF is so great that the proposed classifications do not include all types of NF or are very complex, so it becomes necessary to classify them in a simpler way that allows including all available NF up to the moment. Considering this, NFs can be classified into two large groups considering the size of the nutrient, which can be bulk or nanosized (less than 100 nm). When the nutrient is at nano size, i.e., nanonutrient, it can be applied as single or multielement, or accompanied by other materials to improve its efficiency, which can be organic or inorganic materials (e.g., chitosan and zeolites), or even NMs (e.g., graphene, fullerene, carbon

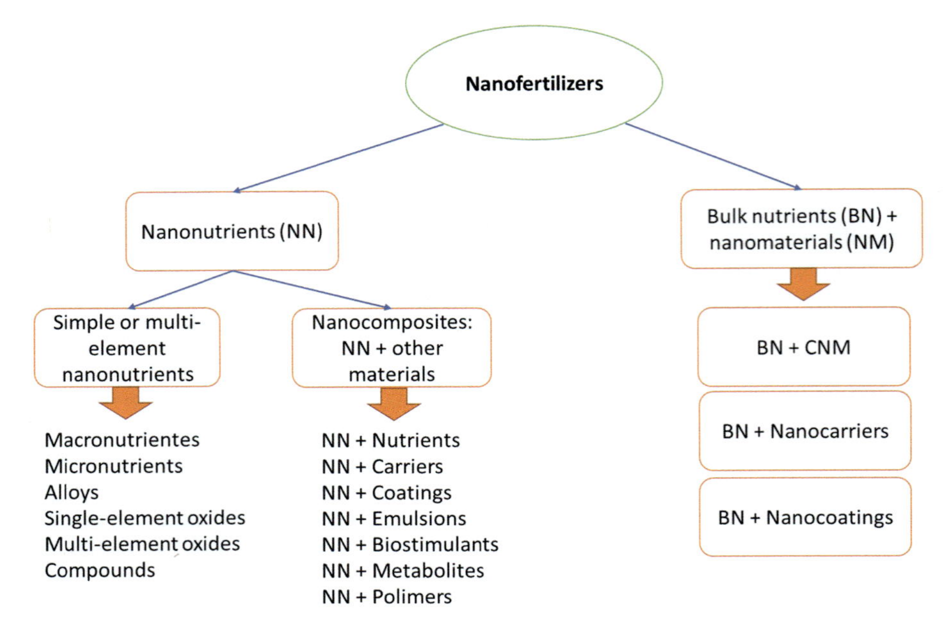

Figure 7.2 Classification of nanofertilizers.

nanotubes, nanochitosan, and nanozeolite). In the case of bulk nutrients, they are accompanied by NMs that can be of organic origin (carbon NMs, nanochitosan, and nanostarch) or inorganic (nanozeolite and nanopolymers) (Fig. 7.2).

7.4.2.1 Controlled release and targeted delivery nanofertilizers

Among the main problems of conventional fertilizers are leaching and volatilization, which results in low efficiency in their use, which is why NF designed with controlled release capacity and directed delivery can solve these problems (Ndaba et al., 2022). This is because NF can be modified in size, shape, chemical composition and surface, in addition to incorporating various organic and inorganic compounds, depending on the needs (Zahra et al., 2022). Controlled-release NF encapsulate nutrients within materials known as carriers or vehicles, and nutrient release occurs when the shell ruptures upon contact with environmental factors such as temperature, pH, and humidity, or by mechanisms of response to stimuli, such as biodegradation or enzyme-mediated degradation (Yadav et al., 2023a) (Fig. 7.3).

Several types of materials have been used as nutrient carriers for crop fertilization, among these are chitosan and zeolites that can be used in bulk or on a nano scale, or some carbon-based NMs such as graphene and nanotubes.

Generally, polymeric materials such as chitosan are used as carriers for fertilizers because they are biodegradable, so they do not represent a risk to the environment (Guo et al., 2018). Chitosan exhibits high film-forming ability with excellent mechanical, antimicrobial, and homeostatic properties (Mohamed & Madian, 2020),

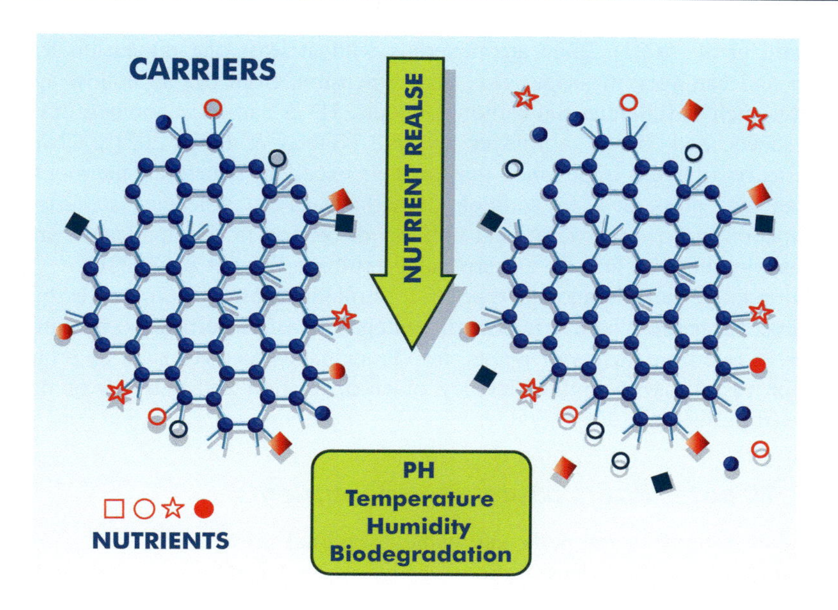

Figure 7.3 Representation of nutrient nanocarriers and the release process due to different factors.

and it has been proven to be an excellent vehicle for the controlled release of fertilizers (Boamah et al., 2023; El-Ganainy et al., 2023; Kongala & Mamidala, 2023).

Zeolites are a type of microporous aluminosilicate minerals that have been used for decades as a soil amendment due to their ability to absorb and retain water, nutrients, and other organic compounds. These occur naturally or can be chemically synthesized, so their porosity can be adjusted to the nanoscale, depending on the application requirements (Sharma et al., 2022; Yadav et al., 2023a). The best-known zeolites that are naturally available include clinoptilolite, mordenite (MOR), erionite, phillipsite, analcime, lind type A (LTA), chabazite (CHA), beta-structured (BEA), sodalite (SOD) among others, which have different pore sizes, ion exchange properties, and bulk densities (Sharma et al., 2022). Synthetic zeolites are prepared using different methodologies, modifying their size, shape, and physicochemical properties, so that they can act as carriers and reservoirs to store and release nutrients over time, providing a more constant supply of these to the plants (Król, 2020; Sharma et al., 2022). In addition, zeolite-based NF can be tailored to the specific nutrient needs of a crop, ensuring that only the necessary nutrients are provided, increasing fertilization efficiency, and reducing cost (Sharma et al., 2022). The application of zeolites can be done through the soil or through the foliar route, and their retention and nutrient capacity depends on factors such as the zeolite structure, the Si/Al ratio, the contact time, the temperature, the pore size and concentration of other added ions (Król, 2020; Sharma et al., 2022).

Carbon-based nanomaterials (CNMs) have the potential to be derived from natural organic raw materials and can therefore be considered eco-friendly

(Goswami et al., 2021). They are materials with at least one dimension less than 100 nm and can vary in shape, size, and dimension, from 0D as hollow spherical structures such as fullerene and carbon nanodots, 1D as carbon nanotubes, 2D as graphene sheets, and 3D as multilayer graphite (Gabris & Ping, 2021). CNM have proven to be of great importance due to their excellent structural, thermophysical, and chemical properties. In addition, they have great molecular stability, good biocompatibility, lower toxicity, and uniform dispersion in the application medium, which is why they can be used as carriers of fertilizers (Zhu et al., 2022).

Additionally, there is another type of controlled-release NF, known as bionanofertilizers. These are characterized by the integration of biofertilizers, prepared with extracts of microorganisms or plants, in NPs or nanostructures such as proteoliposomes or vesicles with high efficiency of encapsulation and delivery of nutrients (Rios et al., 2018).

7.4.3 Nanofertilizers based on nanonutrients

For a plant to develop properly and achieve optimal growth and yield, it requires the addition of specific concentrations of essential nutrients, that is, macronutrients (Ca, K, Mg, Na, P, and S) and micronutrients (Mn, Fe, Ni, Mo, Cu, Zn, B, and Cl) (Veazie et al., 2020). As already mentioned, the inappropriate use of fertilizers that contain these elements is one of the main drivers of environmental devastation, mainly due to their low absorption efficiency by the plant, and their waste through volatilization and leaching (Zahra et al., 2022).

NF cause an increase in nutrient use efficiency and crop yields, reduce soil toxicity, and minimize potential negative effects associated with overdosing (Abdulhameed et al., 2021). Various macro and microelements at the nanometric scale have been evaluated and their efficiency on the performance of various crops has been verified (Table 7.1). Abd El-Azeim et al. (2020) evaluated the efficiency of foliar-applied N, P, and K NPs in the cultivation of *Solanum tuberosum* L. and demonstrated that a greater absorption of N, P, and K was achieved, as well as that the yield parameters of the crop were improved with the application of 50% of N, P, and K NF, that is, half the recommended level of NPK in bulk. Abdulhameed et al. (2021) applied N, P, and K NF to the soil at concentrations of 100, 48, and 30 g of N, P, and K, respectively, and reported an increase in yield and nutrient uptake in *Brassica oleracea* L. plants compared to conventional fertilizers. Zhu et al. (2023) applied 0.5 g L^{-1} of $CaCO_3$ NP to *Prunus persica* L. whit foliar route and observed a higher concentration in the calcium content in the leaf and fruit; in addition, it also increased fruit quality due to the decrease in the Fruit-cracking percentage, compared to bulk calcium application. Du et al. (2019) applied 20 mg L^{-1} of ZnO NP to the *Triticum aestivum* L. plant and reported an increase in yield and grain quality, as well as better assimilation of the micronutrient. Yujing Wang et al. (2019) reported an increase in fresh and dry biomass and the net photosynthesis rate in *Spinacia oleracea* L. leaves with foliar application of 20 mg kg^{-1} Cu NPs. Sabet and Mortazaeinezhad (2018) evaluated the foliar application of FeO_3 NP at a concentration of 1000 mg L^{-1} in *Cuminum cyminum* L. plants and reported an increase in yield and leaf nitrogen efficiency.

Table 7.1 Impact of nanonutrient-based nanofertilizers on different crops.

Group	Nanofertilizer	Concentration	Application route	Vegetal specie	Effects	References
Elements and alloys	N-P-K NPs (hyper feed amino NPs)	2.5 kg ha^{-1}	Sprayed	Wheat (*Triticum aestivum* L.)	Increased spike length, number of grains per spike, weight, straw yield, grain yield.	Abdelsalam et al. (2019)
	P NPs	12.5, 100, and 1000 $\mu g\ mL^{-1}$	Drench	Tomato (*Solanum Lycopersicum* L.)	Increased phosphorus uptake in the plant and increases germination, biomass, and plant growth.	Priyam et al. (2022)
	Ca NPs	0.5 g L^{-1}	Sprayed	Nectarine (*Prunus persica* L.)	Increased the Ca content in peel and leaves, and the calcium pectinate in peel.	Zhu et al. (2023)
	K NPs	20 mg L^{-1}	Sprayed	Wheat (*Triticum aestivum* L.)	Increase in fresh and dry biomass of the plant and root, as well as grain yield and photosynthetic pigments in the leaf.	Sheoran et al. (2021)
	Cu NPs	100 mg L^{-1}	Drench	Strawberry (*Fragaria* × *ananassa*)	Increased the content of Cu, P, K, Mn, and Zn in the leaf.	El-Bialy et al. (2023)
	B NPs	50 mg L^{-1}	Sprayed	Bean (*Phaseolus vulgaris* L.)	Increased fresh and dry biomass, grain yield, and nitrate reductase activity in the leaf.	Márquez-Prieto et al. (2022)

(Continued)

Table 7.1 (Continued)

Group	Nanofertilizer	Concentration	Application route	Vegetal specie	Effects	References
Single-element oxides	ZnO NPs	17 mg L^{-1}	Drench	Lettuce (*Lactuca sativa* L.)	Increased the content of K, Ca, and S in the leaf.	Garza-Alonso et al. (2023)
	CuO NPs	75 mg kg^{-1}	Drench	Sweet potato (*Ipomoea batatas* L.)	Increased yield and carbohydrate accumulation in the tuber.	Bonilla-Bird et al. (2020)
	Fe_2O_3 NPs	25 and 50 mg L^{-1}	Sprayed	Maize (*Zea mays* L.)	Increase nitrate content, and soluble sugars in the grain.	Hasan et al. (2020)
	$CaCO_3$ NPs	5 g L^{-1}	Sprayed	Sweet pepper (*Capsicum annuum* L.)	Increased yield and absorption of N, P, K, Mg, Fe, Zn, Mn, and Cu.	Vidak et al. (2021)
Multielement oxides	Manganese zinc ferrite NPs ($Mn_{0.5}Zn_{0.5}Fe_2O_4$)	20 and 30 mg L^{-1}	Sprayed	Squash (*Cucurbita pepo* L.)	Increased the growth, and content of N, Zn, Fe, and Mn in the fruit.	Shebl et al. (2020)
	Nanohydroxyapatite ($Ca_{10}(PO_4)_6(OH)_2$)	50, 100, 500 and 1000 mg L^{-1}	Sprayed	Pomegranate (*Punica granatum* L.)	Increased the fruit yield, the content of photosynthetic pigments, and carbohydrates in the fruit.	Abdelmigid et al. (2022)

7.4.3.1 Nanofertilizers based on Nanonutrients accompanied by other materials

Some nanonutrients have been applied in conjunction with some other materials, which can be nano-sized or also bulk materials (Table 7.2). It has been common to find the use of hydroxyapatite for the application of nutrients in the literature (Kottegoda et al., 2017; Le et al., 2022; Raguraj et al., 2020); however, other materials such as zeolite (Manikandan & Subramanian, 2016), and CNMs such as graphene (Kabiri et al., 2017), fullerene (Bityutskii et al., 2020), and carbon nanofibers (Ashfaq et al., 2017), have also been used, all with favorable effects on crops. In general, improvements have been observed in the agronomic parameters of the crops, as well as in biomass production and fruit yield. And the use of this type of NF has been evaluated in a variety of crops such as *Polyscias fruticosa*, *Oryza sativa*, *Zea mays*, *Triticum durum*, *Prunus dulcis*, *B. oleracea*, *B. rapa*, *Camellia sinensis*, *Cucumis sativus*, and *Cicer arietinum* (Table 7.2).

7.4.3.1.1 Hydroxyapatite-based nanofertilizers

NF that contain nanohydroxyapatite $Ca_{10}(PO_4)_6(OH)_2$ (nHA) are excellent candidates for use in agronomic applications because they are also a source of calcium and phosphorus and are widely recognized for their biocompatibility and biodegradability (Alamri et al., 2022; Marchiol et al., 2019). The nHA nanohybrids synthesized from ($Ca_3P_2O_8$), have been used successfully for the gradual release of phosphorus (P) and promote the growth of plants of *Punica granatum* L. (Abdelmigid et al., 2022). Mikhak et al. (2017) evaluated the effect of the combination of (nHA) saturated with bulk ammonium sulfate (NH_4^+) and observed that the nHA obtained better phosphorus solubility compared to the commercial triple superphosphate fertilizer. In addition, they evaluated *Matricaria chamomilla* L. plants and reported the increase in fresh and dry biomass, number of branches and flowers, in addition to the increase in the phosphorus content of roots and shoots. Le et al. (2022) applied 0.2% of nHA as a carrier of the commercial fertilizer NPK (15–15-15) bulk sprayed to the leaves and soil around the plant *P. fruticosa* L., and reported an increase in fresh and dry biomass, as well as plant yield.

Ha et al. (2019) evaluated chitosan NPs as KNO_3 NPs carriers and observed a relative release of 10% N and 50% K in the first 48 hours. After 240 hours, the release of nutrients was 66% and 58% of N and K, respectively. Additionally, they report an increase in yield and nutrient content in *Coffea arabica* plants with the application of nanofertilizer foliar.

Manikandan and Subramanian (2016) applied a concentration of 25 kg ha^{-1} of nanozeolite/nanourea (1:1) to the soil for their maize plant cultivation. They reported a higher nitrogen content in the grain, as well as an improvement in root length, dry matter production, and spike.

7.4.3.1.2 Nanofertilizers based on carbon nanomaterials

Graphene oxide (GO) NPs have been used as carriers of NOs of microelements such as Zn and Cu, which showed a greater dissolution and release of microelements in the soil and an improvement in grain yield and assimilation of elements in

Table 7.2 Impact of nanonutrient-based nanofertilizers accompanied by other materials on different crops.

Nanofertilizer	Concentration	Application route	Vegetal specie	Effects	References
Nanohydroxyapatite $Ca_{10}(PO_4)_6(OH)_2$/ NPK NPs	0.2%	Sprayed to the leaves and drenched around the plant	Aralia (*Polyscias fruticosa* L.)	Increased number of branches/ plant, branch length dry matter production, leaf area, and leaf length.	Le et al. (2022)
KH_2PO_4 NPs covered with nanopolymer gel	0.1563 g L^{-1}	Nutrient solution	Rice (*Oryza sativa* L.)	Increased the accumulation of P in shoots and roots. Increased the photosynthetic rate and water-efficient use in plants.	Miranda-Villagómez et al. (2019)
Nanozeolite/nanourea	25 kg ha^{-1}	Drench	Maize (*Z. mays* L.)	Increase the N content in the grain as well as an improvement in the length of the root, production of dry matter, and spike.	Manikandan and Subramanian (2016)
Graphene oxide (GO)/ Zn NPs nano-GO/Cu NPs	20 mg L^{-1}	Drench	Durum wheat (*Triticum durum* L.)	Increased grain yield, promoted the absorption and accumulation of Zn and Cu.	Kabiri et al. (2017)
B (H_3BO_3) and Fe ($FeSO_4$) NPs into nanovegetal vesicles	B (2%) Fe (4%)	Sprayed	Almond (*Prunus dulcisl*)	Increased the concentration of Fe and B in the leaves.	Rios et al. (2020)
Zn NPs into nanovegetal vesicles	500 μM	Sprayed	Broccoli (*Brassica oleracea*) and Pak choi (*Brassica rapa*)	Increase in the concentration of Zn in the leaves.	Rios et al. (2018)

Nanohydroxyapatite ($Ca_{10}(PO_4)_6(OH)_2$/ nanourea)	130 mg L^{-1}	Sprayed	Rice (*O. sativa* L.)	Increased height and NPK content in leaves and grain.	Kottegoda et al. (2017)
Nanohydroxyapatite ($Ca_{10}(PO_4)_6(OH)_2$/ nanourea)	40% N and 6% P_2O_5	Drench	Tea (*Camellia sinensis* L.)	Increased yield and the efficiency of nitrogen and phosphorus in the plant.	Raguraj et al. (2020)
Nanofullerene ($C_{60}(OH)_{22-24}$/$FeSO_4$ NPs	10 mg L^{-1}	Sprayed	Cucumber (*Cucumis sativus* L.)	Suppressed plant Fe deficiency by promoting foliar penetration. Increased Fe accumulation	Bityutskii et al. (2020)
Carbon nanofiber/Cu NPs	10 and 500 μg mL^{-1}	Seed priming	Chickpea (*Cicer arietinum*)	Increased the efficiency and accumulation of Cu in shoots and roots. Increased germination percentage and biomass accumulation.	Ashfaq et al. (2017)
Nanochitosan as carrier of nano-KNO_3	Cs NPs (1%)/ KNO_3 Nps (50 mg L^{-1})	Sprayed	Coffee (*Coffea arabica*)	Increased yield and nutrient content.	Ha et al. (2019)

T. durum plants in comparison with commercial zinc sulfate and copper sulfate fertilizer granules (Kabiri et al., 2017). Bityutskii et al. (2020) cultivated *C. sativus* plants in the absence of Fe and applied foliar NPs of fullerene ($C_{60}OH_{22-24}$) as a carrier of $FeSO_4$ NPs and demonstrated that the penetration of Fe was promoted and the active Fe in the leaves was increased; in addition, they successfully suppressed the plant's Fe deficiency symptoms. Ashfaq et al. (2017) used carbon nanofibers as carriers of Cu NPs (synthesized from Cu $(NO_3)_2$ $3H_2O$) and applied it by seed priming in *C. arietinum*, observing an increase in the germination rate and a greater length of shoots and roots. They also report the controlled release of Cu in aqueous solution, approximately 30% of Cu impregnated in the nanofibers was released within 6 hours after application, subsequently releasing approximately 45% in 15 days.

Rios et al. (2020) studied the potential of membrane vesicles derived from plant material as encapsulating NF of B (H_3BO_3) and Fe ($FeSO_4$) NPs for foliar application in *P. dulcis* and demonstrated an increase of 20% and 87% in the content of Fe and B with respect to plants treated with conventional fertilizers. Rios et al. (2018) applied foliar route to *B. oleracea* and *B. rapa* plants with vesicles containing a 500 μM solution of $ZnSO_4$ NPs which were previously cultivated in the absence of Zn and observed an increase in the foliar Zn concentration almost four times compared to the application of the nutrient in the form of salt for both species.

7.4.4 Nanofertilizers based on bulk nutrients plus nanomaterials

NF based on bulk nutrients accompanied by some NM are less common than those based on nanonutrients; however, they have also proven to be efficient and, above all, have positive effects on the crops evaluated (Table 7.3). Nanochitosan-based NF are probably the most common in this group, mainly due to their wide range of properties, which include biocompatibility, biodegradability, high permeability, cost-effectiveness, nontoxicity, and excellent ability to form films, hydrogels, scaffolds, fibers, and microparticles or NPs. This is due to the deacetylated surfaces that make it easier to handle for many desired modifications (Mujtaba et al., 2020).

Dhlamini et al. (2020) reported a more efficient release of a commercial NPK fertilizer (20:7:3) loaded with chitosan nanoparticles (Cs NPs) (Cs NPs 1%/NPK 40 mg L^{-1}) in vitro conditions; after 24 hours, they were released only 6.3% N, 11.6% P, and 15.1% K compared to fertilizers without Cs NPs that exhibited 89%, 92%, 91% release of N, P, and K, respectively, in the first 12 hours. In addition, it was found that the sum of the nutrient release was less than 25% after 48 hours, and did not exceed 60% after one week. In the greenhouse, the authors observed that the nanoformulated applied to the soil induced *Z. mays* plants with greater height, number of leaves, stem diameter, and chlorophyll content at 10 weeks after sowing, compared to the NPK fertilizer. Kubavat et al. (2020) applied to the commercial fertilizer $K_2S_2O_8$ loaded with Cs NPs and reported a significant increase in the accumulation of fresh and dry biomass of *Z. mays*, and also improved the physical properties of the soil through greater porosity and greater water conductivity. Abd-Elrahman et al. (2023) applied foliar to *Fragaria* × *ananassa* plants nano-Cs

Table 7.3 Impact of nanofertilizers based on bulk nutrients plus nanomaterials on different plant crops.

Nanofertilizer	Concentration	Application route	Vegetal specie	Effects	References
Biochar NPs as carrier of calcium nitrate tetrahydrate ($CaH_8N_2O_{10}$), sodium dihydrogen phosphate dihydrate ($NaH_2PO_4 \cdot 2H_2O$) and potassium sulfate (K_2SO_4) bulk	0.5, 2 and 2 g L^{-1}	Sprayed	Chinese cabbage (*Brassica rapa* L.)	Increased height, stem diameter, fresh and dry biomass.	Yang et al. (2022)
Nanochitosan as a carrier of potassium sulfate (K_2SO_4)bulk	750 and 1000 mg L^{-1}	Soil	Strawberry (*Fragaria* × *ananassa*)	Increased yield and quality parameters of the fruit, total soluble solids TSS, vitamin C, and anthocyanins.	Abd-Elrahman et al. (2023)
Nanozeolite with ammonium sulfate (NH_4^+) bulk	19%	Soil	Chamomile (*Matricaria chamomilla* L.)	Increased fresh and dry biomass, number of flower branches, P content of roots and shoots.	Mikhak et al. (2017)
Nanochitosan as a carrier of commercial fertilizer (NPK, 20:7:3) bulk	40 mg Kg^{-1} of soil	Soil	Maize (*Z. mays*)	Height increase, number of leaves, stem diameter, and photosynthetic pigments.	Dhlamini et al. (2020)
Nanochitosan as carrier of a commercial fertilizer $K_2S_2O_8$ bulk	Cs NPs (0.8%)/ $K_2S_2O_8$ (75%) y	Soil	Maize (Z. mays)	Increase fresh and dry biomass of plants.	Kubavat et al. (2020)
Nanochitosan as carrier of nano-KNO_3	Cs NPs (1%)/KNO_3 Nps (50 mg L^{-1})	Sprayed	Coffee (*C. arabica*)	Increased yield and nutrient content.	Ha et al. (2019)

as a carrier of potassium sulfate K_2SO_4 and reported an increase in vegetative growth parameters, such as plant height, fresh and dry weight, number of leaves, and leaf area per plant.

7.5 Conclusions

The use of NMs and NPs as NF has effectively proven to be a very useful tool to face the great challenges that exist in the nutritional management of agricultural crops. Clearly, the physical and chemical properties of NMs provide significant advantages compared to traditional bulk-applied fertilizers. In the first instance, NF induce positive responses due to the ability of NMs to elicit plants and produce secondary metabolites that will give them the ability to tolerate various stress conditions. However, in addition, NF have proven to be more efficient in their use by increasing the absorption and assimilation capacity of plants, resulting in a lower need to apply fertilizers. This translates directly into savings in the amount of fertilizer applied, making the nutritional management of crops more efficient, and also, reducing the adverse effects derived from excessive applications of traditional fertilizers.

References

Abd El-Azeim, M. M., Sherif, M. A., Hussien, M. S., Tantawy, I. A. A., & Bashandy, S. O. (2020). Impacts of nano- and non-nanofertilizers on potato quality and productivity. *Acta Ecological Sinica*, *40*, 388–397.

Abdelmigid, H. M., Morsi, M. M., Hussien, N. A., Alyamani, A. A., Alhuthal, N. A., & Albukhaty, S. (2022). Green synthesis of phosphorous-containing hydroxyapatite nanoparticles (nHAP) as a novel nano-fertilizer: Preliminary assessment on pomegranate (*Punica granatum* L.). *Nanomaterials*, *12*, 1527.

Abd-Elrahman, S. H., El-Gabry, Y. A. E.-G., Hashem, F. A., Ibrahim, M. F. M., El-Hallous, E. I., Abbas, Z. K., Darwish, D. B. E., Al-Harbi, N. A., Al-Qahtani, S. M., & Taha, N. M. (2023). Influence of nano-chitosan loaded with potassium on potassium fractionation in sandy soil and strawberry productivity and quality. *Agronomy*, *13*, 1126.

Abdelsalam, N. R., Kandil, E. E., Al-Msari, M. A. F., Al-Jaddadi, M. A. M., Ali, H. M., Salem, M. Z. M., & Elshikh, M. S. (2019). Effect of foliar application of NPK nanoparticle fertilization on yield and genotoxicity in wheat (*Triticum aestivum* L.). *The Science of the Total Environment*, *653*, 1128–1139.

Abdulhameed, M. F., Taha, A. A., & Ismail, R. A. (2021). Improvement of cabbage growth and yield by nanofertilizers and nanoparticles. *Environmental Nanotechnology, Monitoring & Management.*, *15*100437.

Akinnawo, S. O. (2023). Eutrophication: Causes, consequences, physical, chemical and biological techniques for mitigation strategies. *Environmental Challenges.*, *12*100733.

Alamri, S., Nafady, N. A., El-Sagheer, A. M., El-Aal, M. A., Mostafa, Y. S., Hashem, M., & Hassan, E. A. (2022). Current utility of arbuscular mycorrhizal fungi and hydroxyapatite nanoparticles in suppression of tomato root-knot nematode. *Agronomy*, *12*, 671.

Albert, S., & Bloem, E. (2023). Ecotoxicological methods to evaluate the toxicity of bio-based fertilizer application to agricultural soils – A review. *The Science of the Total Environment*, *879*163076.

Alzain, M. N., Loutfy, N., & Aboelkassem, A. (2023). Effects of different kinds of fertilizers on the vegetative growth, antioxidative defense system and mineral properties of sunflower plants. *Sustainability*, *15*, 10072.

Arndt, C., Diao, X., Dorosh, P., Pauw, K., & Thurlow, J. (2023). The Ukraine war and rising commodity prices: Implications for developing countries. *Global Food Security*, *36*100680.

Ashfaq, M., Verma, N., & Khan, S. (2017). Carbon nanofibers as a micronutrient carrier in plants: Efficient translocation and controlled release of Cu nanoparticles. *Environmental Science Nano.*, *4*, 138–148.

Avellan, A., Yun, J., Zhang, Y., Spielman-Sun, E., Unrine, J. M., Thieme, J., Li, J., Lombi, E., Bland, G., & Lowry, G. V. (2019). Nanoparticle size and coating chemistry control foliar uptake pathways, translocation, and leaf-to-rhizosphere transport in wheat. *ACS Nano*, *13*, 5291–5305.

Baliyarsingh, B., & Pradhan, C. K. (2023). Prospects of plant-derived metallic nanopesticides against storage pests - A review. *Journal of Agriculture and Food Research*100687. Available from https://doi.org/10.1016/j.jafr.2023.100687.

Bhardwaj, A. K., Arya, G., Kumar, R., Hamed, L., Pirasteh-Anosheh, H., Jasrotia, P., Kashyap, P. L., & Singh, G. P. (2022). Switching to nanonutrients for sustaining agroecosystems and environment: The challenges and benefits in moving up from ionic to particle feeding. *Journal of Nanobiotechnology*, *20*19.

Bityutskii, N. P., Yakkonen, K. L., Lukina, K. A., & Semenov, K. N. (2020). Fullerenol increases effectiveness of foliar iron fertilization in iron-deficient cucumber. *PLoS One*, *15*, e0232765.

Boamah, P. O., Onumah, J., Aduguba, W. O., & Santo, K. G. (2023). Application of depolymerized chitosan in crop production: A review. *International Journal of Biological Macromolecules*, *235*123858.

Bonilla-Bird, N. J., Ye, Y., Akter, T., Valdes-Bracamontes, C., Darrouzet-Nardi, A. J., Saupe, G. B., Flores-Marges, J. P., Ma, L., Hernandez-Viezcas, J. A., Peralta-Videa, J. R., & Gardea-Torresdey, J. L. (2020). Effect of copper oxide nanoparticles on two varieties of sweetpotato plants. *Plant Physiology and Biochemistry: PPB/Societe Francaise de Physiologie Vegetale*, *154*, 277–286.

Carrasco-Correa, E. J., Mompó-Roselló, Ò., & Simó-Alfonso, E. F. (2023). Calcium oxide nanofertilizer as alternative to common calcium products for the improvement of the amount of peel fruit calcium. *Environmental Technology Innovation*, *31*103180.

da Silva Júnior, A. H., Mulinari, J., de Oliveira, P. V., de Oliveira, C. R. S., & Reichert Júnior, F. W. (2022). Impacts of metallic nanoparticles application on the agricultural soils microbiota. *Journal of Hazardous Materials Advance*, *7*100103.

Dhlamini, B., Paumo, H. K., Katata-Seru, L., & Kutu, F. R. (2020). Sulphate-supplemented NPK nanofertilizer and its effect on maize growth. *Materials Research Express*, *7*, 095011.

Dilnawaz, F., Misra, A. N., & Apostolova, E. (2023). Involvement of nanoparticles in mitigating plant's abiotic stress. *Plant Stress*100280. Available from https://doi.org/10.1016/j.stress.2023.100280.

Du, W., Yang, J., Peng, Q., Liang, X., & Mao, H. (2019). Chemosphere comparison study of zinc nanoparticles and zinc sulphate on wheat growth : From toxicity and zinc bioforti fi cation. *Chemosphere*, *227*, 109–116.

El-Bialy, S. M., El-Mahrouk, M. E., Elesawy, T., Omara, A. E.-D., Elbehiry, F., El-Ramady, H., Áron, B., Prokisch, J., Brevik, E. C., & Solberg, S. Ø. (2023). Biological nanofertilizers to enhance growth potential of strawberry seedlings by boosting photosynthetic pigments, plant enzymatic antioxidants, and nutritional status. *Plants*, *12*, 302.

El-Ganainy, S. M., Soliman, A. M., Ismail, A. M., Sattar, M. N., Farroh, K. Y., & Shafie, R. M. (2023). Antiviral activity of chitosan nanoparticles and chitosan silver nanocomposites against alfalfa mosaic virus. *Polymers (Basel)*, *15*, 2961.

Esikova, T. Z., Anokhina, T. O., Abashina, T. N., Suzina, N. E., & Solyanikova, I. P. (2021). Characterization of soil bacteria with potential to degrade benzoate and antagonistic to fungal and bacterial phytopathogens. *Microorganisms*, *9*, 755.

Etesami, H., Fatemi, H., & Rizwan, M. (2021). Interactions of nanoparticles and salinity stress at physiological, biochemical and molecular levels in plants: A review. *Ecotoxicology and Environmental Safety*, *225*112769.

Feregrino-Perez, A. A., Magaña-López, E., Guzmán, C., & Esquivel, K. (2018). A general overview of the benefits and possible negative effects of the nanotechnology in horticulture. *Scientia Horticulturae*, *238*, 126–137.

Fertahi, S., Pistocchi, C., Daudin, G., Amjoud, M., Oukarroum, A., Zeroual, Y., Barakat, A., & Bertrand, I. (2022). Experimental dissolution of biopolymer-coated phosphorus fertilizers applied to a soil surface: Impact on soil pH and P dynamics. *Annals of Agricultural Science.*, *67*, 189–195.

Gabris, M. A., & Ping, J. (2021). Carbon nanomaterial-based nanogenerators for harvesting energy from environment. *Nano Energy*, *90*106494.

Gade, A., Ingle, P., Nimbalkar, U., Rai, M., Raut, R., Vedpathak, M., Jagtap, P., & Abd-elsalam, K. A. (2023). Nanofertilizers : The next generation of agrochemicals for long-term impact on sustainability in farming systems. *Agrochemicals*, *2*, 257–278.

Garza-Alonso, C. A., Juárez-Maldonado, A., González-Morales, S., Cabrera-De la Fuente, M., Cadenas-Pliego, G., Morales-Díaz, A. B., Trejo-Téllez, L. I., Tortella, G., & Benavides-Mendoza, A. (2023). ZnO nanoparticles as potential fertilizer and biostimulant for lettuce. *Heliyon*, *9*e12787.

Geisseler, D., & Scow, K. M. (2014). Long-term effects of mineral fertilizers on soil microorganisms – A review. *Soil Biology & Biochemistry*, *75*, 54–63.

Goswami, A. D., Trivedi, D. H., Jadhav, N. L., & Pinjari, D. V. (2021). Sustainable and green synthesis of carbon nanomaterials: A review. *Journal of Environmental Chemistry & Engineering.*, *9*106118.

Guelfi, D., Nunes, A. P. P., Sarkis, L. F., & Oliveira, D. P. (2022). Innovative phosphate fertilizer technologies to improve phosphorus use efficiency in agriculture. *Sustainability*, *14*, 14266.

Guo, H., White, J. C., Wang, Z., & Xing, B. (2018). Nano-enabled fertilizers to control the release and use efficiency of nutrients. *Current Opinion in Environmental Science & Health*, *6*, 77–83.

Ha, N. M. C., Nguyen, T. H., Wang, S.-L., & Nguyen, A. D. (2019). Preparation of NPK nanofertilizer based on chitosan nanoparticles and its effect on biophysical characteristics and growth of coffee in green house. *Research Chemistry Intermed.*, *45*, 51–63.

Hasan, M., Rafique, S., Zafar, A., Loomba, S., Khan, R., Hassan, S. G., Khan, M. W., Zahra, S., Zia, M., Mustafa, G., Shu, X., Ihsan, Z., & Mahmood, N. (2020). Physiological and anti-oxidative response of biologically and chemically synthesized iron oxide: *Zea mays* a case study. *Heliyon*, *6*e04595.

He, A., Jiang, J., Ding, J., & Sheng, G. D. (2021). Blocking effect of fullerene nanoparticles (nC_{60}) on the plant cell structure and its phytotoxicity. *Chemosphere*, *278*130474.

He, X., Deng, H., & Hwang, H. (2019). The current application of nanotechnology in food and agriculture. *Journal of Food Drug Analytics.*, *27*, 1–21.

Hubbard, J. D., Lui, A., & Landry, M. P. (2020). Multiscale and multidisciplinary approach to understanding nanoparticle transport in plants. *Current Opinion Chemistry Engineering.*, *30*, 135–143.

Huddell, A., Ernfors, M., Crews, T., Vico, G., & Menge, D. N. L. (2023). Science of the total environment nitrate leaching losses and the fate of 15 N fertilizer in perennial intermediate wheatgrass and annual wheat — A field study. *Science of Total Environment.*, *857*159255.

Jat Baloch, M. Y., Zhang, W., Sultana, T., Akram, M., Shoumik, B. A. A., Khan, M. Z., & Farooq, M. A. (2023). Utilization of sewage sludge to manage saline–alkali soil and increase crop production: Is it safe or not? *Environmental Technology Innovation.*, *32*103266.

Jordan, J. T., Singh, K. P., & Cañas-Carrell, J. E. (2018). Carbon-based nanomaterials elicit changes in physiology, gene expression, and epigenetics in exposed plants: A review. *Current Opinion Environ.mental Science Health.*, *6*, 29–35.

Juárez-Maldonado, A., Tortella, G., Rubilar, O., Fincheira, P., & Benavides-Mendoza, A. (2021). Biostimulation and toxicity : The magnitude of the impact of nanomaterials in microorganisms and plants. *Journal of Advanced Research.*, *31*, 113–126.

Junedi, M. A., Mukhopadhyay, R., & Manjari, K. S. (2023). Alleviating salinity stress in crop plants using new engineered nanoparticles (ENPs). *Plant Stress*100184. Available from https://doi.org/10.1016/j.stress.2023.100184.

Kabiri, S., Degryse, F., Tran, D. N. H., da Silva, R. C., McLaughlin, M. J., & Losic, D. (2017). Graphene oxide: A new carrier for slow release of plant micronutrients. *ACS Applications of Material Interfaces*, *9*, 43325–43335.

Khatri-Chhetri, A., Sapkota, T. B., Maharjan, S., Cheerakkollil Konath, N., & Shirsath, P. (2023). Agricultural emissions reduction potential by improving technical efficiency in crop production. *Agricultural Systems*, *207*103620.

Kimbi, S. B., Onodera, S. I., Ishida, T., Saito, M., Tamura, M., Tomozawa, Y., & Nagasaka, I. (2022). Nitrate contamination in groundwater: Evaluating the effects of demographic aging and depopulation in an island with intensive citrus cultivation. *Water (Switzerland)*, *14*.

Komarek, A. M., Drogue, S., Chenoune, R., Hawkins, J., Msangi, S., Belhouchette, H., & Flichman, G. (2017). Agricultural household effects of fertilizer price changes for smallholder farmers in central Malawi. *Agricultural Systems*, *154*, 168–178.

Kong, C., Camps-Arbestain, M., Clothier, B., Bishop, P., & Vázquez, F. M. (2021). Use of either pumice or willow-based biochar amendments to decrease soil salinity under arid conditions. *Environmental Technology Innovation.*, *24*101849.

Kongala, S. I., & Mamidala, P. (2023). Harpin-loaded chitosan nanoparticles induced defense responses in tobacco. *Carbohydrate Polymer Technology Applications.*, *5*100293.

Kottegoda, N., Sandaruwan, C., Priyadarshana, G., Siriwardhana, A., Rathnayake, U. A., Berugoda Arachchige, D. M., Kumarasinghe, A. R., Dahanayake, D., Karunaratne, V., & Amaratunga, G. A. J. (2017). Urea-hydroxyapatite nanohybrids for slow release of nitrogen. *ACS Nano*, *11*, 1214–1221.

Kozłowska, M., Bandurska, H., & Breś, W. (2021). Response of lawn grasses to salinity stress and protective potassium effect. *Agronomy*, *11*, 843.

Król, M. (2020). Natural vs. Synthetic zeolites magdalena. *Crystals*, *10*, 622.

Kubavat, D., Trivedi, K., Vaghela, P., Prasad, K., Vijay Anand, G. K., Trivedi, H., Patidar, R., Chaudhari, J., Andhariya, B., & Ghosh, A. (2020). Characterization of a chitosan-based sustained release nanofertilizer formulation used as a soil conditioner while simultaneously improving biomass production of *Zea mays* L. *Land Degradation and Development.*, *31*, 2734–2746.

Le, T. T. H., Mai, T. T. T., Phan, K. S., Nguyen, T. M., Tran, T. L. A., Dong, T. N., Tran, H. C., Ngo, T. T. H., Hoang, P. H., & Ha, P. T. (2022). Novel integrated nanofertilizers for improving the growth of *Polyscias fruticosa* and asparagus officinalis. *Journal of Nanomaterial.*, *2022*, 1−10.

Li, A., Shi, Z., Yin, Y., Fan, Y., Zhang, Z., Tian, X., Yang, Y., & Pan, L. (2023a). Excessive use of chemical fertilizers in catchment areas raises the seasonal pH in natural freshwater lakes of the subtropical monsoon climate region. *Ecological Indicators.*, *154*110477.

Li, C., Lan, W., Jin, Z., Lu, S., Du, J., Wang, X., Chen, Y., & Hu, X. (2023b). Risk of heavy metal contamination in vegetables fertilized with mushroom residues and swine manure. *Sustainability*, *15*, 10984.

Lv, J., Christie, P., & Zhang, S. (2019). Uptake, translocation, and transformation of metal-based nanoparticles in plants: Recent advances and methodological challenges. *Environmental Science Nano*, *6*, 41−59.

Manikandan, A., & Subramanian, K. S. (2016). Evaluation of zeolite based nitrogen nano-fertilizers on maize growth, yield and quality on inceptisols and alfisols. *International Journal of Plant Soil Science.*, *9*, 1−9.

Marchiol, L., Filippi, A., Adamiano, A., Degli Esposti, L., Iafisco, M., Mattiello, A., Petrussa, E., & Braidot, E. (2019). Influence of hydroxyapatite nanoparticles on germination and plant metabolism of tomato (*Solanum lycopersicum* L.): Preliminary evidence. *Agronomy*, *9*, 161.

Márquez-Prieto, A. K., Palacio-Márquez, A., Sanchez, E., Macias-López, B. C., Pérez-Álvarez, S., Villalobos-Cano, O., & Preciado-Rangel, P. (2022). Impact of the foliar application of potassium nanofertilizer on biomass, yield, nitrogen assimilation and photosynthetic activity in green beans. *Notulae Botanicae. Horti Agrobotanici Cluj-Napoca*, *50*, 12569.

Mikhak, A., Sohrabi, A., Kassaee, M. Z., & Feizian, M. (2017). Synthetic nanozeolite/nano-hydroxyapatite as a phosphorus fertilizer for German chamomile (*Matricaria chamomilla* L.). *Industrial Crops Production.*, *95*, 444−452.

Miranda-Villagómez, E., Trejo-Téllez, L. I., Gómez-Merino, F. C., Sandoval-Villa, M., Sánchez-García, P., & Aguilar-Méndez, M. Á. (2019). Nanophosphorus fertilizer stimulates growth and photosynthetic activity and improves P status in rice. *Journal of Nanomater.*, *2019*, 1−11.

Mohamed, N., & Madian, N. G. (2020). Evaluation of the mechanical, physical and antimicrobial properties of chitosan thin films doped with greenly synthesized silver nanoparticles. *Materials. Today Communication.*, *25*101372.

Mujtaba, M., Khawar, K. M., Camara, M. C., Carvalho, L. B., Fraceto, L. F., Morsi, R. E., Elsabee, M. Z., Kaya, M., Labidi, J., Ullah, H., & Wang, D. (2020). Chitosan-based delivery systems for plants: A brief overview of recent advances and future directions. *International Journal of Biological Macromolecules*, *154*, 683−697.

Ndaba, B., Roopnarain, A., Rama, H., & Maaza, M. (2022). Biosynthesized metallic nanoparticles as fertilizers : An emerging precision agriculture strategy. *Journal of Integrated Agriculture*, *21*, 1225−1242.

Nehela, Y., Mazrou, Y. S. A., Alshaal, T., Rady, A. M. S., El-Sherif, A. M. A., Omara, A. E., Abd El-Monem, A. M., & Hafez, E. M. (2021). The integrated amendment of sodic-saline soils using biochar and plant growth-promoting rhizobacteria enhances maize (*Zea mays* L.) Resilience to water salinity. *Plants*, *10*, 1960.

Neme, K., Nafady, A., Uddin, S., & Tola, Y. B. (2021). Application of nanotechnology in agriculture, postharvest loss reduction and food processing: Food security implication and challenges. *Heliyon*, *7*e08539.

Nisha Raj, S., Anooj, E. S., Rajendran, K., & Vallinayagam, S. (2021). A comprehensive review on regulatory invention of nano pesticides in agricultural nano formulation and food system. *Journal of Molecular Structure*, *1239*130517. Available from https://doi.org/10.1016/j.molstruc.2021.130517.

Periakaruppan, R., Romanovski, V., Thirumalaisamy, S. K., Palanimuthu, V., Sampath, M. P., Anilkumar, A., Sivaraj, D. K., Ahamed, N. A. N., Murugesan, S., Chandrasekar, D., & Selvaraj, K. S. V. (2023). Innovations in modern nanotechnology for the sustainable production of agriculture. *Chemical Engineering*, *7*, 61.

Piash, M. I., Iwabuchi, K., Itoh, T., & Uemura, K. (2021). Release of essential plant nutrients from manure- and wood-based biochars. *Geoderma*, *397*115100.

Priyam, A., Yadav, N., Reddy, P. M., Afonso, L. O. B., Schultz, A. G., & Singh, P. P. (2022). Fertilizing benefits of biogenic phosphorous nanonutrients on *Solanum lycopersicum* in soils with variable pH. *Heliyon*, *8*e09144.

Raguraj, S., Wijayathunga, W. M. S., Gunaratne, G. P., Amali, R. K. A., Priyadarshana, G., Sandaruwan, C., Karunaratne, V., Hettiarachchi, L. S. K., & Kottegoda, N. (2020). Urea—hydroxyapatite nanohybrid as an efficient nutrient source in *Camellia sinensis* (L.) Kuntze (tea). *Journal of Plant Nutrition*, *43*, 2383—2394.

Rahman, Z., & Singh, V. P. (2019). The relative impact of toxic heavy metals (THMs) (arsenic (As), cadmium (Cd), chromium (Cr) (VI), mercury (Hg), and lead (Pb) on the total environment: An overview. *Environmental Monitoring and Assessment*, *191*419.

Rashid, A., Schutte, B. J., Ulery, A., Deyholos, M. K., Sanogo, S., Lehnhoff, E. A., & Beck, L. (2023). Heavy metal contamination in agricultural soil: Environmental pollutants affecting crop health. *Agronomy*, *13*, 1521.

Rios, J. J., García-Ibañes, P., & Carvajal, M. (2018). The use of biovesicles to improve the efficiency of Zn foliar fertilization. *Colloids Surfaces B Biointerfaces*, *18*.

Rios, J. J., Yepes-Molina, L., Martínez-Alonso, A., & Carvajal, M. (2020). Nanobiofertilization as a novel technology for highly efficient foliar application of Fe and B in almond trees. *Royal Socsiety of Open Science*, *7*.

Rop, K., Karuku, G. N., Mbui, D., Michira, I., & Njomo, N. (2018). Annals of agricultural sciences formulation of slow release NPK fertilizer (cellulose-graft-poly (acrylamide)/nano-hydroxyapatite/ soluble fertilizer) composite and evaluating its N mineralization potential. *Annals of Agricultural Science*, *63*, 163—172.

Sabet, H., & Mortazaeinezhad, F. (2018). Yield, growth and Fe uptake of cumin (*Cuminum cyminum* L.) affected by Fe-nano, Fe-chelated and Fe-siderophore fertilization in the calcareous soils. *Journal of Trace Elements in Medicine and Biology: Organ of the Society for Minerals and Trace Elements (GMS)*, *50*, 154—160.

Saffan, M. M., Koriem, M. A., El-Henawy, A., El-Mahdy, S., El-Ramady, H., Elbehiry, F., Omara, A. E.-D., Bayoumi, Y., Badgar, K., & Prokisch, J. (2022). Sustainable production of tomato plants (*Solanum lycopersicum* L.) under low-quality irrigation water as affected by bio-nanofertilizers of selenium and copper. *Sustainability*, *14*, 3236.

Saleem, H., Zaidi, S. J., & Center, N. A. A. (2021). Recent advancements in the nanomaterial application in concrete and its ecological impact. *Materials (Basel)*.

Salo, T., & Salovius-Laurén, S. (2022). Green algae as bioindicators for long-term nutrient pollution along a coastal eutrophication gradient. *Ecological Indicators*, *140*109034.

Saraiva, R., Ferreira, Q., Rodrigues, G. C., & Oliveira, M. (2023). Nanofertilizer use for adaptation and mitigation of the agriculture/climate change dichotomy effects. *Climate*, *11*, 129.

Saritha, G. N. G., Anju, T., & Kumar, A. (2022). Nanotechnology — Big impact: How nanotechnology is changing the future of agriculture? *Journal of Agricultural Food Research*, *10*100457.

Schwab, F., Zhai, G., Kern, M., Turner, A., Schnoor, J. L., & Wiesner, M. R. (2016). Barriers, pathways and processes for uptake, translocation and accumulation of nanomaterials in plants – Critical review. *Nanotoxicology*, *10*, 257–278.

Seleiman, M. F., Almutairi, K. F., Alotaibi, M., Shami, A., Alhammad, B. A., & Battaglia, M. L. (2021). Nano-fertilization as an emerging fertilization technique : Why can modern agriculture benefit from its use ? *Plants*, *2019*, 1–27.

Sharma, V., Javed, B., Byrne, H., Curtin, J., & Forong, T. (2022). Zeolites as carriers of nano-fertilizers : From structures and principles to prospects and challenges. *Applied Nano*, *33*, 163–186.

Shebl, A., Hassan, A. A., Salama, D. M., Abd El-Aziz, M. E., & Abd Elwahed, M. S. A. (2020). Template-free microwave-assisted hydrothermal synthesis of manganese zinc ferrite as a nanofertilizer for squash plant (*Cucurbita pepo* L). *Heliyon*, *6*e03596.

Sheoran, P., Goel, S., Boora, R., Kumari, S., Yashveer, S., & Grewal, S. (2021). Biogenic synthesis of potassium nanoparticles and their evaluation as a growth promoter in wheat. *Plant Gene*, *27*100310.

Silva, L. I. da, Pereira, M. C., Carvalho, A. M. X. de, Buttrós, V. H., Pasqual, M., & Dória, J. (2023). Phosphorus-Solubilizing microorganisms: A key to sustainable agriculture. *Agriculture*, *13*, 462.

Singh, A., Kumar, A., Kumar, R., Sheoran, P., Yadav, R. K., & Sharma, P. C. (2022). Multivariate analyses discern shared and contrasting eco-physiological responses to salinity stress of ziziphus rootstocks and budded trees. *South African Journal of Botany.*, *146*, 573–584.

Swoboda, P., Döring, T. F., & Hamer, M. (2022). Remineralizing soils? The agricultural usage of silicate rock powders: A review. *The Science of the Total Environment*, *807*150976.

Tarafder, C., Daizy, M., Alam, M. M., Ali, M. R., Islam, M. J., Islam, R., Ahommed, M. S., Aly Saad Aly, M., & Khan, M. Z. H. (2020). Formulation of a hybrid nanofertilizer for slow and sustainable release of micronutrients. *ACS Omega*, *5*, 23960–23966.

Tripathi, D. K., Tripathi, A., Shweta., Singh, S., Singh, Y., Vishwakarma, K., Yadav, G., Sharma, S., Singh, V. K., Mishra, R. K., Upadhyay, R. G., Dubey, N. K., Lee, Y., & Chauhan, D. K. (2017). Uptake, accumulation and toxicity of silver nanoparticle in autotrophic plants, and heterotrophic microbes: A concentric review. *Frontiers in Microbiology*, *08*, 1–16.

Tsintskaladze, G., Eprikashvili, L., Mumladze, N., Gabunia, V., & Sharashenidze, T. (2017). Nitrogenous zeolite nanomaterial and the possibility of its application in agriculture. *Annals of Agrarian Science*, *15*, 365–369.

Upadhyay, P. K., Dey, A., Singh, V. K., Dwivedi, B. S., Singh, T., G. A, R., Babu, S., Rathore, S. S., Singh, R. K., Shekhawat, K., Rangot, M., Kumar, P., Yadav, D., Singh, D. P., Dasgupta, D., & Shukla, G. (2023). Conjoint application of nano-urea with conventional fertilizers: An energy efficient and environmentally robust approach for sustainable crop production. *PLoS One*, *18*, e0284009.

Veazie, P., Cockson, P., Henry, J., Perkins-Veazie, P., & Whipker, B. (2020). Characterization of nutrient disorders and impacts on chlorophyll and anthocyanin concentration of *Brassica rapa* var. Chinensis. *Agriculture*, *10*, 461.

Vidak, M., Lazarević, B., Petek, M., Gunjača, J., Šatović, Z., Budor, I., & Carović-Stanko, K. (2021). Multispectral assessment of sweet pepper (*Capsicum annuum* L.) fruit quality affected by calcite nanoparticles. *Biomolecules*, *11*, 832.

Vishwanath, R., & Negi, B. (2021). Conventional and green methods of synthesis of silver nanoparticles and their antimicrobial properties. *Current Research Green Sustainable Chemistry*100205.

Voke, E., Pinals, R. L., Goh, N. S., & Landry, M. P. (2021). In planta nanosensors: Understanding biocorona formation for functional design. *ACS Sensors*, *6*(8), 2802–2814. Available from https://doi.org/10.1021/acssensors.1c01159.

Wang, J., Geng, Y., Zhang, J., Li, L., Guo, F., Yang, S., Zou, J., & Wan, S. (2023a). Increasing calcium and decreasing nitrogen fertilizers improves peanut growth and productivity by enhancing photosynthetic efficiency and nutrient accumulation in acidic red soil. *Agronomy*, *13*, 1924.

Wang, X., Zhu, H., Yan, B., Chen, L., Shutes, B., Wang, M., Lyu, J., & Zhang, F. (2023b). Ammonia volatilization, greenhouse gas emissions and microbiological mechanisms following the application of nitrogen fertilizers in a saline-alkali paddy ecosystem. *Geoderma*, *433*116460.

Wang, Y., Liu, X., Wang, L., Li, H., Zhang, S., Yang, J., Liu, N., & Han, X. (2023c). Effects of long-term application of Cl-containing fertilizers on chloride content and acidification in brown soil. *Sustainability*, *15*, 8801.

Wang, S., Xu, L., & Hao, M. (2022). Impacts of long-term micronutrient fertilizer application on soil properties and micronutrient availability. *International Journal of Environmental Research and Public Health*, *19*, 16358.

Wang, Y., Lin, Y., Xu, Y., Yin, Y., Guo, H., & Du, W. (2019). To copper oxide nanoparticles / microparticles as potential agricultural fertilizer. *Environmental Pollution Bioavailability.*, *31*, 80–84.

Wong, M. H., Misra, R. P., Giraldo, J. P., Kwak, S., Son, Y., Landry, M. P., Swan, J. W., Blankschtein, D., & Strano, M. S. (2016). Lipid exchange envelope penetration (LEEP) of nanoparticles for plant engineering : A universal localization mechanism. *Nano Letters*, *16*, 1161–1172.

Wu, L., Jiang, Y., Zhao, F., He, X., Liu, H., & Yu, K. (2020). Increased organic fertilizer application and reduced chemical fertilizer application affect the soil properties and bacterial communities of grape rhizosphere soil. *Scientific Reports.*, *10*9568.

Xiao, Y., Ma, C., Li, M., Zhangzhong, L., Song, P., & Li, Y. (2023). Interaction and adaptation of phosphorus fertilizer and calcium ion in drip irrigation systems : The perspective of emitter clogging. *Agricultural Water Management.*, *282*108269.

Yadav, A., Yadav, K., & Abd-elsalam, K. A. (2023a). Nanofertilizers : Types, delivery and advantages in agricultural sustainability. *Agrochemicals*, *2*, 296–336.

Yadav, N., Garg, V. K., Chhillar, A. K., & Rana, J. S. (2023b). Recent advances in nanotechnology for the improvement of conventional agricultural systems: A review. *Plant Nano Biology*, *4*100032.

Yan, G., Xing, Y., Wang, J., Zhang, Z., Xu, L., Han, S., Zhang, J., Dai, G., & Wang, Q. (2018). Effects of winter snowpack and nitrogen addition on the soil microbial community in a temperate forest in northeastern China. *Ecological Indicators*, *93*, 602–611.

Yang, R., Shen, J., Zhang, Y., Jiang, L., Sun, X., Wang, Z., Tang, B., & Shen, Y. (2022). The role of biochar nanoparticles performing as nanocarriers for fertilizers on the growth promotion of chinese cabbage (*Brassica rapa* (Pekinensis group)). *Coatings*, *12*, 1984.

Yescas-Coronado, P., Segura-Castruita, M. Á., Chávez-Rodríguez, A. M., Gómez-Leyva, J. F., Martínez-Sifuentes, A. R., Amador-Camacho, O., & González-Medina, R. (2022). Covariables of soil-forming factors and their influence on pH distribution and spatial variability. *Agriculture*, *12*, 2132.

Yildiz, M., Poyraz, İ., Çavdar, A., Özgen, Y., & Beyaz, R. (2021). *Plant responses to salt stress. In:* Plant breeding - Current and future views (pp. 137–144). IntechOpen.

Younis, H., Sha, S., Ehsan, Z., Ishfaq, A., Mehboob, K., Ajaz, M., Hidayat, A., & Muhammad, W. (2023). Radiometric examination of fertilizers and assessment of their health hazards, commonly used in Pakistan. *Nuclear Engineering Technology.*, *55*, 2247–2453.

Zahra, Z., Habib, Z., Hyun, H., & Shahzad, H. M. A. (2022). Overview on recent developments in the design, application, and impacts of nanofertilizers in agriculture. *Sustainability*, *14*, 9397.

Zain, M., Ma, H., Nuruzzaman, M., Chaudhary, S., Nadeem, M., Shakoor, N., Azeem, I., Duan, A., Sun, C., & Ahamad, T. (2023). Nanotechnology based precision agriculture for alleviating biotic and abiotic stress in plants. *Plant Stress*100239. Available from https://doi.org/10.1016/j.stress.2023.100239.

Zhang, M., Liu, Y., Wei, Q., Liu, L., Gu, X., Gou, J., & Wang, M. (2023). Chemical fertilizer reduction combined with biochar application ameliorates the biological property and fertilizer utilization of pod pepper. *Agronomy*, *13*, 1616.

Zhang, R., Li, Y., Zhao, X., Allan Degen, A., Lian, J., Liu, X., Li, Y., & Duan, Y. (2022). Fertilizers have a greater impact on the soil bacterial community than on the fungal community in a sandy farmland ecosystem, Inner Mongolia. *Ecological Indicators*, *140*108972.

Zhao, C., Xu, J., Bi, H., Shang, Y., & Shao, Q. (2023). A slow-release fertilizer of urea prepared via biochar-coating with nano-SiO-starch-polyvinyl alcohol: Formulation and release simulation. *Environmental Technology Innovation*, *32*103264.

Zheng, S., Yin, K., & Yu, L. (2022). Factors influencing the farmer's chemical fertilizer reduction behavior from the perspective of farmer differentiation. *Heliyon*, *8*e11918.

Zhu, L., Chen, L., Gu, J., Ma, H., & Wu, H. (2022). Carbon-based nanomaterials for sustainable agriculture: Their application as light converters, nanosensors, and delivery tools. *Plants*, *11*, 511.

Zhu, M., Yu, J., Wang, R., Zeng, Y., Kang, L., & Chen, Z. (2023). Nano-calcium alleviates the cracking of nectarine fruit and improves fruit quality. *Plant Physiology and Biochemistry: PPB/Societe Francaise de Physiologie Vegetale*, *196*, 370–380.

Zulfiqar, F., Navarro, M., Ashraf, M., Akram, N. A., & Munné-Bosch, S. (2019). Nanofertilizer use for sustainable agriculture: Advantages and limitations. *Plant Science (Shannon, Ireland)*, *289*110270.

Biostimulation of plants with nanocomposites: a new perspective to improve crop production

8

Luz Leticia Rivera-Solís[1], Julia Medrano-Macías[2], Álvaro Morelos-Moreno[3], Zulfiqar Ali Sahito[4] and Adalberto Benavides-Mendoza[2]
[1]Doctorate in Sciences in Protected Agriculture, Universidad Autónoma Agraria Antonio Narro, Saltillo, Mexico, [2]Department of Horticulture, Universidad Autónoma Agraria Antonio Narro, Saltillo, Mexico, [3]CONAHCYT, Universidad Autónoma Agraria Antonio Narro, Saltillo, Mexico, [4]Ministry of Education (MOE), Key Laboratory of Environmental Remediation and Ecosystem Health, College of Environmental and Resources Science, Zhejiang University, Hangzhou, P. R. China

8.1 Introduction

Reducing the ecological impact of agriculture is imperative for sustaining human life. Considering the need to feed a growing population, conventional agricultural practices entail deforestation, overuse of water resources, and heavy reliance on chemical fertilizers and pesticides, which contribute to soil degradation, water pollution, and biodiversity loss. These repercussions threaten food security and exacerbate climate change. Transitioning to sustainable agricultural methods, such as organic farming, agroforestry, and precision agriculture, is essential to preserve ecosystems, ensure long-term agricultural productivity, and maintain the balance necessary for the well-being of current and future generations (Karaca et al., 2023).

Nanotechnology can play a significant role in developing precision agriculture by increasing efficiency and reducing the environmental impact of fertilizers and pesticides. The majority of conventional agrochemicals, including fertilizers, are lost to various environmental compartments through a variety of processes like surface run-off, volatilization, leaching, and air drift. These processes prevent the traditional agrochemicals from reaching the target site in a coordinated manner. The efficiency with which conventional fertilizers use nutrients is poor. For example, for conventional urea, the nitrogen use efficiency is approximately 40% (Chakraborty et al., 2023). Conversely, nanofertilizers ensure better uptake by plant roots or leaf sprays, reducing the amount needed. Controlled release of nutrients is also possible, ensuring plants receive them when needed, optimizing growth, and minimizing wastage. Using nanopesticides enables the targeted delivery of active substances. This reduces the quantities required,

Nanocomposites for Environmental, Energy, and Agricultural Applications. DOI: https://doi.org/10.1016/B978-0-443-13935-2.00008-5

obtaining innocuous food, minimizing pollution, and the development of pesticide-resistant pests. The nanosensors can help to monitor moisture levels, nutrient concentrations, and disease indicators. Data collected can be used to make informed decisions regarding irrigation and agrochemical application planning (Ojeda-Barrios et al., 2020).

Nanocomposites, defined here as multiphase or hybrid materials in which one or more components have at least one nanometric dimension (equal to or less than 100 nm), are emerging as a beneficial class of nanomaterials (NMs) in agriculture. The decrease in size is the driving force behind the notable alterations in the properties and capacities of the nanocomposites, which can be attributed to the high ratio of surface area to volume. Currently, the uses of nanocomposites in the food production and processing industry are more directed toward food packaging and preservation. Regarding applications in the field for crop production, there is an acceptable amount of evidence that indicates the feasibility of its application. However, uses by agribusiness and extensive crop production, although potentially valuable, are still limited (Wypij et al., 2023).

Nanocomposites are obtained by incorporating different fillers or nanofillers, such as metal or metal oxide nanoparticles (NPs), carbon NMs, agrochemicals, natural products (e.g., essential oils), or polymers in a carrier matrix, which in turn can be a bulk material or a NM, for example, chitosan or nanochitosan. Combining two or more materials or NMs in a nanocomposite allows different biostimulation processes in plants—the explanation for the above lies in synergism. As will be explained later, the presence of two or more components in the nanocomposite, with at least one of them having a nanometric dimension, activates different response signaling pathways, synergistically improving the metabolic and biochemical response of the plants (Ahmed et al., 2022; Cabrera-De la Fuente et al., 2020; Chakraborty et al., 2023). Additionally, the different combinations of matrices and materials or NMs allow nanocomposites to be tailored to exhibit properties such as enhanced strength, controlled release, and increased reactivity, making the last two properties particularly valuable for using nanocomposites as crop biostimulants (Camargo et al., 2009).

This chapter aims to describe the uses and potentials of nanocomposites in the biostimulation of crops. To do this, first, there will be an enumeration of the nanocomposites most used in agricultural activity. The biostimulation mechanisms proposed for nanocomposites will be explained in the subsequent section, followed by the enumeration of the diverse applications found in the reviewed literature. The chapter concludes by describing the authors' views on research needs and the possible additional uses of nanocomposites.

8.2 Nanocomposites used in agriculture

In agriculture, nanocomposites are primarily utilized to improve various aspects such as packaging, water treatment, soil improvement, nanosensors, crop protection, and controlled release of fertilizers and pesticides. Some of the widely used nanocomposites in agriculture include the following.

8.2.1 Polymer-based nanocomposites

These are the most widely used in agriculture and the manufacturing of smart packaging for agricultural products. These nanocomposites are comprised of biopolymers or nanobiopolymer (e.g., chitosan, starch, cellulose, alginate, and many others) derived from plants, microbes, and animals, are biodegradable and so ecologically sound, and can be used to develop nanocomposites for applications in the agriculture and food industries. Polymer nanocomposites contain a polymer or nanopolymer matrix, nanofillers that can be nanobiopolymers, metals, nanometals, pesticides, fertilizers, nanofertilizers, or natural extracts, plasticizers, and compatibilizers (Fig. 8.1) (Wypij et al., 2023).

The polymer-based nanocomposites are utilized in fertilizers and pesticides, and nanocomposite-based hydrogels to slow or control the release of fertilizers

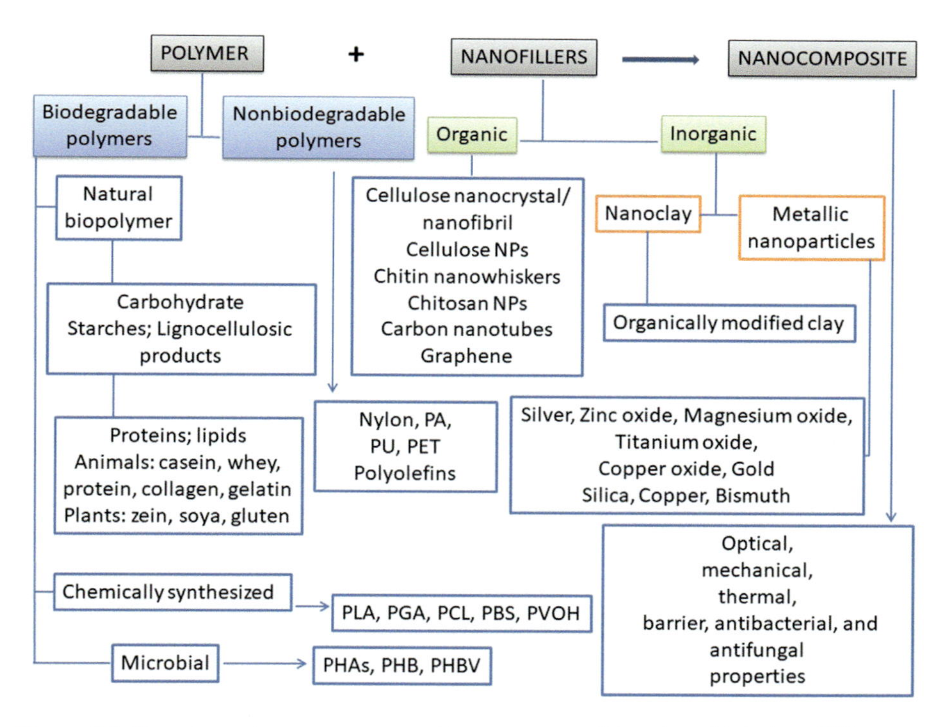

Figure 8.1 Common polymer-nanocomposite components. *PA*, polyamide/nylon; *PBS*, polybutylene succinate; *PCL*, polycaprolactone; *PET*, polyethylene terephthalate; *PGA*, polyglycolic acid; *PHAs*, polyhydroxyalkanoates; *PHB*, polyhydroxybutyrate; *PHBV*, poly(3-hydroxybutyrate-*co*-3-hydroxyvalerate); *PLA*, polylactic acid; *PU*, polyurethane; *PVOH*, polyvinyl alcohol.

Source: From Wypij, M., Trzcińska-Wencel, J., Golińska, P., Avila-Quezada, G. D., Ingle, A. P. & Rai, M. (2023). The strategic applications of natural polymer nanocomposites in food packaging and agriculture: Chances, challenges, and consumers' perception. *Frontiers in Chemistry*, *10*, 1106230. https://doi.org/10.3389/fchem.2022.1106230.

and pesticides and for water retention in soil. They encapsulate the active ingredients and water and release them over time, enhancing the effectiveness and reducing the amount needed. As an example, the nanocomposites of nanochitosan-nanoselenium and nanochitosan-nanosilver showed great therapeutic effectiveness and improvement of tolerance against viral infections in human cells in vitro and pepper plants in the greenhouse (Abdelhamid et al., 2023; El-Ganainy et al., 2023). On the other hand, the nanocomposite hydrogels can release water gradually, which is especially useful in drought-prone areas. As a result, the release of agrochemicals and water is drastically slowed, and their leaching into the soil and aquatic environments and evaporation is lowered. The polymer-based hydrogels can be employed as absorbent agents, slow-release agrochemical agents, and seed-coating components for easy germination by keeping pathogens and pests away from the seeds (Madzokere et al., 2021; Wypij et al., 2023).

8.2.2 Nanobiocomposites

These are made by combining synthetic or natural biopolymers with organic or inorganic NPs. The difference with the previous category is that the polymer is not in a nanometric form and can be natural or synthetic. Non-biodegradable synthetic polymers (Fig. 8.1) can be classified into two main categories: high-density polyethylene and low-density polyethylene (LDPE). Examples of these polymers include ethylene vinyl acetate, polyethylene terephthalate, polyethylene, polypropylene, and polystyrene. On the other hand, the category of synthetic biodegradable biopolymers encompasses various types, such as poly(lactic acid) (PLA), poly-(glycolic acid), poly(lactic-*co*-glycolic acid), poly(butylene succinate) (PBA), polycaprolactone, poly(ethylene adipate), poly(p-dioxanone), and copolymers derived from these materials (Wypij et al., 2023). The proliferation of nondegradable polymers has resulted in critical environmental challenges (e.g., micro- and nanoplastics in water and soil), prompting a growing preference for biodegradable polymers (Sohail et al., 2023).

Nanobiocomposites are used in seed coatings, for example, to protect seeds and plants from pathogens and abiotic stress (Mehta et al., 2021). Natural biopolymers per se can have fungicidal, bactericidal, and biostimulant capacity for crops. Therefore, adding the favorable impact of the nano-biocomposite matrix with the biostimulant contribution of the nanofiller allows for obtaining multifaceted compounds capable of inducing greater tolerance to environments that are not favorable for growth and productivity (Khan, Ijaz, et al., 2023). The nanometric fillers of the nanobiocomposites can be either organic or inorganic, such as nanoclay (such as montmorillonite), carbon NMs (nanotubes or nanodots), natural biopolymers (such as nanochitosan or nanocellulose), biological antimicrobial agents (such as nisin), metal (such as nano-Ag), metal oxides (such as NZnO and $NTiO_2$), and nonmetals (such as silicon or selenium) (Barua et al., 2019; De & Karak, 2019; Othman, 2014).

8.2.3 Nanosensors

They are sensing devices that rely on the distinctive attributes of NMs to distinguish and detect the behavior of materials in a given environment. Nanosensors are often used for soil, water, and crop monitoring. They can comprise carbon-based nanocomposites and metal or metal oxide nanocomposites. These sensors can detect changes in the environment and monitor moisture levels, nutrients, and the presence of pests (Munonde & Nomngongo, 2021).

From the point of view of biostimulation of crops, the most widely used nanocomposites are polymer-based nanocomposites and nanobiocomposites.

When manufacturing nanocomposites, the matrix material typically depends on the desired application and on specific requirements such as biodegradability, non-toxicity, and environmental sustainability. Some commonly used matrix materials for manufacturing nanocomposites for agricultural use are shown in Table 8.1.

In general, for nanocomposites for agricultural applications, there is a preference for using matrix materials that are biodegradable and have minimal impact on the environment.

An interesting fact from the list in Table 8.1 is that practically all the materials used as a matrix (both in bulk and nanoform) per se have a biostimulant impact on plants (Cabrera-De la Fuente et al., 2020; Hernández-Hernández et al., 2018). This is explained by multiple surface charges in the polymers, reactive functional groups, and the release of oligomers and monomers (e.g., oligopeptides, oligoglucans, or oligosaccharines) when the polymer is metabolized in the soil or by the soil microbiome or plant cells (Juárez-Maldonado et al., 2021). The same occurs with the NMs incorporated into the different matrices. These act as biostimulants because the free energy associated with the surface charge of NMs causes a biostimulation effect in plant cells, modifying their metabolism and gene expression (Juárez-Maldonado et al., 2019; Naz et al., 2021).

On the other hand, the NMs commonly used to be added to the carrier matrices of the nanocomposites are listed in Table 8.2.

8.3 Biostimulation mechanisms of nanocomposites

Biostimulation refers to the process by which certain materials, including NMs, can enhance the growth, development, or overall health of plants. The biostimulation of an organism is a phenomenon of modification in metabolic processes that allows more efficient use of environmental resources, greater growth or yield, and greater tolerance to adverse environmental factors.

From a functional point of view, biostimulation has been described as a general biological phenomenon that depends on the interaction between the molecular structures of cells and impulses or stimuli. These impulses or stimuli come from environmental agents (internal or external to the organism) that can be physical, chemical, or biological (Juárez-Maldonado et al., 2019).

Table 8.1 Some common materials that are used as matrices for agricultural nanocomposites.

Material	Use	References
Alginate	Material derived from seaweed. This copolymer is commercially significant because it forms hydrogels, films, and fibers with different ions and components with distinct properties. It is used as a plant biostimulant, in seed coating and seed encapsulation, for the controlled release of nutrients, growth regulators, and pesticides.	Bibi et al. (2019)
β-glucans	Biostimulant and biodegradable polysaccharides with low toxicity, high biodegradability, biocompatibility, and bio-adhesivity. β-glucans can be filled with NPs to obtain biostimulant and antimicrobial materials.	Tsivileva and Perfileva (2022)
Cellulose and cellulose derivatives	Used in creating biodegradable films for agricultural packaging or hydrogels used as soil conditioners and carriers for the slow or controlled release of nutrients and other agrochemicals. It can be used to make composite hydrogels with clays and nanomaterials.	Bauli et al. (2021)
Chitosan or nanochitosan	Biodegradable material used for plant biostimulation and elicitation, slow or controlled release of fertilizers, pesticides, and biomolecules. It is utilized to enhance seed germination, soil amendment, and water purification.	Yu et al. (2021)
3-dimensional hydrogels	Three-dimensional, hydrophilic polymer networks capable of absorbing large amounts of water. They are widely used in agriculture for water retention in soils, as beneficial microorganisms' carriers, and as carriers for controlled-release fertilizers and pesticides.	Singh et al. (2021)
Lignin, nanolignin, and its derivatives	Biodegradable material used for plant biostimulation, antibiotic and antifungal, controlled release of fertilizers or pesticides, enhancing seed germination, soil amendment, and water purification.	Arya et al. (2023)
Poly(lactic acid) (PLA)	PLA is an inexpensive polymer easily recycled into lactic acid. It is used to obtain biodegradable mulch and packaging films, and controlled-release fertilizers.	Ranakoti et al. (2022)

(*Continued*)

Table 8.1 (Continued)

Material	Use	References
Polyvinyl alcohol (PVA)	Biodegradable water-soluble polymer. Used as a biostimulant and for creating hydrogels, films, and coatings for controlled release of agrochemicals and smart packaging for agricultural products.	Kassem et al. (2021)
Protein-based matrices	Materials like soy protein, wheat gluten, and casein are sometimes used as biodegradable matrices, especially for packaging applications.	Jafarzadeh et al. (2022)
Starch-based polymers	Biodegradable material used for agricultural packaging, hydrogels, and as a carrier material for controlled-release fertilizers.	Gamage et al. (2022)
Nanoclays-based matrices	Nanoclays can be used for the controlled delivery of fertilizers, phytohormones, and pesticides. Clays and nanoclays have a substantial surface area and cation exchange capacity, which makes them useful for carrying a variety of substances.	Merino et al. (2020)

Table 8.2 Common nanoparticles or nanomaterials used to be incorporated into polymeric matrices to manufacture nanocomposites for agricultural use.

Material	Use	References
Carbon nanotubes (CNTs)	Carbon nanotubes are tubular carbon NMs known for their large surface area, strength, and electrical conductivity. In agriculture, they can be used as biostimulants, carriers for agrochemicals and fertilizers, water purification, and embedded in polymeric matrices to create sensors for monitoring soil or plant conditions.	Patel et al. (2020)
Chitosan nanoparticles (nanochitosan)	Chitosan is a biodegradable polymer known for its biocompatibility and biostimulant capacity. Additionally, it can be used to encapsulate microorganisms and seeds or for the controlled release of nutrients and pesticides. Chitosan nanoparticles, with a significantly higher specific surface than bulk chitosan, are efficient fillers to ameliorate the properties of nanocomposites.	Garavand et al. (2022)

(*Continued*)

Table 8.2 (Continued)

Material	Use	References
Clay nanoparticles (e.g., montmorillonite, halloysite)	Clay nanoparticles are used to improve the mechanical and barrier properties of polymers. When used in agricultural films or coatings, they can help control the permeability to gases and moisture, which can be advantageous for applications like mulch films or greenhouse covers. Clay nanoparticles dispersed in hydrogels or other polymers allow obtaining nanocomposites for carrying bioactive molecules and fertilizer nanocomposites with greater thermal stability, nutrient release duration, and nutrient release amount significantly improved.	Vejan et al. (2021)
Graphene	In hexagonal lattices, a single layer of carbon atoms is covalently connected to form graphene's two-dimensional structure. Graphene possesses a large surface area, a low weight, and exceptional thermal and mechanical properties. It is used to fabricate nanocomposites using starch, chitosan, cellulose, alginates, gelatin, gluten, zein, soy protein, and PLA. The graphene nanocomposites are used for the adsorption of nutrients and heavy metals, elimination of organic contaminants, agrochemical delivery, and biostimulation under stress conditions.	Dhand et al. (2013)
Metal oxide nanoparticles (e.g., ZnO, TiO_2, and CuO)	These nanoparticles are used for their antimicrobial properties. When embedded into polymers, they can be used to create packaging that keeps agricultural products fresher for longer periods by preventing microbial growth. Additionally, they can be used in coatings to prevent fouling on agricultural equipment. Nanocomposites with metal oxide NPs work as biostimulants, improving metabolism and stress tolerance in crops.	Yu et al. (2021)

(*Continued*)

Table 8.2 (Continued)

Material	Use	References
Nanocellulose	Derived from plant cell walls, nanocellulose can be used to reinforce polymeric matrices, resulting in lightweight, strong, and biodegradable materials. These composites can be used for packaging or as biodegradable pots for seedlings.	Bauli et al. (2021)
Silica nanoparticles (SiO_2)	They are biostimulant materials known for their high absorption and adsorption capacity and can be used as an agrochemical carrier, water treatment, and to improve water retention in soils. When incorporated into polymers, they can enhance mechanical properties and thermal stability, making them ideal for nanosensors and for controlled release systems for fertilizers, pesticides, and gene delivery.	Rajiv et al. (2020)
Silver nanoparticles (Ag NPs)	Due to their antimicrobial properties, silver nanoparticles can be embedded into polymers to create materials that prevent bacterial growth. This can be used in agricultural applications for water purification, protective clothing, or in packaging to increase the shelf life of agricultural products.	Khan, Zahoor et al. (2023)

8.3.1 Biostimulation by nanocomposites

In the case of nanocomposites, the crop biostimulation process occurs due to the matrix material's presence and the NMs carried by the matrix. In the case of nanocomposites for agricultural use, biological polymers are widely used as a matrix, and these materials, such as chitosan and alginate, are very effective biostimulants (Stasińska-Jakubas & Hawrylak-Nowak, 2022). This allows us to conclude that using properly formulated nanocomposites brings additional advantages compared to using individual materials since synergisms are expected to occur between the different materials used in the nanocomposite formulation. An example of the above is the synergistic improvement in the antibacterial activity of the nanocomposites of humic acids and ZnO NPs (Venezia et al., 2023). In the same way, synergism allows for obtaining biostimulation responses in plants of greater diversity and amplitude regarding response and adaptation to environmental challenges (Guha et al., 2020).

What is expected from the interaction between plants and nanocomposites is that surface interactions first occur between the surfaces of the plants and the polymeric material used as a matrix or its degradation products (oligomers and monomers)

(Armentano et al., 2018). This would be the first phase of biostimulation. A second phase (which could occur almost parallel to the previous one) would happen when the NMs contained in the nanocomposite matrix are released and come into contact with plant surfaces. The interaction between both biological responses is the cause of the synergy in the metabolic and physiological behavior of plants exposed to nanocomposites (Yu et al., 2021).

Different reviews have described the responses of plants to the polymeric materials used as matrices in nanocomposites (Mukherjee & Patel, 2020; Mukhtar Ahmed et al., 2020; Shahrajabian et al., 2021; Stasińska-Jakubas & Hawrylak-Nowak, 2022; Sun et al., 2023). In summary, the presence of the matrix polymer (and its oligomers) is perceived by means of receptors located in the cell walls and membranes. Following the perception, a series of events of transduction, signaling, effector action, modification of gene expression, and cellular metabolism occurs (Wang & Lei, 2018).

Compared to the cellular response to nanocomposite matrix polymers, the reaction to NMs is similar in terms of signaling and response events but more complex in terms of perception and response, as it appears to involve a response in two phases (Juárez-Maldonado et al., 2019). In the following paragraphs, a brief explanation of the biostimulation obtained using NMs is presented.

8.3.2 *The first phase of plant cell biostimulation*

The biostimulation process involves modulating plant physiological processes to achieve better outcomes. From a biological point of view, and in particular, from the point of view of biostimulation, the main difference between a material (such as bulk copper or iron) and an NM (such as nanocopper NCu or nanoiron NFe) is that the NMs are very reactive when interacting with cell surfaces, giving rise to responses similar to those observed in situations of stress or exposure to other biostimulants such as humic substances, chitosan, or polysaccharides of fungal or bacterial pathogens (Juárez-Maldonado et al., 2021).

The greater chemical and biological reactivity of NMs results from a characteristic called surface energy. The amount of surface energy of a material is inversely proportional to the size of the particle; in other words, the smaller the particles, the higher the surface energy they will have. As the surface energy increases, the biostimulant capacity also increases because higher surface energy means a stronger interaction with the cell walls, cell membranes, proteins, and nucleic acids of the cells (Cheng et al., 2013; Yu et al., 2023).

NP-cell interactions will change the activity of cell wall proteins and ion transporters and proton pumps in cell membranes, causing metabolism and gene expression changes. As with other biostimulants, the ability of NMs to induce a certain type of response depends on the concentration applied at a certain moment and the dose obtained over time (Anderson et al., 2018; Juárez-Maldonado et al., 2019). Normally, the biological response observed for NMs is biphasic or hormesis, characterized by presenting an inverted U-shaped biological response, positive with a low concentration of NM, followed by a negative response when the concentration of NM increases (Agathokleous et al., 2019). According to different studies, the concentration range for a favorable biological response using NMs goes from 5 to 250 mg L^{-1}. Throughout this spectrum of concentrations, it is possible to observe positive impacts on different response variables of the plants. However, for certain combinations of NMs, plant species, and growth conditions, beneficial effects have been reported with higher concentrations, for example, 500–2000 mg L^{-1} (Afzal et al., 2021; Agrawal et al., 2022; Ali et al., 2021; Liu et al., 2021).

It should be noted that the spectrum of response to NMs applied in concentrations lower than 5 mg L^{-1} has been little studied. However, even such low concentrations are expected to have a biological impact, especially considering long-term or chronic exposures, since, in this case, the effect of the accumulated dose over time (the threshold of biological activity) is the relevant factor, not only the concentration. This long-term dose factor is one of the important points to consider when verifying the potential ecological and health impact of NMs (González-Morales et al., 2020; Lead et al., 2018).

The advantage of the ability of NMs to induce biological reactions when using nanocomposites is that it allows their use for a dual function. For example, a nanofertilizer that contains NFe will have a double impact on crops: it will provide Fe as a nutrient and will additionally serve as a biostimulant; in the same way, a nanocomposite with chitosan and Fe or Zn NPs or a nanocomposite with nanochitosan will have a double impact when applied to the soil, in the nutrient solution or by spraying on the plants: on the one hand it will increase the cation exchange capacity, and it will serve as a complexing agent for essential ions such as Zn^{2+} and Cu^{2+}, whereas the NPs of Fe, Zn, or chitosan will have a biostimulant impact per se, increasing tolerance to biotic or abiotic stress (Verma et al., 2022).

As a result of the synergism mentioned above, there is a potentially enormous number of biological responses to different NMs and mixtures thereof in nanocomposites. In addition, this diversity also results from the vast variety in the composition of the NMs; for example, there are NMs of metals and metal oxides (e.g., NPs of Fe and FeO), NMs of semimetals such as Se and Si, organic NMs (e.g., nanochitosan) and minerals (e.g., nanozeolites and nanoclays). Adding to the above, differences also occur in the same material depending on the size, shape, amount of surface charges, and amount of hydrophobic and hydrophilic sites of the nanostructure or NP. These differences in shape, size, and charges cause different biological responses, so they can be considered different materials from a biological point of view. For example, it has been observed that plants respond differently to NMs of the same composition that present variations in the size or shape of the NPs (Barciela et al., 2023; Juárez-Maldonado et al., 2019; Liu et al., 2021). Considering the above, it can be concluded that the valuable research effort developed so far has only explored a small section of all the opportunities for using nanocomposites as agricultural biostimulants.

On the other hand, studies with different plant species show that the response profile differs, for example, when exposed to NPs of the same composition but with a cubic, spherical, oval, cylindrical, or amorphous shape. The same occurs with the size of the NPs, observing that NPs of the same composition but of different sizes cause different responses in the cultures. The differences observed in the biological response refer to metabolism, gene expression, physiology, chemical or biochemical composition, and, in general, to growth, vigor, yield, or quality of the crop (Juárez-Maldonado et al., 2021).

Another factor that increases the diversity of plant responses to NMs is a phenomenon of the interaction of NMs surface charges with metabolites and proteins found in natural media such as water and soil and in biological fluids such as in the apoplast and the vacuole. The interaction between the NMs and these metabolites and proteins causes the NM to be coated with organic molecules such as humic substances, organic acids, peptides, and proteins by electrostatic attraction. This organic cover is called the corona (Fig. 8.2) (Ahsan et al., 2018). For the above reason, the NMs in nanocomposites, natural environments, or the internal environment

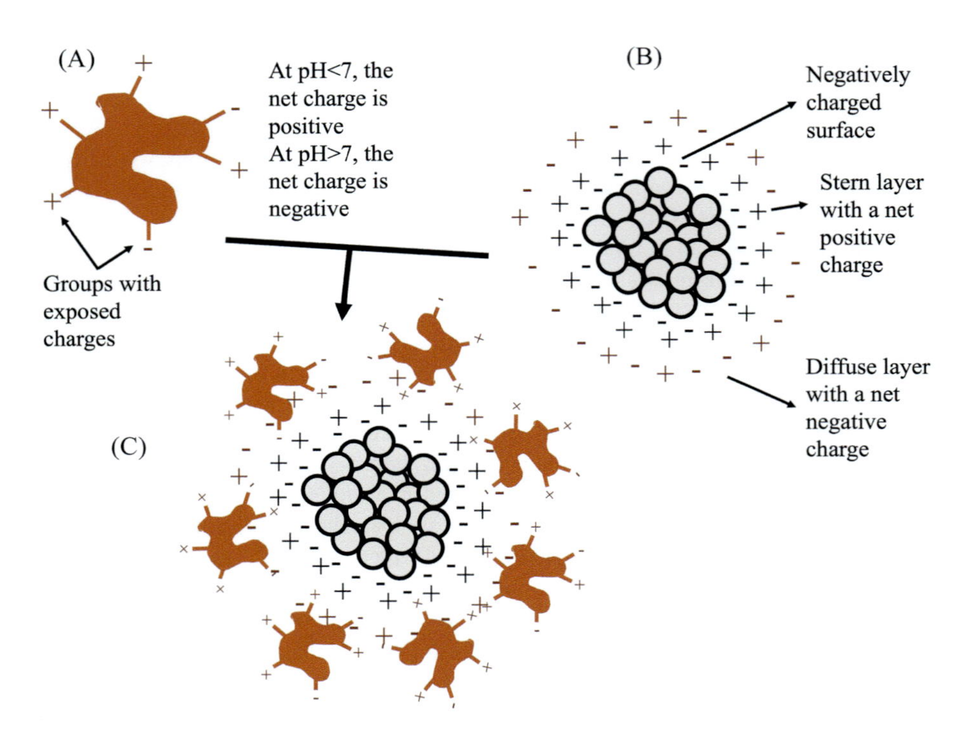

Figure 8.2 Representation of a protein corona formation from a nanomaterial in a natural environment. In (A), a protein is represented whose net surface charge will be positive at pH < 7 (as occurs in the apoplast and most cellular compartments). In addition to its net positive surface charge, the protein has negative and positive charges of exposed groups of the different amino acids that make up the polypeptide sequence. In (B), a pristine nanomaterial is represented whose surface charge is negative. These surface charges unite with positive charges from the chemical compounds in the environment and form a second layer of charges, called the Stern layer, a mixture of mostly positive charges and, therefore, with a net positive charge. A third layer of charges joins the Stern layer, called the diffuse layer, whose net charge is negative. In (C), the electrostatic attraction between the positive surface charges of the proteins and the negative charges of the diffuse layer of the nanomaterial to form a protein corona is represented. The corona will have on its surface the positive surface charge of the proteins and different negatively and positively charged groups. Both sets of charges will have strong interaction with cell surfaces.

of plants (apoplast) or cells (symplast) consist of two parts, an internal one called the nucleus and an external one called the corona. The formation of the corona of the NMs is an instantaneous event since it occurs by the union of opposite sign charges between the surfaces of proteins (or metabolites) with the surface of the NMs (Juárez-Maldonado et al., 2019; Xu et al., 2018).

The corona's composition varies according to the environment where the NMs are found and the exposure time. For nanocomposites with NMs of the same composition but located in different matrices, for example, lignin, chitosan, PLA, or PVA, the corona's composition will be different since the organic components present in each medium are different (Yu et al., 2023). As time passes, the composition of the corona is modified; initially, it forms a corona (called soft corona) whose components are subject to replacement. As time passes, the components become more strongly rooted in the NM surface and form a more stable structure (called hard corona or eco-corona). Some authors call a hard corona a first layer of proteins and metabolites that binds strongly (even irreversibly) to the surface of the NMs, and they call a soft corona a secondary layer of proteins and metabolites that attaches to the primary layer. This soft corona may be subject to a dynamic exchange of components with the environment. In the case of nanocomposites, the soft corona of the NMs is made up of the compound(s) or metabolite(s) used for its functionalization. For the NMs obtained from biological synthesis, the soft corona will be formed by a complex of proteins and metabolites of the cells that synthesize the NMs. However, once the NMs are incorporated into the nanocomposite matrix material, the corona's identity can change again due to interactions with the matrix's components, modifying the NMs' ability to induce biological responses (Da Costa et al., 2021; Kihara et al., 2021, 2019; Milani et al., 2012; Wu et al., 2023; Xu et al., 2018).

The polymers of the nanocomposite matrix and the NMs used as fillers can interact with plant cells and their microbiome, affecting their metabolism, gene expression, and cell signaling pathways. Biostimulation of crops by NMs is a process that occurs in two stages. The first refers to the rapid interaction of the NMs corona's surface charges on the cells' surface charges. That is, the high density of free energy concentrated in the corona of the NM modifies the activity of the integral proteins of the cell membranes and walls, in addition to causing the formation of pores in the membranes that are perceived by the cells as damage molecular patterns (DAMPs). These interfacial processes impact the activity of $NADH_2$/$NADPH_2$ receptors, proton pumps, channels, transporters, and oxidases in cell walls and membranes. The above changes the transmembrane potential and modifies the transport of ions and biomolecules, causing modifications in cell signaling and adjustments in gene expression (Juárez-Maldonado et al., 2019, 2021).

Another mechanism of surface interaction is presented by the formation of DAMPs in the corona of the NMs. Due to molecular distortion, proteins that bind to the corona change their folding pattern. This change in folding is perceived by cells as damage induced by stress. In response, they activate metabolic defense mechanisms and change their gene expression to adjust to the new situation. What happens is a type of priming induced by the NM (Farrera & Fadeel, 2015).

8.3.3 The second phase of plant cell biostimulation

After the interactions of the surfaces, in a second time that goes from minutes to hours or even days, the components of the core of the NMs and their corona are metabolized and released in the rhizosphere, the epidermis, the apoplast or in the cell cytoplasm in the form of ions, peptides, polypeptides and other functional groups that change the cell ionome and metabolome, leading to adjustments in gene expression. In certain cases, the NMs used as fillers or their matrices are not susceptible to being disintegrated by plant cells, or this process occurs slowly, often with the intervention of the plant microbiome (You et al., 2023). Examples of the above are the C NMs, nanoclays, and nanozeolites. In this case, the nanocomposites exert their immediate impact through physicochemical processes carried out by the integral nanostructures. An example of C NMs that can migrate to the chloroplasts and increase the efficiency of solar radiation capture has been described (Giraldo et al., 2014; Mony et al., 2022).

The induction of the above processes using nanocomposites causes modifications in the metabolic and physiological behavior of the crops. Among these changes, the following may be mentioned.

Nanocomposites can produce antioxidants and defense metabolites in plants, which can help mitigate oxidative stress caused by environmental factors like drought, high salinity, or heavy metal toxicity. The increased antioxidant activity can protect plant cells from the damaging effects of reactive species (Chavanke et al., 2022; Haydar et al., 2023). In the same way, the use of biological polymers as matrices of the nanocomposites can activate plant defense responses, making the plants more resistant to pathogens. This could be due to the direct antimicrobial effects of the NMs or their ability to stimulate the plant's defense signaling pathways (Ahmed et al., 2022). NMs can affect gene expression, upregulating the expression of stress response genes and enabling crops to better tolerate adverse conditions. Either by direct interaction with nucleic acids or through free radicals generated in response to the presence of NMs, nanocomposites can influence the synthesis, degradation, or signaling of metabolites, phytoalexins, and plant hormones such as ABA, ethylene, auxins, and gibberellins, which play a central role in plant responses to abiotic or biotic stress (Abideen et al., 2022; Gupta et al., 2019). Nanocomposites can modify the activity of ion channels on plant cell membranes. This could potentially help plants maintain ion homeostasis under stress conditions (Rahimzadeh & Ghassemi-Golezani, 2022).

8.3.4 Other processes of biostimulation

In addition to the previously described processes of direct cellular biostimulation by nanocomposites, other processes indirectly provide biostimulant potential. They are mentioned below.

The enhanced water and nutrient uptake or the controlled release of nutrients or bioactive compounds. NMs and nanocomposites can improve the uptake of nutrients by plants by adsorption of soil solution's ions and soil physical amelioration.

For example, nanochitosan, nanozeolites, or C NMs with high surface area and reactivity can adsorb nutrients and release them near the root zone in a more bioavailable form. Additionally, nanozeolites, nanoclays, CNTs, organic nanocomposites, and various metal oxide NPs can alter the physical and chemical properties of the soil, improving soil structure, moisture retention, and nutrient availability (Ojeda-Barrios et al., 2020).

On the other hand, applying nutrients in the nanometric form using nanofertilizers presents a series of advantages over conventional fertilizers. The greater stability against leaching, precipitation, and volatilization of the nanofertilizers can be mentioned. Regarding nanocomposites, the matrix carriers can be engineered to release their nutrient contents slowly or respond to specific triggers (such as changes in soil moisture or pH), thereby providing a more consistent supply of nutrients to crops and reducing the risk of nutrient leaching and volatilization (Naz et al., 2021). Additionally, nanocomposites based on chitosan (Kumaraswamy et al., 2018), and alginate, cellulose, pectin, and whey protein, among others, are used for encapsulating nutrients or bioactive compounds (Zabot et al., 2022). These nanocomposites can slowly release the encapsulated nutrients or active substances near the plant roots or on the leaf surface, ensuring that the plants receive a steady supply of nutrients. This controlled release can lead to enhanced stress tolerance and more growth yield or quality. The characteristics mentioned above increase the efficiency of nutrient uptake by the plants and less loss of nutrients to the environment and, therefore, less pollution (Zabot et al., 2022).

The improvement in the absorption of nutrients in the crops is also obtained indirectly by the alteration that occurs in the growth and morphology of the root in the presence of NMs and even by the presence of the materials used as matrices in nanocomposites, such as alginate, chitosan, β-glucans, and nanoclays. Certain NMs, such as nanosilica, nanochitosan, and zeolites, can stimulate the root length and the volume of exploration of soil and promote the production of root hairs, which increase the root surface area available for nutrient and water uptake (Agrawal et al., 2022). On the other hand, some NMs, like silica NPs, have been shown to enhance nutrient absorption in plants, potentially by changing root morphology or promoting the development of the root microbiome (Rajput et al., 2021).

Regarding water use efficiency, nanocomposites like hydrogels that contain superabsorbent materials can be used to improve water retention in the soil. These nanocomposites can absorb large quantities of water and plant nutrients and release them slowly, ensuring the plants have a steady water supply, especially in arid or drought-prone areas (Olad et al., 2018). This can enhance the growth and stress tolerance of the plants.

8.4 Applications of nanocomposites as biostimulants

Why is the contribution of biostimulants based on nanocomposites considered potentially relevant to the effort to produce crops under suboptimal and variable environmental conditions? Part of the answer is that the strategies available to mitigate the impact on crops of abiotic stresses are limited. This limitation results from the great complexity of the adaptive mechanisms of plants, with responses mediated by hundreds to thousands of genes, which complicates the efforts carried out through conventional breeding and biotechnological approaches. For this reason, techniques that increase the effectiveness of natural stress tolerance mechanisms, such as biostimulation using nanocomposites, can be beneficial (Teklić et al., 2020).

Table 8.3 contains results obtained in different crops using nanocomposites as biostimulants. Considering the multitude of available studies, it is impossible to include the entire universe of results. The review of Table 8.3 is useful to appreciate the potential usefulness of applying nanocomposites in agricultural activity.

The diversity of applications of nanocomposites can be seen in Table 8.3, directly as a biostimulant to induce tolerance to biotic and abiotic stress or improve the response of plants in their growth environment, in plant nutrition, to improve nutritional and nutraceutical quality, as herbicide, bactericide, or fungicide, to increase water or nutrient use efficiency, as vehicles for beneficial organisms, or to promote the microbiome of the soil or substrate.

The question arises: Of all the possible uses and types of material available (actually manufactured and experimentally tested or potentially designable materials for future use), should some be prioritized for commercial use?

The previous question should be answered in terms of the accessibility and cost of the materials to manufacture the nanocomposite, the ease of use in the field and the expected results, the profitability of its use, the feasibility of scaling it up for commercial production, and the utility in solving the problems faced by agricultural producers and agro-industrial companies. These problems are mainly related to the degradation of agricultural soils or the need to produce on marginal soils, the presence of extreme weather events such as droughts, temperature extremes, and the occurrence of pests and diseases, the greater need for real-time information about the performance of plants in the field and the variables of soil, water, and atmosphere, in addition to the complex commercial and regulatory environment associated with the ecologically and economically sustainable production and distribution of food products.

Considering the above, the question about whether some use or class of material should be prioritized can be answered as follows. The needs of the farmers who produce our food are multiple and require effective and quick responses. The farmers will have different requirements for their climate, soil, and crop. Some farmers will prioritize the efficient use of water, others the efficient use of nutrients, and some others the stress tolerance of their crops. Therefore, the development of multifaceted materials is required, in other words, nanocomposites whose use solves several productive issues at the same time.

Biostimulation is a biological response to environmental stimuli or challenges; it commonly involves a multivariate crop response. A biostimulatory multifaceted response is obtained from the interaction between the nanocomposite matrix and fillers with various receptors, signalers, and gene and metabolic effectors from plant cells. For this reason, the materials to be chosen as matrices and as fillers should be

Table 8.3 Results reported in the literature about the use of nanocomposites as biostimulants.

Crop	Nanocomposite	Conditions	Results	References
Alfalfa	The formulation was prepared by integrating deltamethrin-loaded, nanosized halloysite nanotubes into polyethylene films	Field study	The successful encapsulation of deltamethrin within halloysite nanotubes has been achieved, resulting in an enhancement of the properties of polyethylene (PE) films. Furthermore, it should be noted that films demonstrate a prolonged release mechanism of the active agent for 60 days. The insecticidal efficacy of films was evaluated in a greenhouse using Medicago Sativa plants infested with thrips and aphids. It was found that the nanocomposites exhibited repellent properties towards mature aphids and lethal effects on young aphids and thrips.	Seven et al. (2019)
Arabidopsis thaliana	Carbon nanotubes complexed with chitosan	In vitro and in vivo	In this study, the researchers employed chitosan-complexed single-walled carbon nanotubes, which were used with a lipid exchange envelope penetration mechanism. The objective was to achieve targeted delivery of plasmid DNA to chloroplasts in various plant species without needing external biolistic or chemical assistance. Chloroplast-targeted transgene delivery and transient expression were successfully observed in mature plants of Eruca sativa, Nasturtium officinale, Nicotiana tabacum, and Spinacia oleracea, as well as in isolated Arabidopsis thaliana mesophyll protoplasts.	Kwak et al. (2019)

(*Continued*)

Table 8.3 (Continued)

Crop	Nanocomposite	Conditions	Results	References
Banana tree	Copper oxide nanoparticles in a starch matrix	Banano seedlings	A nanocomposite was produced by employing *Aspergillus niger* AH1 to synthesize copper oxide nanoparticles (CuO NPs), which were then combined with starch. The efficacy of the nanocomposite in enhancing the process of root development and seedling growth in banana plants was assessed. The study's findings indicate that the medium supplemented with 9 mg L^{-1} nanocomposite exhibited the most significant increases in the number of leaves (8.56 leaves), roots formed per explants, and root length (6.40 cm). The chlorophyll content peaked at 9 mg L^{-1} of the nanocomposite, as indicated by the recorded SPAD values of 33.87. The study suggests that CuO NPs and the nanocomposite exhibit significant antifungal activity against *Penicillium expansum*, *Fusarium oxysporum*, and *Rhizoctonia solani*. Notably, the nanocomposite demonstrates a slightly higher efficacy compared to CuO NPs.	Hasanin et al. (2023)

Bermuda grass	Biochar nanocomposite, attapulgite clay, glyphosate, azobenzene, and amino silicon oil	Herbicide	In this study, a nanocomposite with light-responsively controlled-release herbicide particle (LCHP) was synthesized. The LCHP possesses a core-shell structure and was fabricated utilizing a nanocomposite material consisting of biochar, attapulgite (ATP), glyphosate (Gly), azobenzene (AZO), and amino silicon oil (ASO). The AZO molecules undergo trans-cis and cis-trans isomer transition under UV−Vis light radiation, serving as light-driven catalysts to facilitate the release of Gly from LCHP via nanopores. The light-responsively regulated release performance exhibited by LCHP was exceptional, as corroborated by pot studies. Significantly, LCHP showed a favorable adhesion performance on the surface of weed leaves, enhancing weed control measures' effectiveness. In addition, the presence of coexisting ions (CO_3^{2-}, SO_4^{2-}, and Cl^-) and variations in pH exhibited minimal influence on the release of LCHP in aqueous solution, thereby demonstrating the remarkable stability of the nanocomposite. Studies of the water contaminating capacity and the impact on soil nematodes indicated that the nanocomposite is biosafe.	Chen et al. (2018)

(*Continued*)

Table 8.3 (Continued)

Crop	Nanocomposite	Conditions	Results	References
Capsicum annuum	Zinc oxide (ZnO) nanocomposites encapsulated in chitosan	Field application	Applying chitosan-coated ZnO NPs on pepper plantlets cultivated in vitro demonstrated enhanced growth and biomass compared to the uncoated ZnO NPs. The utilization of the nanocomposite resulted in a significant enhancement in the levels of chlorophylls (51%), carotenoids (70%), proline (twofold), proteins (approximately twofold), alkaloids (60.5%), and soluble phenols (40%). The results of the micropropagation test demonstrated that the application of the nanocomposite treatment resulted in a significant enhancement of the organogenesis performance.	Asgari-Targhi et al. (2021)
Cavendish banana	Edible coating of nanochitosan 2% and *Moringa oleifera* 10%	Postharvest	The effect of enriching *Aloe vera* and *M. oleifera* plant extract edible coatings with chitosan nanoparticles on the postharvest quality of "Cavendish" bananas was studied. Fruits were stored at room temperature ($18 \pm 1°C$) for up to 30 days. Sampling was done at a five-day interval, and sensory evaluation was done after 25 days in storage. The results showed that adding chitosan nanoparticles to edible coating significantly impacted the firmness, ethylene rate, respiration, and total phenolic content during storage. *Moringa*-nanochitosan coating reduced weight loss (23%), respiration rate	Odetayo et al. (2022)

			(18 mg kg^{-1} h^{-1}), and ethylene generation (144 μL kg^{-1} h^{-1}) to a more significant extent than *Aloe*-nanochitosan. In addition, the combination treatments preserved fruit quality parameters, such as total soluble solids, fruit firmness, and peel color.	
Chickpea	Carbon nanofibers with Cu nanoparticles 50–100–500 μg mL^{-1}	Seed priming	An aqueous colloidal dispersion of a nanocomposite of carbon nanofibers and Cu NPs was used as a seed priming treatment. The water uptake capacity, germination rate, shoot and root lengths, and chlorophyll and protein contents significantly increased in the plants using the nanocomposite. The Cu NPs and C nanofibers improved the osmotic conditions, resulting in an increase in the water uptake capacity of the seeds, triggering the seed germination and growth at a relatively higher rate than the control. The shoot and root length of the seedlings, chlorophyll content, and the translocation of nutrients from root to shoot increased with increasing concentrations of the materials.	Ashfaq et al. (2017)

(Continued)

Table 8.3 (Continued)

Crop	Nanocomposite	Conditions	Results	References
Chickpea	Nanocomposites chitosan-ZnO, CuO, and Ag	In vitro and in vivo	Chitosan and different nanocomposites (50, 100, and 200 $\mu g\ mL^{-1}$) were evaluated as antifungal agents against *Fusarium oxysporum*, in vitro and in vivo. Among these nanocomposites, chitosan-CuO NPs and chitosan-ZnO NPs were the most effective against *F. oxysporum*. Chitosan nanoparticles and the nanocomposite chitosan-Ag NPs were moderately effective but more efficient than standard fungicides, i.e., copper-oxy-chloride (CuOCl). Based on in-vitro results, 100 $\mu g\ mL^{-1}$ of all nanoformulations was selected for in vivo studies in potted plants. The highest wilt disease reduction was observed using the nanocomposite chitosan-CuO NPs (46.67%), followed by plants treated with chitosan-ZnO (40%) as moderately effective. In comparison, chitosan-Ag NPs and chitosan NPs caused only a 33.33% reduction in wilt incidence. All nanoformulations showed good antifungal efficacy, inhibited the pathogen, and were found to promote chickpea plants' growth compared to control plants.	Kaur et al. (2018)

Chickpea	Cu − Zn/carbon nanofibers with a mixture of polyvinyl alcohol and starch nanocomposite 0.25–4 g kg^{-1}	In pot	A nanocomposite composed of a biodegradable polymeric formulation of PVA-starch was synthesized to facilitate the slow release of carbon nanofibers carrying the Cu-Zn micronutrients. The Cu-Zn-carbon material was dispersed in situ within the PVA-starch blend through polymerization. The efficacy of the nanocomposite was assessed by employing chickpea as a model plant species and administering various doses (0.25, 0.50, 1.0, 2.0, and 4.0 g of nanocomposite kg^{-1} soil) over 30 days. The translocation of the Cu-Zn-carbon from the roots to the shoots was confirmed through elemental analysis. The plants that received treatment exhibited the most significant height, measuring approximately 33 cm, while the control plants only reached a length of roughly 18 cm. The levels of superoxide anion radicals and hydrogen peroxide in the control plants were determined to be 207 ± 3.15 and 272 ± 5.74 nmol g^{-1} plant, respectively. On the other hand, it was observed that plants grown with nanocomposites exhibited concentrations of reactive oxygen species at 129 ± 3.25 and 194 ± 6.47 nmol g^{-1}, respectively.	Kumar et al. (2018)

(*Continued*)

Table 8.3 (Continued)

Crop	Nanocomposite	Conditions	Results	References
Citrus reticulata cv. Shatangju	Mesoporous silica and chitosan nanoparticles loaded with prochloraz (25.4% w/w)	In vitro and postharvest	Silica-chitosan-prochloraz NPs were synthesized by encapsulating the fungicide prochloraz within the pores of mesoporous silica NPs covalently bonded to chitosan. The nanoparticle formulations exhibited a comparatively elevated loading efficiency of prochloraz, with a weight-to-weight ratio of 25.4%. Furthermore, these NPs showed a notable improvement in the light stability of prochloraz. The NPs showed exceptional performance in terms of controlled release behavior. The activity test confirmed that the biological media, specifically citrus fruits, induced the release of pesticides as needed. In contrast to the prochloraz technical emulsifiable concentrate, the utilization of silica-chitosan-prochloraz NPs by preharvest application showed superior antifungal efficacy and a prolonged duration in combating fruit illnesses. The toxicity of the NPs toward the model organism zebrafish showed a reduction of over six times when compared to the toxicity of technical prochloraz.	Liang et al. (2018)

Coffea cane phora	Zn-B nanocomposite with chitosan nanoparticles (20–30 mg kg^{-1})	Greenhouse coffee tree seedlings, foliar application	The preparation of a Zn-B nanofertilizer involved the loading of $ZnSO_4$ and H_3BO_3 onto a chitosan NPs emulsion. The application of nanofertilizer was administered to the foliage of coffee seedlings using varying concentrations of 0, 10, 20, 30, and 40 mg L^{-1}. Using nanofertilizer resulted in increased absorption of Zn, N, and P. The findings indicated a significant enhancement in the coffee's chlorophyll content and photosynthetic activity. Utilizing the Zn-B-nanochitosan nanofertilizer enhanced coffee plant growth in terms of leaf area, plant height, and stem diameter.	Wang and Nguyen (2018)
Eggplant	Chitosan nanocomposites with EDTA and graphene oxide	Greenhouse	The production of nanocomposites consisting of chitosan (CS) or ethylene diamine tetraacetic acid (EDTA) on graphene oxide (GO) was undertaken. The study aimed to evaluate the nematocidal efficacy against *Meloidogyne incognita*, a pathogen responsible for root-knot infection in eggplant. The investigation of the plant immune response involved the quantification of photosynthetic pigments, phenols, and proline concentrations, assessing oxidative stress levels, and evaluating antioxidant enzyme activity. The study's findings reveal that the treatment involving pure GO exhibited the highest mortality rates for second-stage juveniles of Meloidogyne incognita. This was followed by the treatments using GO combined with chitosan (GO-CS)	Attia et al. (2021)

(*Continued*)

Table 8.3 (Continued)

Crop	Nanocomposite	Conditions	Results	References
			and GO combined with EDTA (GO-EDTA). The results obtained from in vivo greenhouse studies demonstrate that the treatment with the highest efficacy in reducing nematodes was GO-CS. This treatment exhibited 85.42%, 75.3%, 55.5%, 87.81%, and 81.32% reductions in the numbers of 2nd juveniles, galls, females, egg masses, and the developmental stage, respectively. The GO-CS treatment exhibited the highest levels of chlorophyll a (104%), chlorophyll b (46%), total phenols (137.5%), and free proline (145.2%). The MDA value was obtained using GO-EDTA (7.22%), and the hydrogen peroxide (H_2O_2) content increased by 47.51% following the application of pure GO at the maximum treatment level. The application of GO-CS resulted in a significant enhancement in the enzymatic activities of catalase (CAT) by 98.3%, peroxidase (POD) by 97.52%, polyphenol oxidase (PPO) by 113.8%, and superoxide dismutase (SOD) by 42.43%.	

Fortunella margarita	Nanocomposite of CuO NPs encapsulated in a biopolymeric coating of chitosan and sodium alginate at 10 mg kg^{-1} (CuO/PEC-10)	Seed priming	The efficacy of a chitosan and sodium alginate complex was evaluated as a biodegradable encapsulation material for the controlled release of CuO NPs. The data gathered confirms spontaneous polymeric aggregation within spherical nanostructures, which have an average diameter of around 300 nm. Furthermore, the data indicates that CuO is encapsulated within the polymeric coating. In addition, the observation was made regarding the considerable deceleration of copper release from the hybrid nanocomposite due to the presence of the biopolymer shell. The study investigated the impact of three different concentrations (10, 50, and 100 mg L^{-1}) of a nanocomposite solution on the germination process of Fortunella margarita Swingle seeds. After a comparative analysis of all treatments, it was observed that the germination rates were highest for the 10 and 50 mg L^{-1} treatments after 31 days, but these differences were not statistically significant. The treatment with a composite concentration of 10 mg L^{-1} exhibited the highest rate of plumule emergence at 80%, substantially different from the rates seen in the other treatments. The combined treatment of 10 mg L^{-1} showed the greatest results in terms of vigor index for both radicles and plumules. The application of nanocomposite treatments positively impacted root extension and the number of secondary roots seen.	Leonardi et al. (2021)

(*Continued*)

Table 8.3 (Continued)

Crop	Nanocomposite	Conditions	Results	References
Maize	Chitosan nanoemulsion loaded with α-terpineol	Shelf life of corn kernels	An α-terpineol nanocomposite encapsulated in chitosan nanoemulsion was synthesized to control fungal aflatoxin B1 and the deterioration of stored maize grain induced by free radicals. The nanocomposite exhibited enhanced antifungal activity against the aflatoxin-secreting strain of *Aspergillus flavus* and 12 other food-borne molds and aflatoxin B1 production at 0.4 and 0.3 $mL\ L^{-1}$, respectively. Further, the nanocomposite inhibited ergosterol synthesis, the aflatoxin enhancer methylglyoxal, and cellular contents release. On the other hand, the nanocomposite showed enhanced radical scavenging activity with IC50 values equivalent to 39.57 and 6.23 $mL\ L^{-1}$ for DPPH and ABTS, respectively. In addition, the nanocomposite completely inhibited aflatoxin B1 production in stored maize samples during *in situ* investigation.	Kumar Chaudhari et al. (2020)

Maize	Seed coat fabricated using nanohydroxyapatite, zinc oxide nanoparticles, and polycaprolactone	Seed cover and priming	The study utilized a biocompatible electrospun polycaprolactone nanofiber as a coating for maize seeds. This coating included synthesized nanohydroxyapatite (P source) and ZnO NPs. The study on pH, conductivity measurements, and metal release demonstrated a regulated release of nutrients from the nanocomposite. During the storage period, the seed was safeguarded by a layer of electrospun porous polymer nanofibrous membrane. This layer facilitates the exchange of water and gases, augmenting the seed's germination potential. The controlled water absorption rate in seeds covered with polymer nanofibers enhanced seed germination during the initial stages, as evidenced by investigations into swelling and imbibition. Applying a polymer nanofiber coating facilitated controlled water absorption and mitigated imbibitional damage. The data presented in the study demonstrates that using the polymeric formulation effectively retards the release of Zn. Based on the findings of statistical analysis, it can be concluded that the consistent availability of nutrients in the vicinity of the seedling significantly influenced the establishment of the plant population and the percentage of successful germination.	Chakkalakkal et al. (2022)

(*Continued*)

Table 8.3 (Continued)

Crop	Nanocomposite	Conditions	Results	References
Onion	Multiwalled carbon nanotubes with Zn 15 $\mu g\ mL^{-1}$	Seed priming	A nanocomposite was fabricated to harness the synergistic effects of ZnO and multiwalled CNTs. The objective was to regulate the provision of micronutrients and promote the growth of onion seeds, with the ultimate aim of substituting chemical fertilizers in onion plants cultivated in arid environments. The nanocomposite exhibited optimal seedling growth and the highest number of cells in telophase at a concentration of 15 $\mu g\ mL^{-1}$. The observed growth trend, under an increase in the concentration of Zn-multiwalled CNTs, did not affect plant growth.	Kumar et al. (2018)
Potato	Chitosan-acacia gum nanocomposite loaded with mancozeb (with 1.0 $mg\ mL^{-1}$ mancozeb) at 1.5 $mg\ L^{-1}$	In vitro and greenhouse	The fungicide mancozeb was incorporated into chitosan-gum acacia polymers to create a nanocomposite with a size of 363.6 nm. The nanocomposite, which had a concentration of 1.0 $mg\ mL^{-1}$ mancozeb, displayed a peak inhibition of $83.8 \pm 0.7\%$ against *Alternaria solani* when used at a concentration of 1.5 $mg\ L^{-1}$. In the case of *Sclerotinia sclerotiorum*, 100% inhibition was observed at concentrations of 1.0 and 1.5 $mg\ L^{-1}$ using the in vitro mycelium inhibition method. The commercial formulation of mancozeb exhibited in vitro inhibitory effects of $84.6 \pm 0\%$ and 100% against both fungi. The disease control efficacy in pathogen-treated plants using the	Kumar, Duhan et al. (2022)

			nanocomposite was 64.6 ± 5.0% and 60.2 ± 1.4% against early blight and stem rot diseases, respectively. At the specified concentration, the nanocomposite exhibited a lower level of cytotoxicity compared to the commercially available mancozeb.	
Potato	Selenium nanocomposites based on polymeric matrices of arabinogalactan (5.92% Se), and carrageenan (3.67% Se)	Seed priming and field assay	Chemically synthesized Se nanocomposites were utilized as nanopriming agents for potato tubers in a field study conducted over multiple years. Additionally, these nanocomposites were employed in a germination study with soybeans, and their in vitro impact on the phytopathogen *Pectobacterium carotovorum* was assessed. The nanocomposites were based on natural polymeric matrices of arabinogalactan, carrageenan, and starch. The findings of field experiments indicate that applying arabinogalactan and carrageenan nanocomposites as preplant treatments for tubers resulted in a noteworthy augmentation in both the stem count and tuber yield per plant in potato plants. However, it should be noted that the observed increase in arabinogalactan nanocomposite was limited to specific years, and the positive impact of both arabinogalactan nanocomposite and carrageenan nanocomposite on tuber weight was only observed in a single year. The influence of all nanocomposites increased in the proportion of marketable seed tubers within the yield structure. Furthermore, it was demonstrated that the nanocomposites	Perfileva et al. (2023)

(Continued)

Table 8.3 (Continued)

Crop	Nanocomposite	Conditions	Results	References
			stimulate the biomass of soybean seedling roots. Applying *P. carotovorum* during seed biopriming resulted in the activation of antioxidant enzymes and a subsequent reduction in the levels of diene conjugates, which exhibited a significant increase. The study demonstrated the antibacterial properties of arabinogalactan nanocomposite against *P. carotovorum*, which decreased bacterial growth, biofilm formation, and cellular dehydrogenase activity.	
Rice	Chitosan-Mg nanocomposite (Mg 100 μg mL^{-1})	Field culture and in vitro	The bactericidal activity of chitosan-iron nanocomposites was evaluated through in vitro and in vivo experiments against *Xanthomonas oryzae* pv. *oryzae* (Xoo), the causative agent of bacterial leaf blight (BLB) disease in rice. Furthermore, the influence of nanocomposites on the endophytic microbiome of healthy and BLB-diseased rice was assessed using a high-throughput sequencing methodology. The nanocomposites exhibited a spherical morphology with an average diameter of 86 nm. The results of in vitro antibacterial assays demonstrated that the nanocomposites exhibited a significant inhibitory effect on the pathogen's biological activities at a concentration of 250 mg L^{-1} compared to the control group. A greenhouse experiment	Ahmed et al. (2022)

			investigated the impact of foliar exposure to nanocomposites at a concentration of 250 mg L^{-1} on bacterial leaf blight disease incidence. The results showed that the nanocomposite treatment significantly reduced the disease incidence by 67.1%. This reduction was attributed to the modulation of antioxidant enzymes, specifically superoxide dismutase, peroxidase, and ascorbate peroxidase, which exhibited 49.2%, 38.8%, and 53.4%, respectively. Furthermore, applying nanocomposites also positively impacted the rice plants' photosynthesis efficiency. This was evident through increased production of total chlorophyll (43.2%) and carotenoids (60.0%). Additionally, the nutritional profile of the rice plants was improved compared to the untreated diseased control. Moreover, the disease resistance response induced by nanocomposites was associated with an upregulation of defense-related genes, including OsPRs, OsSOD, and OsAPX, in rice plants. The utilization of high-throughput sequencing techniques yielded findings indicating that applying nanocomposite amendment reduced the relative abundance of Xanthomonas bacteria by approximately 87.5%. This reduction was observed by altering the bacterial community associated with both the phyllosphere (the aerial parts of the rice plant) and the endophytic regions of the plant's	

(*Continued*)

Table 8.3 (Continued)

Crop	Nanocomposite	Conditions	Results	References
			roots. Furthermore, applying beneficial native microorganisms (BNCs) has enhanced the diversity of bacteria in healthy and diseased plants. Noticeable enhancements in the comparative prevalence of advantageous bacteria were observed in plants afflicted with disease and those in good health following treatment with BNCs.	
Rice	Polysaccharide-Se nanocomposite	Rice seedlings	Crude polysaccharides from spent mushroom substrate were extracted for preparing a polysaccharide-selenium nanocomposite. The results revealed that the nanocomposite exhibited great free radical scavenging ability. Rice seedlings treated with AaPs-SeNPs showed significant enhancements in growth characteristics compared to control, and foliar application exhibited a better growth-promoting effect than root application. Moreover, adding the nanocomposite enhanced rice seedlings' growth performance and antioxidant enzyme activities, and the absorption efficiency of essential nutrients N, P, K, Fe, Zn, and Mn was also improved.	Peng et al. (2023)

Tomato	Nanoparticles with 1.0 mg mL^{-1} of mancozeb in chitosan-acacia gum biopolymers	Pot assay and in vitro assay	The process involved the incorporation of mancozeb into chitosan-gum acacia biopolymers to create a nanoparticulate matrix with a diameter ranging from 322.2 ± 0.9 to 403.7 ± 0.7 nm. The nanocomposite containing 1.0 mg mL^{-1} mancozeb (CSGA-1.0) demonstrated a significant inhibitory effect of 85.2 ± 0.7% on *Alternaria alternata* at a concentration of 0.5 mg L^{-1}. In the mycelium inhibition method, the inhibition of *Stemphylium lycopersici* by CSGA-1.0 was measured to be 62.1 ± 0.7%. The antimicrobial action of nanocomposites was observed in pot greenhouse environments. The release of mancozeb from the nanoformulations occurred after ten hours, which can be attributed to the diffusion and relaxation of the polymer matrix. In contrast, the commercial mancozeb exhibited a release time of only 2 hours. Although drug-loaded conjugated nanoparticles exhibit comparable antifungal activities, their release rate is lower, resulting in reduced toxicity compared to commercial mancozeb.	Kumar, Nain, et al. (2022)

(Continued)

Table 8.3 (Continued)

Crop	Nanocomposite	Conditions	Results	References
Toadskin melon	Ag-chitosan nanocomposites	Edible coating	Ag-chitosan nanocomposites were integrated into chitosan coatings, which were subsequently evaluated for their impact on the quality of fresh-cut melon over 13 days at a temperature of 5°C. The application of coating treatments resulted in a decrease in the respiration rate of fresh-cut melon. Applying chitosan derived from red claw crayfish, including red claw crayfish-derived Ag-chitosan nanocomposites, exhibited a reduced respiratory rate level throughout the storage period compared to the other coatings. The prevention of softening during extended storage periods was achieved by applying a coating composed of chitosan derived from red claw crayfish, including nanocomposites of Ag-chitosan extracted from red claw crayfish. The coating treatments did not significantly affect the color, soluble solids content, sucrose, glucose and fructose, pH, TA, and citric and malic acids. In addition, it was observed that the application of a coating containing chitosan derived from red claw crayfish, including nanocomposites of red claw crayfish-derived Ag-chitosan, resulted in the highest concentration of total vitamin C after a 13-day storage period at a temperature of 5°C, surpassing the other coating treatments.	Ortiz-Duarte et al. (2019)

Tomato	Chitosan nanoparticles loaded with vanillin and cinnamaldehyde	Greenhouse and in vitro	Composites were synthesized through the process of intercalating chitosan into sodium montmorillonite, incorporating chitosan with polyaniline, and combining chitosan/polyaniline/exfoliated montmorillonite as carriers for the antioxidants vanillin and cinnamaldehyde. The researchers conducted an in vitro assay to investigate the antifungal activity of the substance against *Fusarium oxysporum* and *Pythium debaryanum*. The results demonstrated a significant inhibitory effect on the linear growth of these pathogens, even at lower concentrations. The results of the greenhouse assay demonstrated that the application of chitosan-polyaniline-cinnamaldehyde and chitosan-sodium montmorillonite-cinnamaldehyde treatments on seedlings resulted in a reduction of both disease index and disease incidence parameters for both pathogens. Furthermore, these treatments exhibited the ability to promote the growth of tomato seedlings when compared to untreated-infected controls.	Elsherbiny et al. (2022)

(*Continued*)

Table 8.3 (Continued)

Crop	Nanocomposite	Conditions	Results	References
Tomato	Hydroxyapatite nanoparticles loaded on carboxymethylcellulose	Seed priming	The effects of hydroxyapatite nanoparticles, which were stabilized using carboxymethylcellulose, were assessed on tomato plants' germination, seedling growth, and metabolism. The germination percentage in *S. lycopersicum* remains unaffected by the escalation of nanocomposite concentrations. However, it was observed that root elongation experienced a significant enhancement as a result of this treatment. Tomato plants cultivated using hydroponic techniques and exposed to the nanocomposite have exhibited no signs of phytotoxicity. In conclusion, it has been determined that the nanocomposite exhibited negligible toxicity towards the model plant. Furthermore, it has demonstrated potential utility as a provider of phosphorus and carrier for various other elements and molecules.	Marchiol et al. (2019)

Tomato	Nanocomposite of hydrogel-nano natural carbon (starch base) in three levels 0%, 0.3%, and 0.6%	Greenhouse	Hydrogel-nano natural char composite (reinforced starch-based hydrogels with natural char NPs) was applied to tomato plants to study the nutritional and morphological responses. Additionally, the study examined the effects of this composite on certain soil biological properties under water-deficit stress conditions. The composite was tested at three different levels: 0%, 0.3%, and 0.6% (w/w). The water-deficit stress was induced by subjecting the plants to three levels of water-holding capacity: 50%, 75%, and 85%. The various nanocomposite and water deficit stress levels significantly influenced the plant morpho-nutritional indices and soil microbial traits. The assay results indicated a decrease in all measured parameters under water-deficit stress conditions. Nevertheless, utilizing nanocomposite materials has been found to mitigate the detrimental impacts of water-deficit stress on the growth and development of tomatoes. The extent of the reactions to the nanocomposite treatment varied based on the concentration of the applied nanocomposite and the severity of stress. The most notable improvements were observed in the growth of tomato plants, with an increase ranging from 22% to 45%, as well as in the nutritional indices, specifically the concentrations of phosphorus (P), iron (Fe), and zinc (Zn), which showed an increase ranging from 16% to 29%. These positive effects were observed when using a 0.3%	Nassaj-Bokharaei et al. (2021)

(Continued)

Table 8.3 (Continued)

Crop	Nanocomposite	Conditions	Results	References
			hydrogel nanocomposite and maintaining 85% water-holding capacity. Furthermore, applying a 0.6% hydrogel nanocomposite and maintaining 75% water-holding capacity resulted in a significant increase in soil respiration rate, with a 61% increase, and the microbial population's size showed an 89% increase.	
Tomato	Natural carbon nanoparticles with starch-*g*-poly(styrene-*co*-butyl acrylate) latex	Tomato plants	Slow-release fertilizer nanocomposites were developed by employing a polymeric coating that is both bio-based and waterborne. This coating consists of starch-*g*-poly(styrene-*co*-butylacrylate) latex, further strengthened by including natural char nanoparticles. A series of nanocomposite samples were fabricated by applying latex formulations containing varying quantities of char nanoparticles onto urea granules. Subsequently, the release rates of urea from these samples were assessed. The study's findings indicate that including char nanofiller and temperature reduction can potentially extend the duration of urea release. Nanocomposite samples' efficacy was compared to urea granules and a commercially available latex coating (poly(styrene-co-butyl acrylate)) regarding their impact on tomato production within a greenhouse setting. The study's findings confirmed that including char	Salimi et al. (2021)

			NPs in the coating formulation resulted in enhanced plant growth and a decrease in the rate of nitrogen release. The efficiencies of all the synthesized nanocomposite samples were superior to that of the latex-coated urea, thus confirming the significant role of starch in the coating formulations. In a sample with a concentration of 1 wt.% of char NPS, the nitrogen content and nitrate levels in the aboveground portions of tomato plants exhibited an increase of 21.1% and a decrease of 42.1%, respectively, compared to the original urea. Significantly, the utilization of a formulation containing 1% char NPs has been observed to improve nitrogen use efficiency, nitrogen agronomic efficiency, and nitrogen apparent recovery fraction by 68.3%, 49.9%, and 29.7%, respectively, compared to urea granules.	

(*Continued*)

Table 8.3 (Continued)

Crop	Nanocomposite	Conditions	Results	References
Tomato	Polyhydroxyalkanoate (PHA) nanocomposites and calcium phosphate nanoparticles (Ca-P-NPs)	Drench application on tomato plants	Biodegradable polymer nanocomposites were synthesized using polyhydroxyalkanoate and calcium phosphate NPs to deliver P in soil or substrate. Their performance was compared to a conventional P source ($CaHPO_4$) by conducting experiments on tomato plants. The efficacy of the nanocomposites was evaluated by quantifying plant biomass, fruit yield and quality, tissue elemental and chlorophyll content, and enzymatic biomarkers. Using phosphorus in the leachate served as a proxy for assessing runoff. The nanocomposites undergo biodegradation due to microbial processes occurring in the soil, thereby regulating the release of P during the early phases of plant development. The nanocomposites exhibited comparable effects on plant performance as the conventional P source while demonstrating a substantial reduction in soil phosphorus loss by more than 80%.	Sigmon et al. (2021)

Wheat	NPK-loaded chitosan-poly(methacrylic acid) NPs	Pot assay	The efficacy of a chitosan nanocomposite encapsulating nitrogen N, P, and K was evaluated in wheat plants administered via foliar spray. The study's findings indicated that nanocomposites were effectively absorbed by the surfaces of leaves and stomata and subsequently transported through the phloem tissues. The application of nanocomposites-NPK to wheat plants cultivated in sandy soil resulted in notable enhancements in the harvest index, crop index, and mobilization index of the measured wheat yield variables in comparison to the control variables of wheat plants treated with standard nonfertilized and conventionally fertilized NPK. The duration of the life cycle of wheat plants that were fertilized with nanoparticles was found to be shorter when compared to wheat plants that were fertilized conventionally. Specifically, the nano-fertilized wheat plants exhibited a reduction of 23.5% in their life cycle, with a period of 130 days from sowing to yield production, in contrast to the 170 days observed in the conventional-fertilized wheat plants.	Abdel-Aziz et al. (2016)

(*Continued*)

Table 8.3 (Continued)

Crop	Nanocomposite	Conditions	Results	References
Wheat	Chitosan and P and ZnO nanocomposites	Seed priming and nanofertilizer	The release stability and impact on wheat plants of nanocomposites consisting of tripolyphosphate-chitosan and tripolyphosphate-chitosan-ZnO nanofertilizers were examined. Utilizing nanocomposites resulted in a substantial decrease in the overall phosphorus leaching over 72 hours. Specifically, the nanocomposites exhibited 91% and 97% reductions compared to a conventional fertilizer, monoammonium phosphate (MAP). The nanofertilizers showed a significant decrease in cumulative phosphorus leaching after 72 hours, with 84% and 95% lower values than the control tripolyphosphate. Adding ZnO to tripolyphosphate-chitosan resulted in a 65% increase in effectiveness in reducing P leaching compared to tripolyphosphate-chitosan alone. Compared to the control treatment (MAP), the application of tripolyphosphate-chitosan-ZnO resulted in a statistically significant increase in wheat plant height, with a mean increase of 33.0%. Compared to MAP, applying tripolyphosphate-chitosan and tripolyphosphate-chitosan-ZnO resulted in a notable increase in wheat grain yield, with increments of 21% and 30%, respectively. It is	Dimkpa et al. (2023)

			worth mentioning that the utilization of tripolyphosphate-chitosan-ZnO resulted in a substantial reduction of shoot phosphorus (P) levels. Specifically, the reduction percentages were 35.5%, 47%, and 45% compared to MAP, tripolyphosphate, and tripolyphosphate-chitosan, respectively. The release of zinc over 72 hours from tripolyphosphate-chitosan-ZnO nanoparticles was significantly lower compared to the control group of ZnO nanoparticles. The average release of zinc (Zn) from the nanocomposite tripolyphosphate-chitosan-ZnO was found to be 34.7 mg L^{-1} and 0.065 mg L^{-1}, respectively. Notably, the release of Zn from ZnO NPs was 534 times higher than the other components. The zinc concentration in grains was found to be significantly higher in the tripolyphosphate-chitosan treatment treatment than in the treatment involving MAP. The application of tripolyphosphate-chitosan resulted in a notable increase in the mobilization of potassium (K), sulfur (S), magnesium (Mg), and calcium (Ca) from the soil to the plant. This phenomenon enhances the overall nutritional composition and reinforces the role of chitosan in facilitating nutrient mobilization.	

(*Continued*)

Table 8.3 (Continued)

Crop	Nanocomposite	Conditions	Results	References
Wheat	Ce-Mn ferrite nanocomposite at 100 mg kg^{-1}	Seed priming	A nanocomposite was synthesized to harness the beneficial effects of Mn, Fe, and Ce NPs on crop growth. The nanocomposite was administered to wheat L. seeds at 100, 250, and 500 mg L^{-1} concentrations for 10 days. The germination rate exhibited a notable increase of 15%. Furthermore, the exposure of seedlings to a concentration of 100 mg L^{-1} resulted in a significant augmentation of total chlorophyll levels by 61% and carotenoid levels by 38%. A notable enhancement of 14% in the absorption of Fe and Mn in the aerial parts of the plants, coupled with an 18% increase in overall productivity, was observed. A direct relationship was observed when quantifying the dose-dependent effect of the nanocomposite on the total content of superoxide dismutase and peroxidase.	Zarinkoob et al. (2021)

Zoysia matrella	Nanocomposite of attapulgite, NH_4HCO_3, amino silicone oil, PVA, and glyphosate	Herbicide	A herbicide with temperature-responsive controlled-release properties was formulated by incorporating attapulgite, NH_4HCO_3, amino silicon oil, poly(vinyl alcohol) (PVA), and glyphosate into a nanocomposite material. The core of the mixture consists of attapulgite-NH_4HCO_3-glyphosate, while the shell is composed of amino silicon oil and PVA. Adenosine triphosphate (ATP) exhibits a micro/nano network structure characterized by porosity, enabling it to bind numerous glycine (Gly) molecules effectively. The quantity of pores can be effectively regulated by temperature. In the context of this discussion, it is observed that the PVA shell exhibits a propensity to undergo dissolution in an aqueous solution when subjected to elevated temperatures. This characteristic enables the effective regulation of Gly release. Utilizing the nanocomposite has demonstrated its ability to significantly mitigate glyphosate loss during simulated rainfall events, thereby enhancing the efficacy of weed control measures. The hydrophobic amino silicone oil in the herbicide exhibits notable stability when dissolved in an aqueous solution, maintaining its integrity for a minimum duration of three months.	Chi et al. (2017)

selected from among those capable of acting as DAMPs or microbe/pathogen associated molecular patterns (M/PAMPs) (e.g., chitosan, alginate, or nanozeolites), in addition to improving the retention of water and nutrients in the soil or substrate or well directly provide nutrients to plants (e.g., hydrogels filled with NPs of Fe, Zn, Se, and Co).

8.5 Conclusion, trends, and possibilities

The questions raised in the preceding section allow us to establish a scenario for the conclusions, trends, and possibilities regarding the use of nanocomposites in agricultural activity (Iavicoli et al., 2017; Mittal et al., 2020; Sampathkumar et al., 2020).

Biostimulation of plants with nanocomposites represents a revolutionary and promising approach to sustainable agriculture. Through the efficient delivery of nutrients, enhancement of plant stress tolerance, and harvest quality and resilience, it has the potential to significantly improve crop yields while reducing environmental impacts. However, at the moment, there are limitations to be resolved to ensure the wide use of nanocomposites in agriculture.

As mentioned before, developing tailored solutions is necessary: future developments could lead to the creation of tailor-made nanocomposites designed for specific crops or environments. Additionally, as precision agriculture evolves, integrating nanocomposite-based biostimulation with real-time monitoring systems can allow for more targeted and efficient applications of biostimulants.

Another limitation is the lack of comprehensive understanding and research data on how the components of nanocomposites (particularly the NMs and nanofertilizers) interact with plants, microbiomes, and soil over the long term. Related to the above, concerns about the health and safety of ecosystems, trophic chains, and humans (farmers and consumers) cannot be overlooked. Little information about NMs' dynamics and long-term impact is available, especially in soil microbiomes and the microbiomes of different organisms. Assessing the toxicity and environmental impact of nanocomposites is crucial for regulatory compliance. Conducting comprehensive toxicity studies and environmental risk assessments can be complex and resource-intensive.

Regarding the regulatory challenges: Regulatory entities worldwide are grappling with how to assess, regulate, and monitor nanotechnology (including nanocomposites) applications in agriculture. The diverse composition, properties, and applications of nanocomposites make it challenging to devise a one-size-fits-all regulatory approach, and the need to balance innovation with potential environmental and health risks creates further challenges.

Partially related to the above, there is public skepticism about using nanotechnology in agriculture, often due to fears about potential environmental and health impacts. Creating awareness among farmers, agronomists, and other stakeholders about nanocomposites' potential applications and advantages is vital for market acceptance. Educating end-users about using, handling, and disposing of nanocomposites can help address concerns or misconceptions.

On the other hand, while nanotechnology has made great strides, the cost of producing nanocomposites and other NMs at a scale large enough to be practical for agricultural use remains a significant barrier. This also includes challenges related to manufacturing processes, quality control, and the practicalities of delivering and using these materials on the farm efficiently and effectively. Demonstrating the tangible benefits of nanocomposites in agriculture is crucial for market acceptance, and communicating these benefits effectively to potential users is essential for driving adoption.

The previous points indicate areas that require investment of talent, resources, and time to solve them. Surely new areas of opportunity in research on nanocomposites will emerge in the coming years.

Regarding the challenges of lab-based research toward the industrial implementation of nanocomposites in agriculture, one of the primary challenges in implementing lab-based research findings into industrial applications is the scale-up process. The transition from small-scale laboratory experiments to large-scale production involves numerous complexities that must be addressed. Some key scale-up challenges include:

1. **Uniform Dispersion:** Achieving uniform dispersion of NPs within the polymer matrix is crucial for obtaining consistent properties in nanocomposites. In lab-based research, researchers often have better control over dispersion due to smaller sample sizes and specialized equipment. However, achieving the same level of dispersion at an industrial scale can be challenging due to factors like increased volume, mixing limitations, and processing conditions.
2. **Processing Techniques:** Lab-scale experiments often utilize techniques that are not easily scalable to industrial production. For example, researchers may use specialized equipment or manual processes that are not feasible for large-scale manufacturing. Developing scalable processing techniques that can maintain the desired properties of nanocomposites is a significant challenge.
3. **Cost Considerations:** Scaling up production introduces cost considerations that may not have been relevant during lab-based research. Factors such as raw material costs, equipment expenses, energy consumption, and waste management must be carefully evaluated to ensure the economic viability of industrial implementation. The cost-effectiveness of nanocomposites compared to existing alternatives is a significant consideration for market adoption. If the cost of implementing nanocomposites outweighs the perceived benefits, it may hinder their widespread adoption in agriculture.
4. **Standardization and Certification:** Industrial implementation requires adherence to standardized protocols and certification processes. Developing standardized testing methods specific to nanocomposites in agriculture is an ongoing challenge. Additionally, obtaining certifications from regulatory bodies adds another layer of complexity to the industrial implementation process.

References

Abdel-Aziz, H. M. M., Hasaneen, M. N. A., & Omer, A. M. (2016). Nano chitosan-NPK fertilizer enhances the growth and productivity of wheat plants grown in sandy soil. *Spanish Journal of Agricultural Research*, *14*(1), 0902. Available from https://doi.org/10.5424/sjar/2016141-8205, http://revistas.inia.es/index.php/sjar/article/view/8205.

Abdelhamid, A. E., Ahmed, E. H., Awad, H. M., & Ayoub, M. M. H. (2023). Synthesis and cytotoxic activities of selenium nanoparticles incorporated nano-chitosan. *Polymer Bulletin*. Available from https://doi.org/10.1007/s00289-023-04768-8, https://doi.org/10.1007/s00289-023-04768-8.

Abideen, Z., Hanif, M., Munir, N., & Nielsen, B. L. (2022). Impact of nanomaterials on the regulation of gene expression and metabolomics of plants under salt stress. *Plants*, *11* (5), 691. Available from https://doi.org/10.3390/plants11050691, https://www.mdpi.com/2223-7747/11/5/691.

Afzal, S., Aftab, T., & Singh, N. K. (2021). Impact of zinc oxide and iron oxide nanoparticles on uptake, translocation, and physiological effects in *Oryza sativa* L. *Journal of Plant Growth Regulation*. Available from https://doi.org/10.1007/s00344-021-10388-1, https://doi.org/10.1007/s00344-021-10388-1.

Agathokleous, E., Feng, Z. Z., Iavicoli, I., & Calabrese, E. J. (2019). The two faces of nanomaterials: A quantification of hormesis in algae and plants. *Environment International*, *131*105044. Available from https://doi.org/10.1016/j.envint.2019.105044, http://www.sciencedirect.com/science/article/pii/S0160412019321610.

Agrawal, S., Kumar, V., Kumar, S., & Kumar Shahi, S. (2022). Plant development and crop protection using phytonanotechnology: A new window for sustainable agriculture. *Chemosphere*, *299*134465. Available from https://doi.org/10.1016/j.chemosphere.2022.134465, https://www.sciencedirect.com/science/article/pii/S0045653522009584.

Ahmed, T., Noman, M., Jiang, H., Shahid, M., Ma, C., Wu, Z., Nazir, M. M., Ali, M. A., White, J. C., Chen, J., & Li, B. (2022). Bioengineered chitosan-iron nanocomposite controls bacterial leaf blight disease by modulating plant defense response and nutritional status of rice (*Oryza sativa* L.). *Nano Today*, *45*101547. Available from https://doi.org/10.1016/j.nantod.2022.101547, https://www.sciencedirect.com/science/article/pii/S174801322200175X.

Ahsan, S. M., Rao, C. M., Ahmad, M. F., Saquib, Q., Faisal, M., Al-Khedhairy, A. A., & Alatar, A. A. (2018). *Nanoparticle-protein interaction: The significance and role of protein corona cellular and molecular toxicology of nanoparticles* (pp. 175–198). Cham: Springer International Publishing. Available from https://doi.org/10.1007/978-3-319-72041-8_11.

Ali, S. S., Al-Tohamy, R., Koutra, E., Moawad, M. S., Kornaros, M., Mustafa, A. M., Mahmoud, Y. A.-G., Badr, A., Osman, M. E. H., Elsamahy, T., Jiao, H., & Sun, J. (2021). Nanobiotechnological advancements in agriculture and food industry: Applications, nanotoxicity, and future perspectives. *Science of The Total Environment*, *792*148359. Available from https://doi.org/10.1016/j.scitotenv.2021.148359, https://www.sciencedirect.com/science/article/pii/S0048969721034306.

Anderson, A. J., McLean, J. E., Jacobson, A. R., & Britt, D. W. (2018). CuO and ZnO nanoparticles modify interkingdom cell signaling processes relevant to crop production. *Journal of Agricultural and Food Chemistry*, *66*(26), 6513–6524. Available from https://doi.org/10.1021/acs.jafc.7b01302, https://doi.org/10.1021/acs.jafc.7b01302.

Armentano, I., Puglia, D., Luzi, F., Arciola, C. R., Morena, F., Martino, S., & Torre, L. (2018). Nanocomposites based on biodegradable polymers. *Materials*, *11*(5), 795. Available from https://doi.org/10.3390/ma11050795, https://www.mdpi.com/1996-1944/11/5/795.

Arya, A., Ahamad, A., Kumar, P., & Chandra, A. (2023). A review on lignin based nanocomposites: Fabrication, characterization and application. *Materials Today: Proceedings*. Available from https://doi.org/10.1016/j.matpr.2023.03.569, https://www.sciencedirect.com/science/article/pii/S2214785323016371.

Asgari-Targhi, G., Iranbakhsh, A., Oraghi Ardebili, Z., & Hatami Tooski, A. (2021). Synthesis and characterization of chitosan encapsulated zinc oxide (ZnO) nanocomposite and its biological assessment in pepper (*Capsicum annuum*) as an elicitor for in vitro tissue culture applications. *International Journal of Biological Macromolecules, 189*, 170–182. Available from https://doi.org/10.1016/j.ijbiomac.2021.08.117, https://www.sciencedirect.com/science/article/pii/S0141813021017773.

Ashfaq, M., Verma, N., & Khan, S. (2017). Carbon nanofibers as a micronutrient carrier in plants: Efficient translocation and controlled release of Cu nanoparticles. *Environmental Science: Nano, 4*(1), 138–148. Available from https://doi.org/10.1039/C6EN00385K, https://pubs.rsc.org/en/content/articlelanding/2017/en/c6en00385k.

Attia, M. S., El-Sayyad, G. S., Abd Elkodous, M., Khalil, W. F., Nofel, M. M., Abdelaziz, A. M., Farghali, A. A., El-Batal, A. I., & El Rouby, W. M. A. (2021). Chitosan and EDTA conjugated graphene oxide antinematodes in eggplant: Toward improving plant immune response. *International Journal of Biological Macromolecules, 179*, 333–344. Available from https://doi.org/10.1016/j.ijbiomac.2021.03.005, https://www.sciencedirect.com/science/article/pii/S0141813021005225.

Barciela, P., Carpena, M., Li, N.-Y., Liu, C., Jafari, S. M., Simal-Gandara, J., & Prieto, M. A. (2023). Macroalgae as biofactories of metal nanoparticles; biosynthesis and food applications. *Advances in Colloid and Interface Science, 311*102829. Available from https://doi.org/10.1016/j.cis.2022.102829, https://www.sciencedirect.com/science/article/pii/S0001868622002317.

Barua, S., Gogoi, S., Khan, R., Karak, N., & Karak, N. (2019). *Chapter 8 - Silicon-based nanomaterials and their polymer nanocomposites. Nanomaterials and Polymer Nanocomposites* (pp. 261–305). Elsevier. Available from https://www.sciencedirect.com/science/article/pii/B9780128146156000084.

Bauli, C. R., Lima, G. F., de Souza, A. G., Ferreira, R. R., & Rosa, D. S. (2021). Eco-friendly carboxymethyl cellulose hydrogels filled with nanocellulose or nanoclays for agriculture applications as soil conditioning and nutrient carrier and their impact on cucumber growing. *Colloids and Surfaces A: Physicochemical and Engineering Aspects, 623*126771. Available from https://doi.org/10.1016/j.colsurfa.2021.126771, https://www.sciencedirect.com/science/article/pii/S0927775721006403.

Bibi, A., Rehman, S.-ur, & Yaseen, A. (2019). Alginate-nanoparticles composites: Kinds, reactions and applications. *Materials Research Express, 6*(9)092001. Available from https://doi.org/10.1088/2053-1591/ab2016, https://doi.org/10.1088/2053-1591/ab2016.

Cabrera-De la Fuente, M., Ortega-Ortiz, H., Juárez-Maldonado, A., Sandoval-Rangel, A., González-Morales, S., Cadenas-Pliego, G., & Benavides-Mendoza, A. (2020). Use of chitosan-polyacrylic acid (CS-PAA) complex, chitosan-polyvinyl alcohol (CS-PVA) and chitosan hydrogels in greenhouses as a carrier for beneficial elements, nanoparticles, and microorganisms. *Acta Horticulturae* (1296), 1153–1160. Available from https://doi.org/10.17660/ActaHortic.2020.1296.146, https://www.actahort.org/books/1296/1296_146.htm.

Camargo, P. H. C., Satyanarayana, K. G., & Wypych, F. (2009). Nanocomposites: Synthesis, structure, properties and new application opportunities. *Materials Research, 12*, 1–39. Available from https://doi.org/10.1590/S1516-14392009000100002, http://www.scielo.br/j/mr/a/53qXWM7k3BwVR74PZ8YGS9t/?lang = en.

Chakkalakkal, N. D., Thomas, M., Chittillapilly, P. S., Sujith, A., & Anjali, P. D. (2022). Electrospun polymer nanocomposite membrane as a promising seed coat for controlled release of agrichemicals and improved germination: Towards a better agricultural prospect. *Journal of Cleaner Production, 377*134479. Available from https://doi.org/10.1016/j.jclepro.2022.134479, https://www.sciencedirect.com/science/article/pii/S0959652622040513.

Chakraborty, R., Mukhopadhyay, A., Paul, S., Sarkar, S., & Mukhopadhyay, R. (2023). Nanocomposite-based smart fertilizers: A boon to agricultural and environmental sustainability. *Science of The Total Environment, 863*160859. Available from https://doi.org/10.1016/j.scitotenv.2022.160859, https://www.sciencedirect.com/science/article/pii/S0048969722079621.

Chavanke, S. N., Penna, S., & Dalvi, S. G. (2022). β-Glucan and its nanocomposites in sustainable agriculture and environment: An overview of mechanisms and applications. *Environmental Science and Pollution Research, 29*(53), 80062–80087. Available from https://doi.org/10.1007/s11356-022-20938-z, https://doi.org/10.1007/s11356-022-20938-z.

Chen, C., Zhang, G., Dai, Z., Xiang, Y., Liu, B., Bian, P., Zheng, K., Wu, Z., & Cai, D. (2018). Fabrication of light-responsively controlled-release herbicide using a nanocomposite. *Chemical Engineering Journal, 349*, 101–110. Available from https://doi.org/10.1016/j.cej.2018.05.079, https://www.sciencedirect.com/science/article/pii/S1385894718308933.

Cheng, L.-C., Jiang, X., Wang, J., Chen, C., & Liu, R.-S. (2013). Nano–bio effects: Interaction of nanomaterials with cells. *Nanoscale, 5*(9), 3547–3569. Available from https://doi.org/10.1039/C3NR34276J, https://pubs.rsc.org/en/content/articlelanding/2013/nr/c3nr34276j.

Chi, Y., Zhang, G., Xiang, Y., Cai, D., & Wu, Z. (2017). Fabrication of a temperature-controlled-release herbicide using a nanocomposite. *ACS Sustainable Chemistry & Engineering, 5*(6), 4969–4975. Available from https://doi.org/10.1021/acssuschemeng.7b00348, https://doi.org/10.1021/acssuschemeng.7b00348.

Da Costa, L. S., Khan, L. U., Franqui, L. S., de Souza Delite, F., Muraca, D., Martinez, D. S. T., & Knobel, M. (2021). Hybrid magneto-luminescent iron oxide nanocubes functionalized with europium complexes: Synthesis, hemolytic properties and protein corona formation. *Journal of Materials Chemistry B, 9*(2), 428–439. Available from https://doi.org/10.1039/D0TB02454F, https://pubs.rsc.org/en/content/articlelanding/2021/tb/d0tb02454f.

De, B., & Karak, N. (2019). *Chapter 7 - Carbon dots and their polymeric nanocomposites. Nanomaterials and polymer nanocomposites* (pp. 217–260). Elsevier. Available from https://www.sciencedirect.com/science/article/pii/B9780128146156000072.

Dhand, V., Rhee, K. Y., Ju Kim, H., & Ho Jung, D. (2013). A comprehensive review of graphene nanocomposites: Research status and trends. *Journal of Nanomaterials, 2013*e763953. Available from https://doi.org/10.1155/2013/763953, https://www.hindawi.com/journals/jnm/2013/763953/?utm_source = google&utm_medium = cpc&utm_campaign = HDW_MRKT_GBL_SUB_ADWO_PAI_DYNA_JOUR_IJELC_X0000&gclid = CjwKCAjwue6hBhBVEiwA9YTx8PwAjjHMeCwKyT2WJAdx8pQIE_Sn59ev_ia3SXBo34q1V-z1UBvnhxoCQrYQAvD_BwE.

Dimkpa, C. O., Deng, C., Wang, Y., Adisa, I. O., Zhou, J., & White, J. C. (2023). Chitosan and zinc oxide nanoparticle-enhanced tripolyphosphate modulate phosphorus leaching in soil. *ACS Agricultural Science & Technology, 3*(6), 487–498. Available from https://doi.org/10.1021/acsagscitech.3c00054, https://doi.org/10.1021/acsagscitech.3c00054.

El-Ganainy, S. M., Soliman, A. M., Ismail, A. M., Sattar, M. N., Farroh, K. Y., & Shafie, R. M. (2023). Antiviral activity of chitosan nanoparticles and chitosan silver nanocomposites against alfalfa mosaic virus. *Polymers, 15*(13), 2961. Available from https://doi.org/10.3390/polym15132961, https://www.mdpi.com/2073-4360/15/13/2961.

Elsherbiny, A. S., Galal, A., Ghoneem, K. M., & Salahuddin, N. A. (2022). Novel chitosan-based nanocomposites as ecofriendly pesticide carriers: Synthesis, root rot inhibition and growth management of tomato plants. *Carbohydrate Polymers, 282*119111. Available from https://doi.org/10.1016/j.carbpol.2022.119111, https://www.sciencedirect.com/science/article/pii/S0144861722000157.

Farrera, C., & Fadeel, B. (2015). It takes two to tango: Understanding the interactions between engineered nanomaterials and the immune system. *European Journal of Pharmaceutics and Biopharmaceutics, 95*, 3–12. Available from https://doi.org/10.1016/j.ejpb.2015.03.007, https://www.sciencedirect.com/science/article/pii/S0939641115001381.

Gamage, A., Thiviya, P., Mani, S., Ponnusamy, P. G., Manamperi, A., Evon, P., Merah, O., & Madhujith, T. (2022). Environmental properties and applications of biodegradable starch-based nanocomposites. *Polymers, 14*(21), 4578. Available from https://doi.org/10.3390/polym14214578, https://www.mdpi.com/2073-4360/14/21/4578.

Garavand, F., Cacciotti, I., Vahedikia, N., Rehman, A., Tarhan, Ö., Akbari-Alavijeh, S., Shaddel, R., Rashidinejad, A., Nejatian, M., Jafarzadeh, S., Azizi-Lalabadi, M., Khoshnoudi-Nia, S., & Jafari, S. M. (2022). A comprehensive review on the nanocomposites loaded with chitosan nanoparticles for food packaging. *Critical Reviews in Food Science and Nutrition, 62*(5), 1383–1416. Available from https://doi.org/10.1080/10408398.2020.1843133, https://doi.org/10.1080/10408398.2020.1843133.

Giraldo, J. P., Landry, M. P., Faltermeier, S. M., McNicholas, T. P., Iverson, N. M., Boghossian, A. A., Reuel, N. F., Hilmer, A. J., Sen, F., Brew, J. A., & Strano, M. S. (2014). Plant nanobionics approach to augment photosynthesis and biochemical sensing. *Nature Materials, 13*(4), 400–408. Available from https://doi.org/10.1038/nmat3890, https://www.nature.com/articles/nmat3890.

González-Morales, S., Parera, C. A., Juárez-Maldonado, A., la Fuente, M. C. D., Benavides-Mendoza, A., Husen, A., & Jawaid, M. (2020). *Chapter 13 - The ecology of nanomaterials in agroecosystems. Nanomaterials for Agriculture and Forestry Applications* (pp. 313–355). Elsevier. Available from http://www.sciencedirect.com/science/article/pii/B9780128178522000135.

Guha, T., Gopal, G., Kundu, R., & Mukherjee, A. (2020). Nanocomposites for delivering agrochemicals: A comprehensive review. *Journal of Agricultural and Food Chemistry, 68*(12), 3691–3702. Available from https://doi.org/10.1021/acs.jafc.9b06982, https://doi.org/10.1021/acs.jafc.9b06982.

Gupta, G. S., Kumar, A., & Verma, N. (2019). Bacterial homoserine lactones as a nanocomposite fertilizer and defense regulator for chickpeas. *Environmental Science: Nano, 6*(4), 1246–1258. Available from https://doi.org/10.1039/C9EN00199A, https://pubs.rsc.org/en/content/articlelanding/2019/en/c9en00199a.

Hasanin, M., Hashem, A. H., Lashin, I., & Hassan, S. A. M. (2023). In vitro improvement and rooting of banana plantlets using antifungal nanocomposite based on myco-synthesized copper oxide nanoparticles and starch. *Biomass Conversion and Biorefinery, 13*(10), 8865–8875. Available from https://doi.org/10.1007/s13399-021-01784-4, https://doi.org/10.1007/s13399-021-01784-4.

Haydar, M. S., Ali, S., Mandal, P., Roy, D., Roy, M. N., Kundu, S., Kundu, S., & Choudhuri, C. (2023). Fe–Mn nanocomposites doped graphene quantum dots alleviate salt stress of Triticum aestivum through osmolyte accumulation and antioxidant defense. *Scientific Reports, 13*(1)11040. Available from https://doi.org/10.1038/s41598-023-38268-6, https://www.nature.com/articles/s41598-023-38268-6.

Hernández-Hernández, H., González-Morales, S., Benavides-Mendoza, A., Ortega-Ortiz, H., Cadenas-Pliego, G., & Juárez-Maldonado, A. (2018). Effects of chitosan–PVA and Cu nanoparticles on the growth and antioxidant capacity of tomato under saline stress. *Molecules*, *23*(1), 178. Available from https://doi.org/10.3390/molecules23010178, https://www.mdpi.com/1420-3049/23/1/178.

Iavicoli, I., Leso, V., Beezhold, D. H., & Shvedova, A. A. (2017). Nanotechnology in agriculture: Opportunities, toxicological implications, and occupational risks. *Toxicology and Applied Pharmacology*, *329*, 96–111. Available from https://doi.org/10.1016/j.taap.2017.05.025, https://www.sciencedirect.com/science/article/pii/S0041008X17302314.

Jafarzadeh, S., Forough, M., Amjadi, S., Javan Kouzegaran, V., Almasi, H., Garavand, F., & Zargar, M. (2022). Plant protein-based nanocomposite films: A review on the used nanomaterials, characteristics, and food packaging applications. *Critical Reviews in Food Science and Nutrition*, 1–27. Available from https://doi.org/10.1080/10408398.2022.2070721, https://doi.org/10.1080/10408398.2022.2070721.

Juárez-Maldonado, A., Ortega-Ortíz, H., Morales-Díaz, A. B., González-Morales, S., Morelos-Moreno, Á., Cabrera-De la Fuente, M., Sandoval-Rangel, A., Cadenas-Pliego, G., & Benavides-Mendoza, A. (2019). Nanoparticles and nanomaterials as plant biostimulants. *International Journal of Molecular Sciences*, *20*(1), 162. Available from https://doi.org/10.3390/ijms20010162, https://www.mdpi.com/1422-0067/20/1/162.

Juárez-Maldonado, A., Tortella, G., Rubilar, O., Fincheira, P., & Benavides-Mendoza, A. (2021). Biostimulation and toxicity: The magnitude of the impact of nanomaterials in microorganisms and plants. *Journal of Advanced Research*, *31*, 113–126. Available from https://doi.org/10.1016/j.jare.2020.12.011, https://www.sciencedirect.com/science/article/pii/S2090123220302617.

Karaca, M., Ince, A. G., Farooq, M., Gogoi, N., & Pisante, M. (2023). *Chapter 8 - Revisiting sustainable systems and methods in agriculture. Sustainable agriculture and the environment* (pp. 195–246). Academic Press. Available from https://www.sciencedirect.com/science/article/pii/B978032390500800004X.

Kassem, I., Ablouh, E.-H., El Bouchtaoui, F.-Z., Kassab, Z., Khouloud, M., Sehaqui, H., Ghalfi, H., Alami, J., & El Achaby, M. (2021). Cellulose nanocrystals-filled poly (vinyl alcohol) nanocomposites as waterborne coating materials of NPK fertilizer with slow release and water retention properties. *International Journal of Biological Macromolecules*, *189*, 1029–1042. Available from https://doi.org/10.1016/j.ijbiomac.2021.08.093, https://www.sciencedirect.com/science/article/pii/S0141813021017529.

Kaur, P., Duhan, J. S., & Thakur, R. (2018). Comparative pot studies of chitosan and chitosan-metal nanocomposites as nano-agrochemicals against fusarium wilt of chickpea (*Cicer arietinum* L.). *Biocatalysis and Agricultural Biotechnology*, *14*, 466–471. Available from https://doi.org/10.1016/j.bcab.2018.04.014, http://www.sciencedirect.com/science/article/pii/S1878818118300707.

Khan, F., Ijaz, M., Akhlaq, A., Nawaz, S., Munawar, J., Rashid, E. U., Castro, G. R., Nadda, A. K., Nguyen, T. A., Sharma, S., & Bilal, M. (2023). *16 - Applications of nanomaterials to build a sustainable agriculture system. Nanomaterials for bioreactors and bioprocessing applications* (pp. 427–453). Elsevier. Available from https://www.sciencedirect.com/science/article/pii/B9780323917827000138.

Khan, S., Zahoor, M., Sher Khan, R., Ikram, M., & Islam, N. U. (2023). The impact of silver nanoparticles on the growth of plants: The agriculture applications. *Heliyon*, *9*(6) e16928. Available from https://doi.org/10.1016/j.heliyon.2023.e16928, https://linkinghub.elsevier.com/retrieve/pii/S240584402304135X.

Kihara, S., Köper, I., Mata, J. P., & McGillivray, D. J. (2021). Reviewing nanoplastic toxicology: It's an interface problem. *Advances in Colloid and Interface Science*,

*288*102337. Available from https://doi.org/10.1016/j.cis.2020.102337, https://www.sciencedirect.com/science/article/pii/S0001868620306060.

Kihara, S., van der Heijden, N. J., Seal, C. K., Mata, J. P., Whitten, A. E., Köper, I., & McGillivray, D. J. (2019). Soft and hard interactions between polystyrene nanoplastics and human serum albumin protein Corona. *Bioconjugate Chemistry*, *30*(4), 1067−1076. Available from https://doi.org/10.1021/acs.bioconjchem.9b00015, https://doi.org/10.1021/acs.bioconjchem.9b00015.

Kumar, R., Ashfaq, M., & Verma, N. (2018). Synthesis of novel PVA−starch formulation-supported Cu−Zn nanoparticle carrying carbon nanofibers as a nanofertilizer: Controlled release of micronutrients. *Journal of Materials Science*, *53*(10), 7150−7164. Available from https://doi.org/10.1007/s10853-018-2107-9, https://doi.org/10.1007/s10853-018-2107-9.

Kumar, R., Duhan, J. S., Manuja, A., Kaur, P., Kumar, B., & Kumar Sadh, P. (2022). Toxicity assessment and control of early blight and stem rot of *Solanum tuberosum* L. by mancozeb-loaded chitosan−gum acacia nanocomposites. *Journal of Xenobiotics*, *12* (2), 74−90. Available from https://doi.org/10.3390/jox12020008, https://www.mdpi.com/2039-4713/12/2/8.

Kumar, R., Nain, V., & Duhan, J. S. (2022). An ecological approach to control pathogens of *Lycopersicon esculentum* L. by slow release of mancozeb from biopolymeric conjugated nanoparticles. *Journal of Xenobiotics*, *12*(4), 329−343. Available from https://doi.org/10.3390/jox12040023, https://www.mdpi.com/2039-4713/12/4/23.

Kumar, V., Sachdev, D., Pasricha, R., Maheshwari, P. H., & Taneja, N. K. (2018). Zinc-supported multiwalled carbon nanotube nanocomposite: A synergism to micronutrient release and a smart distributor to promote the growth of onion seeds in arid conditions. *ACS Applied Materials & Interfaces*, *10*(43), 36733−36745. Available from https://doi.org/10.1021/acsami.8b13464, https://doi.org/10.1021/acsami.8b13464.

Kumaraswamy, R. V., Kumari, S., Choudhary, R. C., Pal, A., Raliya, R., Biswas, P., & Saharan, V. (2018). Engineered chitosan based nanomaterials: Bioactivities, mechanisms and perspectives in plant protection and growth. *International Journal of Biological Macromolecules*, *113*, 494−506. Available from https://doi.org/10.1016/j.ijbiomac.2018.02.130, https://www.sciencedirect.com/science/article/pii/S0141813017338412.

Kumar Chaudhari, A., Singh, A., Kumar Singh, V., Kumar Dwivedy, A., Das, S., Grace Ramsdam, M., Dkhar, M. S., Kayang, H., & Kishore Dubey, N. (2020). Assessment of chitosan biopolymer encapsulated α-Terpineol against fungal, aflatoxin B1 (AFB1) and free radicals mediated deterioration of stored maize and possible mode of action. *Food Chemistry*, *311*126010. Available from https://doi.org/10.1016/j.foodchem.2019.126010, https://www.sciencedirect.com/science/article/pii/S0308814619321533.

Kwak, S.-Y., Lew, T. T. S., Sweeney, C. J., Koman, V. B., Wong, M. H., Bohmert-Tatarev, K., Snell, K. D., Seo, J. S., Chua, N.-H., & Strano, M. S. (2019). Chloroplast-selective gene delivery and expression in planta using chitosan-complexed single-walled carbon nanotube carriers. *Nature Nanotechnology*, *14*(5), 447−455. Available from https://doi.org/10.1038/s41565-019-0375-4, https://www.nature.com/articles/s41565-019-0375-4.

Lead, J. R., Batley, G. E., Alvarez, P. J. J., Croteau, M.-N., Handy, R. D., McLaughlin, M. J., Judy, J. D., & Schirmer, K. (2018). Nanomaterials in the environment: Behavior, fate, bioavailability, and effects—An updated review. *Environmental Toxicology and Chemistry*, *37*(8), 2029−2063. Available from https://doi.org/10.1002/etc.4147, https://setac.onlinelibrary.wiley.com/doi/abs/10.1002/etc.4147.

Leonardi, M., Caruso, G. M., Carroccio, S. C., Boninelli, S., Curcuruto, G., Zimbone, M., Allegra, M., Torrisi, B., Ferlito, F., & Miritello, M. (2021). Smart nanocomposites of chitosan/alginate nanoparticles loaded with copper oxide as alternative nanofertilizers. *Environmental Science: Nano*, *8*(1), 174–187. Available from https://doi.org/10.1039/D0EN00797H, https://pubs.rsc.org/en/content/articlelanding/2021/en/d0en00797h.

Liang, Y., Fan, C., Dong, H., Zhang, W., Tang, G., Yang, J., Jiang, N., & Cao, Y. (2018). Preparation of MSNs-chitosan@prochloraz nanoparticles for reducing toxicity and improving release properties of prochloraz. *ACS Sustainable Chemistry & Engineering*, *6*(8), 10211–10220. Available from https://doi.org/10.1021/acssuschemeng.8b01511, https://doi.org/10.1021/acssuschemeng.8b01511.

Liu, Y., Xiao, Z., Chen, F., Yue, L., Zou, H., Lyu, J., & Wang, Z. (2021). Metallic oxide nanomaterials act as antioxidant nanozymes in higher plants: Trends, meta-analysis, and prospect. *Science of The Total Environment*, *780*146578. Available from https://doi.org/10.1016/j.scitotenv.2021.146578, https://www.sciencedirect.com/science/article/pii/S0048969721016466.

Madzokere, T. C., Murombo, L. T., & Chiririwa, H. (2021). Nano-based slow releasing fertilizers for enhanced agricultural productivity. *Materials Today: Proceedings*, *45*, 3709–3715. Available from https://doi.org/10.1016/j.matpr.2020.12.674, https://www.sciencedirect.com/science/article/pii/S2214785320403918.

Marchiol, L., Filippi, A., Adamiano, A., Degli Esposti, L., Iafisco, M., Mattiello, A., Petrussa, E., & Braidot, E. (2019). Influence of hydroxyapatite nanoparticles on germination and plant metabolism of tomato (*Solanum lycopersicum* L.): Preliminary evidence. *Agronomy*, *9*(4), 161. Available from https://doi.org/10.3390/agronomy9040161, https://www.mdpi.com/2073-4395/9/4/161.

Mehta, M. R., Mahajan, H. P., & Hivrale, A. U. (2021). Green synthesis of chitosan capped-copper nano biocomposites: Synthesis, characterization, and biological activity against plant pathogens. *BioNanoScience*, *11*(2), 417–427. Available from https://doi.org/10.1007/s12668-021-00823-8, https://doi.org/10.1007/s12668-021-00823-8.

Merino, D., Tomadoni, B., Salcedo, M. F., Mansilla, A. Y., Casalongué, C. A., Alvarez, V. A., Kharissova, O. V., Torres Martínez, L. M., & Kharisov, B. I. (2020). *Nanoclay as carriers of bioactive molecules applied to agriculture. Handbook of nanomaterials and nanocomposites for energy and environmental applications* (pp. 1–22). Cham: Springer International Publishing. Available from https://doi.org/10.1007/978-3-030-11155-7_62-1.

Milani, S., Baldelli Bombelli, F., Pitek, A. S., Dawson, K. A., & Rädler, J. (2012). Reversible versus irreversible binding of transferrin to polystyrene nanoparticles: Soft and hard Corona. *ACS Nano*, *6*(3), 2532–2541. Available from https://doi.org/10.1021/nn204951s, https://doi.org/10.1021/nn204951s.

Mittal, D., Kaur, G., Singh, P., Yadav, K., & Ali, S. A. (2020). Nanoparticle-based sustainable agriculture and food science: Recent advances and future outlook. *Frontiers in Nanotechnology*, *2*. Available from https://doi.org/10.3389/fnano.2020.579954, https://www.frontiersin.org/articles/10.3389/fnano.2020.579954/full.

Mony, C., Kaur, P., Rookes, J. E., Callahan, D. L., Eswaran, S. V., Yang, W., & Manna, P. K. (2022). Nanomaterials for enhancing photosynthesis: Interaction with plant photosystems and scope of nanobionics in agriculture. *Environmental Science: Nano*, *9*(10), 3659–3683. Available from https://doi.org/10.1039/D2EN00451H, https://pubs.rsc.org/en/content/articlelanding/2022/en/d2en00451h.

Mukherjee, A., & Patel, J. S. (2020). Seaweed extract: Biostimulator of plant defense and plant productivity. *International Journal of Environmental Science and Technology*, *17*(1), 553–558. Available from https://doi.org/10.1007/s13762-019-02442-z, https://doi.org/10.1007/s13762-019-02442-z.

Mukhtar Ahmed, K. B., Khan, M. M. A., Siddiqui, H., & Jahan, A. (2020). Chitosan and its oligosaccharides, a promising option for sustainable crop production—A review. *Carbohydrate Polymers, 227*115331. Available from https://doi.org/10.1016/j.carbpol.2019.115331, https://www.sciencedirect.com/science/article/pii/S0144861719309981.

Munonde, T. S., & Nomngongo, P. N. (2021). Nanocomposites for electrochemical sensors and their applications on the detection of trace metals in environmental water samples. *Sensors, 21*(1), 131. Available from https://doi.org/10.3390/s21010131, https://www.mdpi.com/1424-8220/21/1/131.

Nassaj-Bokharaei, S., Motesharezedeh, B., Etesami, H., & Motamedi, E. (2021). Effect of hydrogel composite reinforced with natural char nanoparticles on improvement of soil biological properties and the growth of water deficit-stressed tomato plant. *Ecotoxicology and Environmental Safety, 223*112576. Available from https://doi.org/10.1016/j.ecoenv.2021.112576, https://www.sciencedirect.com/science/article/pii/S0147651321006886.

Naz, M., Benavides-Mendoza, A., & Husen, A. (2021). *Nanofertilizers as tools for plant nutrition and plant biostimulation under adverse environment plant performance under environmental stress: Hormones, biostimulants and sustainable plant growth management* (pp. 387–415). Cham: Springer International Publishing. Available from https://doi.org/10.1007/978-3-030-78521-5_15.

Odetayo, T., Sithole, L., Shezi, S., Nomngongo, P., Tesfay, S., & Ngobese, N. Z. (2022). Effect of nanoparticle-enriched coatings on the shelf life of *Cavendish bananas. Scientia Horticulturae, 304*111312. Available from https://doi.org/10.1016/j.scienta.2022.111312, https://www.sciencedirect.com/science/article/pii/S0304423822004332.

Ojeda-Barrios, D. L., Morales, I., Juárez-Maldonado, A., Sandoval-Rangel, A., Fuentes-Lara, L. O., Benavides-Mendoza, A., Srivastava, A. K., & Hu, C. (2020). *Importance of nanofertilizers in fruit nutrition. Fruit crops* (pp. 497–508). Elsevier. Available from http://www.sciencedirect.com/science/article/pii/B9780128187326000356.

Olad, A., Zebhi, H., Salari, D., Mirmohseni, A., & Reyhani Tabar, A. (2018). Slow-release NPK fertilizer encapsulated by carboxymethyl cellulose-based nanocomposite with the function of water retention in soil. *Materials Science and Engineering: C, 90*, 333–340. Available from https://doi.org/10.1016/j.msec.2018.04.083, https://www.sciencedirect.com/science/article/pii/S0928493117325535.

Ortiz-Duarte, G., Pérez-Cabrera, L. E., Artés-Hernández, F., & Martínez-Hernández, G. B. (2019). Ag-chitosan nanocomposites in edible coatings affect the quality of fresh-cut melon. *Postharvest Biology and Technology, 147*, 174–184. Available from https://doi.org/10.1016/j.postharvbio.2018.09.021, https://www.sciencedirect.com/science/article/pii/S0925521418304290.

Othman, S. H. (2014). Bio-nanocomposite materials for food packaging applications: Types of biopolymer and nano-sized filler. *Agriculture and Agricultural Science Procedia, 2*, 296–303. Available from https://doi.org/10.1016/j.aaspro.2014.11.042, https://www.sciencedirect.com/science/article/pii/S2210784314000436.

Patel, D. K., Kim, H.-B., Dutta, S. D., Ganguly, K., & Lim, K.-T. (2020). Carbon nanotubes-based nanomaterials and their agricultural and biotechnological applications. *Materials, 13*(7), 1679. Available from https://doi.org/10.3390/ma13071679, https://www.mdpi.com/1996-1944/13/7/1679.

Peng, S.-ying, Yan, J., Li, M., Yan, Z.-xuan, Wei, H.-yu, Xu, D.-jun, & Cheng, X. (2023). Preparation of polysaccharide-conjugated selenium nanoparticles from spent mushroom substrates and their growth-promoting effect on rice seedlings. *International Journal of*

*Biological Macromolecules*126789. Available from https://doi.org/10.1016/j.ijbiomac.2023.126789, https://www.sciencedirect.com/science/article/pii/S0141813023036863.

Perfileva, A. I., Kharasova, A. R., Nozhkina, O. A., Sidorov, A. V., Graskova, I. A., & Krutovsky, K. V. (2023). Effect of nanopriming with selenium nanocomposites on potato productivity in a field experiment, soybean germination and viability of *Pectobacterium carotovorum*. *Horticulturae*, *9*(4), 458. Available from https://doi.org/10.3390/horticulturae9040458, https://www.mdpi.com/2311-7524/9/4/458.

Rahimzadeh, S., & Ghassemi-Golezani, K. (2022). Biochar-based nutritional nanocomposites altered nutrient uptake and vacuolar H + -pump activities of dill under salinity. *Journal of Soil Science and Plant Nutrition*, *22*(3), 3568–3581. Available from https://doi.org/10.1007/s42729-022-00910-z, https://doi.org/10.1007/s42729-022-00910-z.

Rajiv, P., Chen, X., Li, H., Rehaman, S., Vanathi, P., Abd-Elsalam, K. A., Li, X., & Abd-Elsalam, K. A. (2020). *Chapter 18 - Silica-based nanosystems: Their role in sustainable agriculture. Multifunctional hybrid nanomaterials for sustainable agri-food and ecosystems* (pp. 437–459). Elsevier. Available from https://www.sciencedirect.com/science/article/pii/B9780128213544000182.

Rajput, V. D., Minkina, T., Feizi, M., Kumari, A., Khan, M., Mandzhieva, S., Sushkova, S., El-Ramady, H., Verma, K. K., Singh, A., van Hullebusch, E. D., Singh, R. K., Jatav, H. S., & Choudhary, R. (2021). Effects of silicon and silicon-based nanoparticles on rhizosphere microbiome, plant stress and growth. *Biology*, *10*(8), 791. Available from https://doi.org/10.3390/biology10080791, https://www.mdpi.com/2079-7737/10/8/791.

Ranakoti, L., Gangil, B., Mishra, S. K., Singh, T., Sharma, S., Ilyas, R. A., & El-Khatib, S. (2022). Critical review on polylactic acid: Properties, structure, processing, biocomposites, and nanocomposites. *Materials*, *15*(12), 4312. Available from https://doi.org/10.3390/ma15124312, https://www.mdpi.com/1996-1944/15/12/4312.

Salimi, M., Motamedi, E., Safari, M., & Motesharezadeh, B. (2021). Synthesis of urea slow-release fertilizer using a novel starch-g-poly(styrene-co-butylacrylate) nanocomposite latex and its impact on a model crop production in greenhouse. *Journal of Cleaner Production*, *322*129082. Available from https://doi.org/10.1016/j.jclepro.2021.129082, https://www.sciencedirect.com/science/article/pii/S0959652621032716.

Sampathkumar, K., Tan, K. X., & Loo, S. C. J. (2020). Developing nano-delivery systems for agriculture and food applications with nature-derived polymers. *iScience*, *23*(5) 101055. Available from https://doi.org/10.1016/j.isci.2020.101055, https://www.sciencedirect.com/science/article/pii/S2589004220302406.

Seven, S. A., Tastan, Ö. F., Tas, C. E., Ünal, H., Ince, İ. A., & Menceloglu, Y. Z. (2019). Insecticide-releasing LLDPE films as greenhouse cover materials. *Materials Today Communications*, *19*, 170–176. Available from https://doi.org/10.1016/j.mtcomm.2019.01.015, https://www.sciencedirect.com/science/article/pii/S2352492818303994.

Shahrajabian, M. H., Chaski, C., Polyzos, N., Tzortzakis, N., & Petropoulos, S. A. (2021). Sustainable agriculture systems in vegetable production using chitin and chitosan as plant biostimulants. *Biomolecules*, *11*(6), 819. Available from https://doi.org/10.3390/biom11060819, https://www.mdpi.com/2218-273X/11/6/819.

Sigmon, L. R., Adisa, I. O., Liu, B., Elmer, W. H., White, J. C., Dimkpa, C. O., & Fairbrother, D. H. (2021). Biodegradable polymer nanocomposites provide effective delivery and reduce phosphorus loss during plant growth. *ACS Agricultural Science & Technology*, *1*(5), 529–539. Available from https://doi.org/10.1021/acsagscitech.1c00149, https://doi.org/10.1021/acsagscitech.1c00149.

Singh, N., Agarwal, S., Jain, A., & Khan, S. (2021). 3-Dimensional cross linked hydrophilic polymeric network "hydrogels": An agriculture boom. *Agricultural Water Management*,

*253*106939. Available from https://doi.org/10.1016/j.agwat.2021.106939, https://www.sciencedirect.com/science/article/pii/S0378377421002043.

Sohail, M., Urooj, Z., Noreen, S., Baig, M. M. F. A., Zhang, X., & Li, B. (2023). Micro- and nanoplastics: Contamination routes of food products and critical interpretation of detection strategies. *Science of The Total Environment*, *891*164596. Available from https://doi.org/10.1016/j.scitotenv.2023.164596, https://www.sciencedirect.com/science/article/pii/S0048969723032175.

Stasińska-Jakubas, M., & Hawrylak-Nowak, B. (2022). Protective, biostimulating, and eliciting effects of chitosan and its derivatives on crop plants. *Molecules*, *27*(9), 2801. Available from https://doi.org/10.3390/molecules27092801, https://www.mdpi.com/1420-3049/27/9/2801.

Sun, W., Shahrajabian, M. H., Petropoulos, S. A., & Shahrajabian, N. (2023). Developing sustainable agriculture systems in medicinal and aromatic plant production by using chitosan and chitin-based biostimulants. *Plants*, *12*(13), 2469. Available from https://doi.org/10.3390/plants12132469, https://www.mdpi.com/2223-7747/12/13/2469.

Teklić, T., Parađiković, N., Špoljarević, M., Zeljković, S., Lončarić, Z., & Lisjak, M. (2020). Linking abiotic stress, plant metabolites, biostimulants and functional food. *Annals of Applied Biology*, *178*(2), 169–191. Available from https://doi.org/10.1111/aab.12651, https://onlinelibrary.wiley.com/doi/abs/10.1111/aab.12651.

Tsivileva, O. M., & Perfileva, A. I. (2022). Mushroom-derived novel selenium nanocomposites' effects on potato plant growth and tuber germination. *Molecules*, *27*(14), 4438. Available from https://doi.org/10.3390/molecules27144438, https://www.mdpi.com/1420-3049/27/14/4438.

Vejan, P., Khadiran, T., Abdullah, R., & Ahmad, N. (2021). Controlled release fertilizer: A review on developments, applications and potential in agriculture. *Journal of Controlled Release*, *339*, 321–334. Available from https://doi.org/10.1016/j.jconrel.2021.10.003, https://www.sciencedirect.com/science/article/pii/S0168365921005320.

Venezia, V., Verrillo, M., Gallucci, N., Di Girolamo, R., Luciani, G., D'Errico, G., Paduano, L., Piccolo, A., & Vitiello, G. (2023). Exploiting bioderived humic acids: A molecular combination with ZnO nanoparticles leads to nanostructured hybrid interfaces with enhanced pro-oxidant and antibacterial activity. *Journal of Environmental Chemical Engineering*, *11*(1)108973. Available from https://doi.org/10.1016/j.jece.2022.108973, https://www.sciencedirect.com/science/article/pii/S2213343722018462.

Verma, K. K., Song, X.-P., Joshi, A., Tian, D.-D., Rajput, V. D., Singh, M., Arora, J., Minkina, T., & Li, Y.-R. (2022). Recent trends in nano-fertilizers for sustainable agriculture under climate change for global food security. *Nanomaterials*, *12*(1), 173. Available from https://doi.org/10.3390/nano12010173, https://www.mdpi.com/2079-4991/12/1/173.

Wang, S.-L., & Nguyen, A. D. (2018). Effects of Zn/B nanofertilizer on biophysical characteristics and growth of coffee seedlings in a greenhouse. *Research on Chemical Intermediates*, *44*(8), 4889–4901. Available from https://doi.org/10.1007/s11164-018-3342-z, https://doi.org/10.1007/s11164-018-3342-z.

Wang, Y.-P., & Lei, Q.-Y. (2018). Metabolite sensing and signaling in cell metabolism. *Signal Transduction and Targeted Therapy*, *3*(1), 1–9. Available from https://doi.org/10.1038/s41392-018-0024-7, https://www.nature.com/articles/s41392-018-0024-7.

Wu, L., Fu, F., Wang, W., Wang, W., Huang, Z., Huang, Y., Pan, X., & Wu, C. (2023). Plasma protein corona forming upon fullerene nanocomplex: Impact on both counterparts. *Particuology*, *73*, 26–36. Available from https://doi.org/10.1016/j.partic.2022.04.006, https://www.sciencedirect.com/science/article/pii/S1674200122000803.

Wypij, M., Trzcińska-Wencel, J., Golińska, P., Avila-Quezada, G. D., Ingle, A. P., & Rai, M. (2023). The strategic applications of natural polymer nanocomposites in food packaging and agriculture: Chances, challenges, and consumers' perception. *Frontiers in Chemistry, 10*1106230. Available from https://doi.org/10.3389/fchem.2022.1106230, https://www.frontiersin.org/articles/10.3389/fchem.2022.1106230.

Xu, X., Mao, X., Wang, Y., Li, D., Du, Z., Wu, W., Jiang, L., Yang, J., & Li, J. (2018). Study on the interaction of graphene oxide-silver nanocomposites with bovine serum albumin and the formation of nanoparticle-protein corona. *International Journal of Biological Macromolecules, 116*, 492–501. Available from https://doi.org/10.1016/j.ijbiomac.2018.05.043, https://www.sciencedirect.com/science/article/pii/S0141813018312303.

You, Y., Kerner, P., Shanmugam, S., & Khodakovskaya, M. V. (2023). Emerging investigator series: Differential effects of carbon nanotubes and graphene on the tomato rhizosphere microbiome. *Environmental Science: Nano, 10*(6), 1570–1584. Available from https://doi.org/10.1039/D2EN01026G, https://pubs.rsc.org/en/content/articlelanding/2023/en/d2en01026g.

Yu, J., Wang, D., Geetha, N., Khawar, K. M., Jogaiah, S., & Mujtaba, M. (2021). Current trends and challenges in the synthesis and applications of chitosan-based nanocomposites for plants: A review. *Carbohydrate Polymers, 261*117904. Available from https://doi.org/10.1016/j.carbpol.2021.117904, https://www.sciencedirect.com/science/article/pii/S0144861721002915.

Yu, Y., Dai, W., & Luan, Y. (2023). Bio- and eco-corona related to plants: Understanding the formation and biological effects of plant protein coatings on nanoparticles. *Environmental Pollution, 317*120784. Available from https://doi.org/10.1016/j.envpol.2022.120784, https://www.sciencedirect.com/science/article/pii/S0269749122019984.

Zabot, G. L., Schaefer Rodrigues, F., Polano Ody, L., Vinícius Tres, M., Herrera, E., Palacin, H., Córdova-Ramos, J. S., Best, I., & Olivera-Montenegro, L. (2022). Encapsulation of bioactive compounds for food and agricultural applications. *Polymers, 14*(19), 4194. Available from https://doi.org/10.3390/polym14194194, https://www.mdpi.com/2073-4360/14/19/4194.

Zarinkoob, A., Esmaeilzadeh Bahabadi, S., Rahdar, A., Hasanein, P., & Sharifan, H. (2021). Ce-Mn ferrite nanocomposite promoted the photosynthesis, fortification of total yield, and elongation of wheat (*Triticum aestivum* L.). *Environmental Monitoring and Assessment, 193*(12), 800. Available from https://doi.org/10.1007/s10661-021-09506-z, https://doi.org/10.1007/s10661-021-09506-z.

Improving fruit quality and bioactive compounds in plants: new trends using nanocomposites

9

Fabián Pérez-Labrada and Antonio Juárez-Maldonado
Department of Botany, Autonomous Agrarian University Antonio Narro, Saltillo, Coahuila, Mexico

9.1 Introduction

Today, consumers are more aware of the food they eat, and above all, the quality of the fruits, be it their physicochemical characteristics or the content of bioactive compounds. The quality of the fruit ranges from visual, physical, and sensory characteristics, as well as nutraceutical properties (Campos Alencar Oldoni et al., 2022; Mikulic-Petkovsek et al., 2021). In the case of bioactive compounds, it is well known that fruits and vegetables are an important source, since from these, one can obtain carbohydrates, acids, minerals and trace elements, polyphenols, vitamins, amino acids, aromatic compounds, carotenoids, fibers, and phytosterols, among other phytochemicals (Oyedele et al., 2020; Ruiz Rodríguez et al., 2020; El-Ramady et al., 2022). Particularly, the consumption of bioactive compounds, also known as phytochemicals, is extremely important for human health since their intake is related to preventing diseases caused by oxidative stress, such as cardiovascular, cancer, neurodegenerative, cognitive, obesity, diabetes, and others (Adefegha, 2018; Granato et al., 2020). Hence, the consumption of fruits with high quality and content of bioactive compounds plays a crucial role in the prevention of malnutrition and the reduction of the risk of noncommunicable diseases.

The ripening and senescence of the fruit are inevitable and irreversible physiological processes during which the biochemical characteristics and the physiology of these organs are altered and influence various quality parameters (Tang et al., 2020). However, these same physiological processes result in the loss of fruit quality for various reasons, such as lipid peroxidation, an increase in reactive oxygen species (ROS), an increase in the enzymatic activity that consumes the structural polysaccharides, and respiration rate (Nasr et al., 2021). It is for these reasons that it becomes a fundamental need to develop new tools that allow controlling the characteristics related to the quality of the fruit to maintain or even improve them.

Nanotechnology is a tool that has been used in agriculture since it is considered that it can promote sustainability because it is used in crop production and protection in different aspects such as nanofertilizers, nano pesticides, and nanobiosensors (Usman et al., 2020). The applications of nanotechnology in agriculture can

Nanocomposites for Environmental, Energy, and Agricultural Applications. DOI: https://doi.org/10.1016/B978-0-443-13935-2.00009-7

increase tolerance to stress conditions and increase the efficiency of fertilizers and pesticides, in addition to other applications such as nanosensors for smart agriculture, making it a very important tool (Kumari et al., 2023). According to the Nanotechnology Products Database website, there are currently 259 products for agricultural applications, which are produced by 95 companies and distributed in 29 countries. Of these products, fertilizers are the ones that include the most products (120), while the rest are distributed among animal husbandry, plant protection, plant breeding, and soil improvement (https://product.statnano.com/industry/agriculture; accessed on September 4, 2023).

Nanotechnology can be the tool that is needed to improve the quality of the fruit since it is known that nanomaterials (NMs, materials with at least one dimension on a nanometric scale of 1−100 nm) can induce the activation of the antioxidant defense through the induction of ROS (Venzhik and Deryabin, 2023). In addition, these NMs can also stimulate secondary metabolism and therefore induce the production of secondary metabolites (Selvakesavan et al., 2023). This will result in the production of antioxidant compounds and secondary metabolites, that is, phytochemicals, which will translate into higher-quality foods and a higher content of bioactive compounds that will be beneficial to human health. Both the antioxidant defense system and secondary metabolism are processes that are closely linked; in addition, some of the secondary metabolites are also antioxidant compounds.

The use of NMs has shown promising results since they promote the antioxidant defense system that allows for the reduction and elimination of ROS and induces membrane integrity (Nasr et al., 2021). This is quite relevant since firmness is considered the main quality attribute of the fruit (Wang 2023a,b,c). In addition, it has been shown that the application of NMs induces the activation of the antioxidant defense system and secondary metabolism, which gives rise to the production of antioxidant compounds and secondary metabolites (Ji et al., 2023; Pejam et al., 2021; Zahedi et al., 2023).

The application of NMs during the development of the crop has shown positive effects on the quality of the fruit due to changes in the metabolism and physiology of the plants (Azizkhani et al., 2023; García-López et al., 2019; Liu 2023a,b). Another strategy that has been followed is the application of NMs and nanocomposites during the postharvest of the fruit, where very promising results have been observed both in the physicochemical characteristics of the fruit and in the content of bioactive compounds. The application of the NMs through the use of covers or a solution on the fruits has proven to be effective in controlling various problems that affect the quality of the fruit, such as weight loss, growth of pathogenic organisms, ripening, discoloration, and browning (Arroyo et al., 2020; Hernández-López et al., 2020; Melo et al., 2020; Taheri et al., 2020).

This chapter aims to analyze the impact of nanotechnology through the use of NMs and nanocomposites to improve fruit quality and the content of bioactive compounds. For this, the mechanisms through which NMs act on plants and fruits will be analyzed, as well as the application routes, and the different strategies followed to achieve the positive results observed.

9.2 Fruit quality and bioactive compounds

Fruits and vegetables are the main foods for human consumption, so their production depends directly on the demand of the population, and as the population continues to grow, more and more food is required (Maringgal et al., 2020). In addition to this, the quality of food has become essential for consumers, not only due to its physicochemical characteristics but also the content of bioactive compounds associated with human health. The quality of the fruit then consists of visual, physical, sensory, and nutraceutical properties, whose constitutive parameters, for example, esthetic appearance, weight, size, firmness, aroma, flavor, texture, nutrient content, and bioactive compounds (Campos Alencar Oldoni et al., 2022; Mikulic-Petkovsek et al., 2021), are essential to achieving a longer shelf life and promoting nutritional intake.

9.2.1 Fruit quality parameters

The main characteristics that determine the quality of the fruits are the flavor quality parameters (total soluble solids, soluble sugars, organic acids, and sugar/acid content ratio), external quality parameters (color, lightness, chroma, hue angle, green/red shade, blue/yellow shade, and color index), and parameters that determine storage quality, that is, firmness (Lu et al., 2021). Soluble sugars and organic acids are primary products of photosynthesis and function as energy stores in plants that are necessary for the construction of cellular components, in addition to being quality indicators and precursors of aromatic compounds and signaling molecules (Vulk et al., 2023). While the total soluble solids indicate the proportion of solids (sugars, acids, and other minor components) dissolved in a solution (Lu et al., 2021). Therefore, these parameters must be considered at all times to maintain the quality of the fruits.

The loss of firmness or softening in the fruits is a typical physiological characteristic of maturation and senescence and is due to the deterioration or damage of the cuticular membrane, a process that is related to the production of ethylene as well as the solubilization and degradation of the pectin by the action of oxidase enzymes (Kuang et al., 2023; Li et al., 2021; Posé et al., 2019). This process results in the loss of water and therefore weight, in addition to the fact that it can lead to the proliferation of pathogenic microorganisms since the appropriate conditions are developed for this (Huang et al., 2020). That is why firmness is considered the main quality attribute of the fruit, and together with juiciness, they are the most important textural properties for consumers (Wang, 2023a,b,c). Thus, maintaining the firmness of the fruits is essential for their quality.

9.2.2 Bioactive compounds and human health

Fruits and vegetables are an important source of nutrients such as carbohydrates, acids, minerals, polyphenols, vitamins, amino acids, aromatic compounds, carotenoids, fibers, and phytosterols, among other phytochemicals; hence, consumption plays a crucial role in the prevention of malnutrition and in reducing the risk of noncommunicable diseases (El-Ramady et al., 2022; Oyedele et al., 2020; Ruiz Rodríguez et al., 2020).

The bioactive compounds or phytochemicals result from the primary and secondary metabolism of the plants; however, in some cases, there are plant species that represent the main sources of these, such as the tomato fruit that provides β-carotene and lycopene, the chili fruit that provides capsaicin, and garlic that provides allicin and alliin, among others (Ali et al., 2021; Sharma et al., 2021). Their consumption is important due to the positive effects they have on human health when they are included in the regular diet, since they are related to preventing diseases caused by oxidative stress, such as cardiovascular, cancer, neurodegenerative, cognitive, obesity, diabetes, and others (Adefegha, 2018; Granato et al., 2020). Some of the most studied bioactive compounds in relation to the prevention of noncommunicable diseases are phenolic compounds, carotenoids, long-chain unsaturated fatty acids, vitamins, glutathione, and organic acids (Amengual, 2019; Calder, 2018; Delgado et al., 2019; Dwivedi et al., 2020; Fenech et al., 2019).

9.2.3 Loss of fruit quality and bioactive compounds

The ripening and senescence of the fruits are inevitable and irreversible physiological processes, during which the biochemical characteristics and the physiology of these organs are altered and influence various quality parameters of the fruit such as firmness, color, aroma, sugar content, and organic acids, among others (Tang et al., 2020).

However, due to the perishable nature of the fruits, serious quality deterioration and losses will inevitably occur due to a variety of reasons during the preharvest and postharvest stages, so it is estimated that one-third of the fresh fruit produced does not reach consumers in optimal conditions (Chen et al., 2020). In this sense, the loss of fruit quality is generally caused in following ways: (i) a reduction in the resistance and permeability of cell membranes (lipid peroxidation); (ii) an increase in ROS, mainly H_2O_2, $O_2^{\bullet-}$, and radicals (hydroxyl); (iii) an increase in the enzymatic activity that consumes the structural polysaccharides (hemicellulose and cell wall pectins) that increase the total soluble solids, as well as by alterations in the titratable acidity (% of organic acids) that are converted into sugar in respiration; and (iv) due to the increase in the respiration rate (depending on the fruit, climacteric, or nonclimacteric) (Nasr et al., 2021).

It has been reported that postharvest losses of crops can range from 10% to 40% due to mechanical, microbial, and physiological damage as the main causes (Bose et al., 2021; Mossa et al., 2021). In general, most fruits are not marketed with a water loss of 5%–10% of their initial weight, as this leads to a serious reduction in quality and wilting. The loss of water that occurs through transpiration depends on the characteristics of the epidermis, which acts as a critical barrier to resist the movement of water from the fruit tissue toward the outer surface (Wei et al., 2019). The excessive softening derived from the loss of water results in an increase in susceptibility to pathogens and leads to significant economic losses and postharvest deterioration (Posé et al., 2019). In addition to causing marketable weight loss, it promotes browning, loss of firmness, texture, flavor, accelerated senescence, and membrane disintegration (Lufu et al., 2020).

Browning is an important physiological disorder that causes economic losses and is due to the enzyme polyphenol oxidase (PPO). This enzyme is characterized by containing copper, which, under the presence of oxygen, hydroxylates monophenols to o-diphenols, and subsequently these are oxidized to o-quinones, finally causing brown, red, and black pigments (Karimirad et al., 2018; Lo'ay and Ameer, 2019; Queiroz et al., 2008).

Mechanical damage such as cuticular cracking, pitting, and bruising can lead to serious damage to the epidermis and excessive loss of water in the fruit, but it can also increase the production of ethylene and favor contamination with pathogenic microorganisms (Hussein et al., 2020; Jansasithorn et al., 2014). These microorganisms can also affect the cosmetic, organoleptic, and nutritional characteristics of the fruits (Sharma et al., 2020). Postharvest diseases cause a loss of 10–30% of total crop yield and occur during handling, transport, storage, and marketing (Mossa et al., 2021). In addition, infected fruits can be risky for human health due to the fact that some pathogens, such as *Penicillium expansum*, produce toxic secondary metabolites such as patulin and citrinin, as in the case of apple fruits (Yu et al., 2020).

9.3 Nanotechnology and nanomaterials

Nanotechnology encompasses the use of materials with at least one dimension on a nanometric scale called NMs, which, unlike bulk materials, possess other properties due to their very small size (1–100 nm). Among these properties, the large surface/volume ratio stands out, which results in a greater amount of surface free energy and greater reactivity (Ma et al., 2023; Venzhik and Deryabin, 2023). These characteristics of NMs can impact plants at different levels and therefore lead to a variety of responses, from positive to negative (Rajpal et al., 2023).

NMs can be applied directly to the leaves or to the roots of the plants, and once they enter the plants through different pathways (stomata, cuticle, trichomes, root hairs, etc.), they can be transported to other organs through the xylem and phloem (Rajpal et al., 2023). Inside the plants, the NMs can move through plant organ structures, through the cell wall, and traverse the lipid membranes of the cells and the organelles (Hubbard et al., 2020). Once they enter the cells, NMs interact with the various organelles and induce positive effects on the plants by bringing about changes at genetic, biochemical, anatomical, and physiological levels, however; at high concentrations, they can induce negative effects (Rajpal et al., 2023).

9.3.1 Antioxidant defense system and secondary metabolism in plants

Plants commonly keep their antioxidant defense system active to maintain redox homeostasis with the use of antioxidants (compounds that work as electron donors) that can be enzymatic or nonenzymatic, to neutralize ROS and inhibit the oxidation process, preventing damage at the cellular level and damage to the different cellular structures (Fujita and Hasanuzzaman, 2022; Rajput et al., 2021; Sarraf et al., 2022). Superoxide dismutase (SOD), catalase (CAT), ascorbate peroxidase (APX), glutathione peroxidase (GPX), and peroxidase (POX) are some enzymatic antioxidants, while ascorbic acid, glutathione, phenolic acids, alkaloids, flavonoids, and carotenoids are nonenzymatic antioxidants (Fujita and Hasanuzzaman, 2022).

ROS are interesting compounds in plants since they have a double function and can cause oxidative stress at high concentrations, while at proper concentrations (commonly low), they act as signaling molecules and stimulate the production of beneficial compounds in plants like antioxidants, secondary metabolites, abscisic acid, salicylic acid, or jasmonic acid (Sarraf et al., 2022; Venzhik and Deryabin, 2023). The secondary metabolites of plants are products that are not essential for survival but provide strength and the ability to adapt to changes in the environment (biotic and abiotic stresses) (Selvakesavan et al., 2023).

9.3.2 Impact of nanomaterials on secondary metabolites and antioxidants

When NMs come into contact with different materials that can be of biotic or abiotic origin, they can absorb organic components or biomolecules and form a corona (Juárez-Maldonado et al., 2019; Liu, 2023a,b). The composition of the corona becomes important since perception with cellular receptors and physiological responses depend on the attached molecules (Francia et al., 2019; Liu, 2023a,b; Prakash and Deswal, 2020). Furthermore, the first interaction between NMs and plants occurs through the surface of the corona and cell walls and/or membranes, which modifies the activity of receptors, transporters, and proteins, triggering a series of cellular responses (Juárez-Maldonado et al., 2021; Liu, 2023a,b; Rajpal et al., 2023).

When the NMs enter the cell, they can interact with different organelles such as the chloroplast, mitochondria, peroxisome, and nucleus, which can induce changes at genetic, biochemical, anatomical, and physiological levels. This, in turn, can result in positive responses for the plant, but at high concentrations, NMs can induce oxidative stress (through the production of ROS), and adverse results may be observed (Liu, 2023a,b; Rajpal et al., 2023). ROS not only cause oxidative stress but, in adequate concentrations (usually low), act as signaling molecules and stimulate the production of a variety of compounds like abscisic acid, salicylic acid, jasmonic acid, ethylene, nitric oxide, and brassinosteroids, which can modulate secondary metabolism (Sarraf et al., 2022; Selvakesavan et al., 2023; Venzhik and Deryabin, 2023) (Fig. 9.1).

Both the antioxidant defense system and secondary metabolism are processes that are closely linked, so it is common to observe a greater activity of antioxidant enzymes while secondary metabolites accumulate (Lala, 2021). For this reason, it has been observed that the application of NMs induces positive responses both in

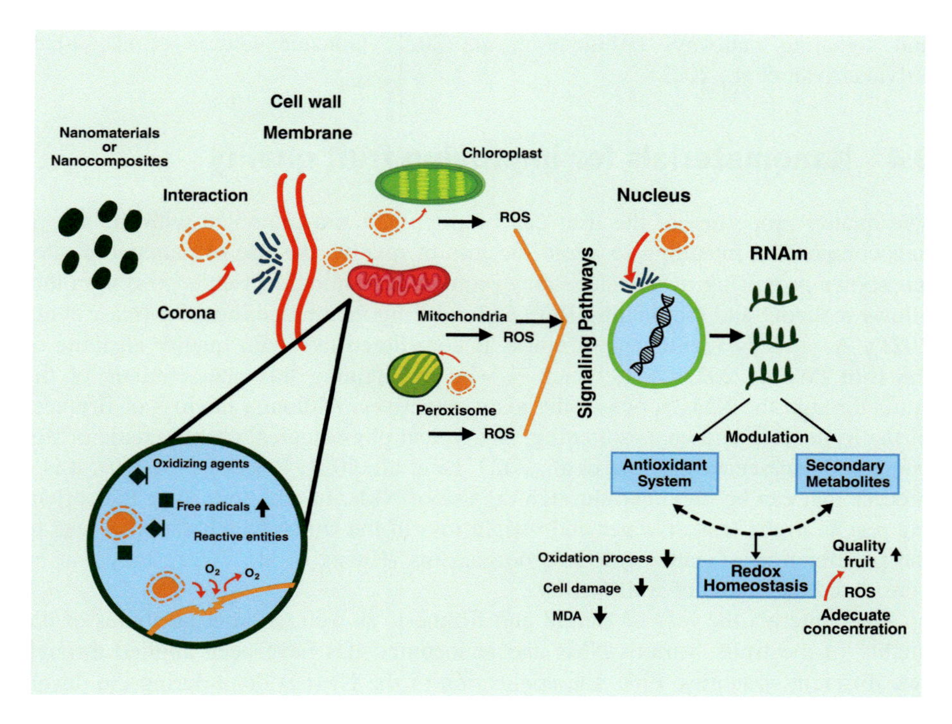

Figure 9.1 Induction of antioxidants and secondary metabolites by nanocomposites and nanomaterials. The first interaction between nanomaterials (NMs) and plants occurs through the surface of the corona and cell walls and/or membranes, which modifies the activity of receptors, transporters and proteins, triggering a series of cellular responses. Moreover, when NMs enter the cell they can interact with different organelles such as the chloroplast, mitochondria, peroxisome and nucleus, which can induce changes at genetic, biochemical, anatomical, and physiological levels, this in turn can result in positive responses for the plant, but at high concentrations, NMs can induce oxidative stress (through the production of reactive oxygen species) and adverse results may be observed.

the antioxidant defense system and in the secondary metabolism of plants (Abideen et al., 2022; Fu et al., 2023; Sarraf et al., 2022).

The ability of NMs to produce ROS can be achieved through two routes and is due to the high reactivity of their surface caused by their very small size: (i) oxidizing agents and free radicals bind to their surface and form more reactive entities; or (ii) due to changes in the chemistry of the surface, size alterations, conduction defects, and modifications of the electronic properties arise that create reactive surface sites that, when interacting with O_2, can exchange electrons and generate $O_2^{\bullet -}$ (Vardakas et al., 2022).

In the case of secondary metabolism, it has been documented through metabolomic studies in plants that the application of NMs modifies the concentration of secondary metabolites through the modulation of reactive species, gene expression, and signaling pathways (Abideen et al., 2022; González-García et al., 2023; Selvakesavan et al., 2023).

9.4 Nanomaterials for improving fruit quality

The quality properties of the fruit can be improved and promoted with the use of nanocomposites. Intending to avoid the loss of quality, the use of nanocomposites has shown promising results since they promote the antioxidant defense system that allows reducing and eliminating ROS and inducing membrane integrity (Nasr et al., 2021). As already mentioned, firmness is considered the main quality attribute of the fruit (Wang, 2023a,b,c), which is why maintaining this characteristic of the fruits through the NMs is one of the main objectives. Although the loss of firmness of the fruits, also known as softening, is a typical physiological characteristic of ripening and senescence (Kuang et al., 2023; Li et al., 2021; Posé et al., 2019), it is a process that can be regulated through the use of NMs. In addition, since the softening results in the loss of water and weight loss of the fruits, which can also lead to the proliferation of pathogenic microorganisms (Huang et al., 2020), they can be considered indicators of fruit quality.

To counteract the loss of weight and firmness, as well as other attributes of the quality of the fruits, various NMs and nanocomposites have been applied through two different strategies. First, the application of the NMs is done during the development of the crop to modify the physiology of the plant and the fruits and thus improve the quality of the fruits. The second strategy is based on the application of NMs, or nanocomposites, after harvest directly to the fruits, to maintain or improve quality during postharvest.

9.4.1 Application of nanomaterials in crops to increase fruit quality

It has been observed that the application of some NMs such as Ca, Zn, and chitosan in the crops has managed to avoid the loss of weight of the fruits, thus avoiding softening, and therefore, the quality of the fruits is maintained for a longer time (González-Saucedo et al., 2019; Ranjbar et al., 2018; Rasouli and Saba, 2018). Also, the application of NMs such as zinc oxide (ZnO) and chitosan has improved the quality of the fruits through a decrease in the growth of pathogenic organisms (Luksiene et al., 2020; Rasouli and Saba, 2018). In any case, this shows that the use of NMs is an effective tool to maintain or improve fruit quality when applied during crop development (Table 9.1).

9.4.2 Application of nanomaterials in postharvest to increase fruit quality

The second strategy, based on the application of the NMs or nanocomposites directly to the fruits after harvest, that is, postharvest, has proven to be very

Table 9.1 Impacts of nanomaterials and nanocomposites application during crop development on fruit quality.

Nanomaterial	Fruit	Effect	References
Chitosan nanocoating (CuO NP/ZnO NP)	Guava (*Psidium guajava* L.)	Reduced weight loss, respiration rate, maturation rate, and ethylene effects.	Kalia et al. (2021)
Chitosan (0.5%) + nano-ZnO	Oranges (*Citrus* × *sinensis* L.)	Reduced total soluble solids, total acidity, weight loss, and loss of firmness.	Dulta et al. (2022)
Multi-walled carbon nanotubes	Grapes (*Vitis vinifera* L.)	Induced stability of firmness, soluble sugar, titratable acidity, and pH.	Sha et al. (2022)
Nanocalcium	Apple (*Malus domestica* L.)	Increased firmness and titratable acidity. Reduction in weight loss, total soluble solids, and internal browning.	Ranjbar et al. (2018)
Nano-Chelate Zn	Apple (*Malus domestica* L.)	Decreased browning index and total color changes. Greater firmness. Lower polyphenol oxidase activity.	Rasouli and Saba (2018)
NPs composed of chitosan	Bell pepper (*Capsicum annuum* L.)	Reduced weight loss and change of color. Microbiological activity was reduced.	González-Saucedo et al. (2019)
Selenium NPs	Jalapeño Pepper *Capsicum annuum* L.	Increased fruit firmness.	Sariñana-Navarrete et al. (2023)
Selenium and copper nanofertilizer	Tomato (*Solanum lycopersicum* L.)	Increased fruit firmness.	Saffan et al. (2022)
ZnO NPs photoactivated	Strawberries (*Fragaria* × *ananassa* Duch.)	Reduced incidence of *Botrytis cinerea*.	Luksiene et al. (2020)

efficient in maintaining fruit quality. Here, the application of the NMs or nanocomposites is done either through the use of covers or through a solution added with the NMs commonly applied through the immersion of the fruits. The main effects observed in the fruits with the use of this strategy are to avoid weight loss (Zambrano-Zaragoza et al., 2020), decrease or suppress the growth of pathogenic organisms (Lavinia et al., 2020; Vieira et al., 2020), retard ripening (Arroyo et al., 2020), reduce discoloration or maintain the color (Esyanti et al., 2019; Pina-Barrera et al., 2019), and reduce browning (Lo'ay and Ameer, 2019) (Table 9.2).

9.5 Nanomaterials for improving bioactive compounds in fruits

The use of NMs to improve the content of bioactive compounds in fruits has proven to be an effective tool. However, it is important to explain the mechanisms through which these positive effects are induced. These mechanisms are basically related to the ability of NMs to induce responses in the antioxidant defense system and in the primary and secondary metabolism of the fruits, which ultimately influences the content of bioactive compounds and nutrients (Fig. 9.2).

9.5.1 Induction of the antioxidant defense system

The increase in the quality of the fruit may be due to an increase in the metabolism of the foliar tissue, which will serve as a source of biocompounds during the stages of flowering, fruit set, and development. In this sense, in the tomato foliar tissue, it was documented that there was increased SOD, CAT, and APX activity when exposed to ZnO NPs (Wang et al., 2018). The use of chitosan-based Schiff metal complexes (CH, CH-Fe, CH-Cu, and CH-Zn) in pomegranate (cv. Malase Saveh) promotes a higher content of chlorophyll *a* (28%), chlorophyll *b* (29.5%), chlorophyll $a + b$ (28.6%), carotenoids (85.7%), and high levels of SOD (35.3%), APX (56.0%) in leaves, as well as an increase in abscisic acid (25.1%) and indole-3-acetic acid (40.5%) (Zahedi et al., 2023). In tomato, the use of Cu nanoclusters (NC; 3.0 ± 0.5 nm; 1 mg kg^{-1}) promotes the enzymes SOD, CAT, and POX in addition to improving photosynthesis and carbohydrate content in leaves and higher fruit production (Wang et al., 2021a). In the same way, ZnO NPs increase the activity of nitrate reductase, soluble phenols, the activity of phenylalanine ammonia-lyase, CAT, and POX in tomato leaves (Pejam et al., 2021). Likewise, silver NP in tomato seedlings (10, 20, or 30 mg L^{-1} in hydroponic media for 7 days) promoted the content of H_2O_2 and MDA as well as the content of anthocyanins, CAT, POX, and higher expression of defense genes ethylene-inducing xylanase, POX 51 (*POX*), and phenylalanine ammonia-lyase (*PAL*) in leaves (Noori et al., 2020).

The use of nanocomposites (nano-TiO_2) in litchi improved the pollen germination rate and pollen tube length (Huang et al., 2022). Similarly, $MnFe_2O_4$ nanocomposites (10 mg L^{-1}) in tomatoes increased pollen activity and ovule size

Table 9.2 Postharvest application of nanomaterials and nanocomposites in fruits to improve fruit quality.

Nanomaterial	Fruit	Effect	References
Sodium alginate film with Ag NPs	Carrots (*Daucus carota* L.), Pears (*Pyrus communis* L)	Reduced weight loss.	Fayaz et al. (2009)
Agar-Ag NPs film	Lime (*Citrus aurantifolium* L.), Apple (*Pyrus malus* L.)	Reduced weight loss. Increased antimicrobial capacity.	Gudadhe et al. (2014)
Dipping of Solid lipid nanoparticles + xanthan gum	Guava (*Psidium guajava* L.)	Reduced weight loss and preserved the best quality.	Zambrano-Zaragoza et al. (2013)
Chitosan–surfactant nanostructure	Tomato (*Solanum lycopersicum* L.)	Lower loss of firmness, respiration, and weight loss. Decrease titratable acidity. Increase in soluble solids were delayed.	Mustafa et al. (2014)
Coating Carboxymethylcellulose-Ag NPs and guar gum-Ag NPs.	Kinnow (*Citrus reticulata* L.)	Lower incidence of fungal diseases.	Shah et al. (2015)
Chitosan/nanosilica	Loquat (*Eriobotrya japonica* Lindl.)	Delayed internal browning. Reduced weight loss.	Song et al. (2016)
Ag NPs coating	Cherry tomatoes (*Solanum lycopersicum* L.)	Reduced weight loss and increased total soluble solids.	Gao et al. (2017)
Solid lipid NPs from candeuba wax and xanthan gum	Guava (*Psidium guajava* L.)	Lower respiration rate. Reduced weight loss.	García-Betanzos et al. (2017)
Nano chitosan	Peach (*Prunus persica* L.)	Reduced weight loss and improved fruit firmness.	Ali and Toliba (2018)
Edible NPs coatings and films	Apples (*Malus domestica* L.), red grapes (*Vitis vinifera* L.), tomatoes (*Solanum lycopersicum* L.) and weet green pepper (*Capsicum annuum* L.)	Reduced weight loss, total soluble solids, firmness loss.	Bakhy et al. (2018)

(*Continued*)

Table 9.2 (Continued)

Nanomaterial	Fruit	Effect	References
ZnO NPs coating	Fig (*Ficus carica* L.)	Delayed maturation, slowed weight loss, reduced color change, increased firmness, and inhibited the growth of microorganisms.	Lakshmi et al. (2018)
Carrageenan-ZnO NPs coating	Mango (*Mangifera indica* L. cv. Gedong Gincu)	Increased antimicrobial activity. It stood firm. Delayed discoloration.	Meindrawan et al. (2018)
Chitosan NPs	Grapes (*Vitis labrusca* L.)	Delayed grape ripening, reduced weight loss, soluble solids and sugar content. Inhibited microorganism growth.	Castelo Branco Melo et al. (2018)
Coating thymol nanoemulsions incorporated in quinoa protein/chitosan edible films	Cherry tomatoes (*Solanum lycopersicum* L.)	Decreased growth of *Botrytis cinerea.*	Robledo et al. (2018)
Chitosan NPs	Cavendish banana (*Musa acuminata* AAA group)	Slower skin discoloration.	Esyanti et al. (2019)
Nanocomposite coating (soybean protein isolate + cinnamaldehyde + ZnO NPs)	Banana (Musaceae)	Reduced weight loss and maintained firmness. Reduced content of total soluble sugars and increased titratable acidity.	Li et al. (2019)
Ca NPs	Mango (*Mangifera indica* L.)	Reduced internal browning index.	Lo'ay and Ameer (2019)
Multisystem coating based on polymeric nanocapsules containing *Thymus vulgaris* L. essential oil	Grapes (*Vitis vinifera* L)	Maintained characteristics of color, firmness, titratable acidity, and TSS for longer time.	Pina-Barrera et al. (2019)
Ag NPs coating	Loquat (*Eriobotrya japonica* Lindl.)	Reduced weight loss, total soluble solids, and sugars.	Ali et al. (2020)

Coating of chitosan + nano-ZnO	Guava (*Psidium guajava* L.)	Inhibition of rot, delayed the ripening process, and reduced weight loss.	Arroyo et al. (2020)
Chitosan + nano-SiO_2	Blueberry (*Vaccinium myrtillus* L.)	Inhibited microbial populations.	Eldib et al. (2020)
Nanocurcumin and nanorosemarinic acid	French basil (*Ocimum basilicum* L.)	Reduced fresh weight loss.	Hammam and Shoala (2020)
Chitosan NPs with α-pineno	Bell peppers (*Capsicum annuum* L.)	Reduced weight loss and preserved physicochemical quality. Inhibited *Alternaria alternata*.	Hernández-López et al. (2020)
Coating chitosan NPs	Bell pepper (*Capsicum annuum* L.)	Reduced weight loss. Inhibited pathogens.	Hu et al. (2020)
Nanocomposite coating (chitosan combined with ZnO NPs)	Papaya (*Carica papaya* L.)	Suppression of microbial growth.	Lavinia et al. (2020)
Natamycin-loaded zein nanoparticles stabilized by carboxymethyl chitosan	Hongjia strawberries (*Fragaria × ananassa* Duch.)	Reduction of the occurrence of rot, mildew, and gray mold.	Lin et al. (2020)
Edible coatings chitosan NPs	Strawberries (*Fragaria × ananassa* Duch.)	Reduced weight loss and maintained firmness.	Martínez-González et al. (2020)
Edible coating of ZnO NPs	Mango (*Mangifera indica* L.)	Maturation delay, weight loss slowdown, and increased firmness.	Malek et al. (2020)
Edible coating of chitosan NPs	Strawberry (*Fragaria × ananassa* Duch.)	Antifungal action. Reduced weight loss, total soluble solids, maturity index, and moisture loss.	Melo et al. (2020)
Nano-net of Cu-chitosan NPs	Tomato (*Solanum lycopersicum* L.)	Reduced microbial decomposition. Maintained fruit firmness.	Meena et al. (2020)

(*Continued*)

Table 9.2 (Continued)

Nanomaterial	Fruit	Effect	References
Nano chitosan	Strawberry (*Fragaria* × *ananassa* Duch.)	Preserved quality index. Reduced weight loss and maintained firmness.	Nguyen and Nguyen (2020)
$CaCl_2$ + nano chitosan	Strawberry (*Fragaria* × *ananassa* Duch.)	Reduced weight loss.	Nguyen et al. (2020)
Ag NPs	Apricot (*Prunus armeniaca* L.)	Reduced weight loss.	Shahat et al. (2020a)
Chitosan nanoparticles	Strawberry (*Fragaria* × *ananassa* Duch.)	Reduced weight loss.	Shahat et al. (2020b)
Chitosan coated iron oxide nanoparticles	Peach (*Prunus persica* L.)	Reduced weight loss. Inhibited microbial growth on the fruit surface.	Saqib et al. (2020)
Heracleum persicum essential oil-loaded chitosan NPs	Red sweet bell pepper (*Capsicum annuum* L.)	Reduced change of color, loss of firmness and weight.	Taheri et al. (2020)
Coating of Ag NPs + hydroxypropylmethylcellulose	Papaya (*Carica papaya* L.)	Reduced weight loss, incidence, and severity of *Colletotrichum gloeosporioides*.	Vieira et al. (2020)
Chitosan/nano-TiO_2 nanocomposite coating	Mango (*Mangiferaindica* L.)	Reduced decomposition index. Increased firmness.	Xing et al. (2020)
Beeswax solid lipid NPs nanocoatings + xanthan gum and propylene glycol	Strawberry (*Fragaria* × *ananassa* Duch. cv. camarosa)	Reduced weight loss, decay rate, and loss of firmness.	Zambrano-Zaragoza et al. (2020)

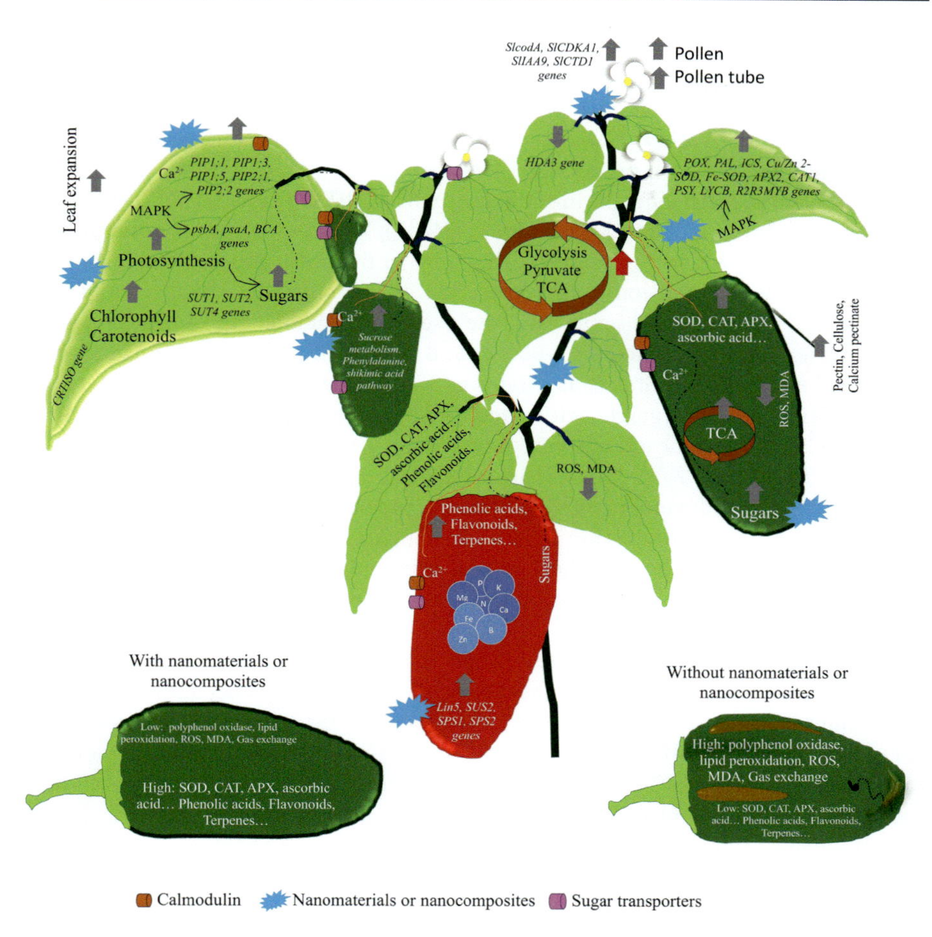

Figure 9.2 Impact of nanomaterials and nanocomposites on the content of bioactive compounds in the fruit. The interaction between NMs and plants, specifically at the cellular level, occurs when NMs enter the cell and come into contact with different organelles such as the chloroplast, mitochondria, peroxisome, and nucleus, which can induce changes at the genetic level, biochemical, anatomical, and physiological. These interactions modify the activity of receptors, transporters, and proteins, triggering a series of cellular responses that will modulate the antioxidant defense system and secondary metabolism, resulting in the production of enzymatic and nonenzymatic antioxidant compounds, as well as secondary metabolites.

(Yue et al., 2022), which can lead to a higher fruit set rate and therefore higher fruit production per plant. In a study where copper NPs (10–30 nm) were used on cucumber plants (*Cucumis sativus*), a high expression of the Cu-Zn SOD gene was documented in the leaves (Mosa et al., 2018). While using selenium-engineered nanocomposites (Se ENMs, 75 $\mu g\ kg^{-1}$), higher expression of the genes *SlcodA*,

SlCDKA1, *SlIAA9*, and *SlCTD1* (related to flower enlargement, cell separation, and expansion) was found in the apical tissue, in turn improving the production of cherry tomato fruit, which increased its diameter by 12.4% (Cheng et al., 2022a). Similarly, the use of CeO_2 nanocomposites (10 mg kg^{-1}) accelerated early flowering and increased tomato production. The same study reported a promotion in the relative expression of aquaporin-inducing genes (*PIP1;3* and *PIP1;5*), synthesis of the waxy layer (*CER6*, *SHN2*, and *THM27*), and synthesis of salicylic acid (*PAL5* and *ICS*), which can lead to efficiency in the absorption of water and calcium in the fruit in addition to the protection of biocomposites by the waxy layer (Feng et al., 2022). In the same way, the use of this nanocomposite in cucumber induced the synthesis of indole-3-acetic acid, which allowed early flowering and optimized translocation of carbohydrates as well as the promotion of floral metabolism (amino acids, organic acids, and flavonoids) and fruit production (Feng et al., 2023). The same authors report an increase in the relative expression of *PIP1;1*, *PIP2;1*, and *PIP2;2* genes in the leaves. Likewise, it is reported that Zn NPs can alter fruit development since Zn is a precursor of tryptophan, required in the biosynthesis of indoleacetic acid, which in turn participates in cell division and elongation. In addition to the fact that this nutrient participates in the photosynthetic process, this alteration in the hormonal load can trigger a better quality of the fruit (Farooq et al., 2023).

In barley, it is reported that applications of iron oxide NPs cause an increase in the expression of the genes *psbA* (photosystem II reaction center protein A), *BCA* (β-carbonic anhydrase), and *psaA* (photosystem I P700 chlorophyll A apoprotein A1) in 98-, 25-, and 9-fold change, in addition to a higher expression of the stress marker gene *HSP90.1*. These data suggest an improvement in the photosynthetic machinery promoting growth and fruit quality, thanks to efficiency in the generation of sugars (Tombuloglu et al., 2020). Likewise, the use of micronutrients in the form of NPs (ZnO, MnO_2, and MoO_3) significantly improves photosynthetic pigments (Salama et al., 2022). In this sense, the use of fluorescent carbon dots doped with nitrogen with blue fluorescence seems to act as artificial antennae, improving the collection and conversion of light, the supply of electrons, and eventually the photosynthetic process, increasing the synthesis of sugars and the yield of the fruit (Wang et al., 2021b). Similarly, carbon quantum dots functionalized with putrescine (10 mg L^{-1}) and cerium oxide NPs (25, 50, and 100 mg L^{-1}) on grapes improve photosynthetic pigments, chlorophyll fluorescence, and promote proline content, as well as the activity of CAT, APX, GPX, and SOD, reducing MDA and levels of H_2O_2 (Gohari et al., 2021a,c). Taken together, these data suggest that improvements in the photosynthetic apparatus and antioxidant system can promote fruit quality.

In addition, the use of synergistic fertilizer based on nanocarbon and calcium nanocarbonate in wheat increased the expression of the genes *TaNRT2.2* and *TaNRT2.3* (coding for nitrate transporter) and *TaGS1* and *TaGS2* (genes that code for the enzyme glutamine synthase). Derived from this, it was possible to streamline the assimilation of nitrogen and, therefore, the synthesis of proteins, which may belong to the antioxidant system of the fruit (Yang et al., 2023). In the case of the spraying of $MnFe_2O_4$ nanocomposites (10 mg L^{-1}) in tomatoes, it allows for an increase in the chlorophyll content and the expression of ferredoxin, psaA, and psbA in leaves. It also induces an increase in the efficiency of photosynthesis, probably because it is an electron donor. Similarly, there was a positive regulation of *SUT1* and *SUT2*, genes related to sucrose synthesis (Yue et al., 2022). The expression of the genes related to the synthesis of sucrase and their heterologous genes (*SlSUT1*, *SlSUT2*, and *SlSUT4*) suggests that the transport of sucrose may be more efficient from the leaf to the flowers, allowing its accumulation in the fruit and improving its quality (Wang et al., 2022).

The promotion of fruit quality can also be stimulated by lanthanum oxide (La_2O_3) with different sizes (10–20 nm) and surface modifications (citrate, polyvinylpyrrolidone, and polyethylene glycol) applied at a rate of 20–200 mg kg^{-1} or mg L^{-1}. The use of these nanocomposites in the seed and plant of *Cucumis sativus* for the control of *Fusarium oxysporum* (Schl.) f. sp. *cucumerinum* Owen causes an increase in the soluble sugars, phenols, and total amino acids of the fruit (Luo et al., 2023). The same authors point out that at the transcriptomic and metabolomic level, La_2O_3, when interacting with calmodulin, activates acquired systemic resistance (related to salicylic acid) and promotes the expression of antioxidant genes, indirectly improving fruit quality.

9.5.2 Primary metabolites in the fruit

The application of NMs and/or nanocomposites to plants has shown that primary metabolism can be modified, which translates into changes in the primary metabolites of the fruits (Fig. 9.3). The spraying of nanomicronutrients (Mn, Fe, and Zn, at 50 mg L^{-1}) on snap bean (*Phaseolus vulgaris* L.) caused an increase in the crude protein content of the pod (Marzouk et al., 2019). Some nanofertilizers significantly modified the content of ash, protein, fat, carbohydrates, and energy in strawberry fruits (BARI Strawberry-1) (Rahman et al., 2021a), while a mixed foliar applied nanofertilizer (NP of Zn, Fe, and Cu) induced an increase in the protein and fiber content of the fruit (Rahman et al., 2021b).

The application of NPs of Se and Cu in the substrate of tomato plants increases the content of vitamin C, glutathione, firmness, total soluble solids, and titratable acidity (Hernández-Hernández et al., 2019). In cherry tomato fruit, selenium nanocomposites (75 μg kg^{-1}) improve the metabolism of carbohydrates and amino acids (glycine, serine, lysine, alanine, and valine), which leads to the synthesis of ascorbic acid, glutathione, indole-3-acetic acid, jasmonic acid, and salicylic acid (Cheng et al., 2022a). While CeO_2 nanocomposites (10 mg kg^{-1}) generate an increase in the content of malic acid, fructose, ascorbic acid, coumaric acid, glutamate, L-phenylalanine, and salicylic acid in tomato (Feng et al., 2022), they also promote the production of vitamins and sugars in cucumber fruit (Feng et al., 2023), and the development of waxy layers in tomato fruit (Feng et al., 2022). A mixed nanofertilizer (NPs of Zn, Fe, and Cu) increases the ascorbic acid content, accompanied by a reduction in DPPH activity and an increase in antioxidant capacity ($ABTS^+$) (Rahman et al., 2021b). In contrast, biogenic CuO NPs (synthesized with green tea) applied to *Lactuca sativa* reduced APX, but were able to increase

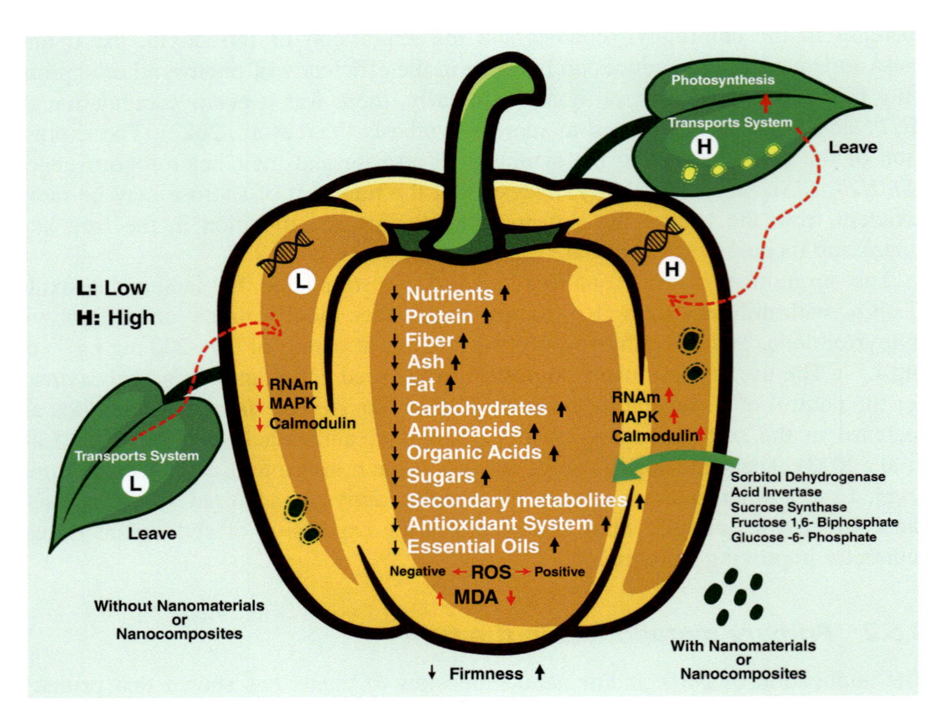

Figure 9.3 Impact of NMs and nanocomposites in primary and secondary metabolism, and the nutrient content of the fruits. The application of NMs and/or nanocomposites to plants modifies the primary metabolism, but also has a promoting effect on secondary metabolism, by stimulating the content of terpenes or isoprenoids, carotenes and xanthophylls, this results in an increase in the content of biocompounds (primary and secondary). In addition to this, the application of NMs can significantly increase the nutritional level of the plant through the accumulation of nutrients such as N, P, K, Mg, Ca, Fe, Cu, Zn, Mn, and Se in the fruit.

nitric oxide levels (R-S-nitrosothiols, RSNO) (Kohatsu et al., 2021). On the other hand, the use of nanoenabled Zn (nano-Zn at 200 $\mu g\ g^{-1}$) in strawberry *Fragaria* × *ananassa* Duch cv. Sweet Charlie increased fruit set, yield, firmness, total soluble solids, reducing sugars, total sugars, and ascorbic acid (Saini et al., 2021). Similar results are reported in tomato fruit when applying Mn_3O_4 and ZnO nanocomposites, since an increase in sugar content was observed by 118% and 111%, respectively (Cantu et al., 2022). In apple fruits (*Malus domestica* L. cv. Golden Delicius), it is reported that Zn NPs induce an increase in the concentration of total sugars (92%), reducing sugars (35%), and nonreducing sugars (Montaño-Herrera et al., 2022). This may be due to a stimulation of sorbitol dehydrogenase, acid invertase, and sucrose synthase activity by Zn (Zhang et al., 2014). Se (applied as Se NPs) increased carbohydrate metabolism by activating the enzyme fructose 1,6-biphosphatase in pomegranate fruits (Zahedi et al., 2019).

In litchi, the use of nano-TiO_2 in the bud stage promotes the vitamin C content (Huang et al., 2022), while the application of multi-walled carbon nanotubes (25 and 50 mg L^{-1}) in 'Flame Seedless' grapes prior to harvest induces stability of firmness, soluble sugar, titratable acid, and pH in the skin of the fruit, as well as an increase in the activity of POX, CAT, SOD, and APX enzymes, decreasing the content of MDA and excessive accumulation of ROS in the pulp (Sha et al., 2022). The increase in enzyme activity appears to be due to multi-walled carbon nanotubes being taken up by parenchymal cells in the peel and through the epidermal cell layer in the pulp, generating oxidative stress (Sha et al., 2022). Likewise, an increase in the content of glucose-6-phosphate, phenylalanine, and ascorbic acid in tomato fruit was documented when applying nanocomposites of $MnFe_2O_4$ (10 mg L^{-1}) (Yue et al., 2022). While $Fe_7(PO_4)_6$ nanocomposite regulates sucrose metabolism, the shikimic acid pathway, and phenylalanine synthesis in tomato fruit (Wang et al., 2022).

The suspension of nano-biochar (1, 3, and 5%) at the beginning of flowering of *Citrus reticulata* L. improved fruit set, size, fruit weight, total dissolved solids, vitamin C content, and lower pH. This response may be due to the fact that nano-biochar induces cell expansion and greater photosynthetic capacity (Shani et al., 2023). A similar response is reported in watermelon, where applying NPs of α- and γ-Fe_2O_3 (400 ppm) promoted the water and vitamin C content of the fruit (Li et al., 2020). Likewise, an increase in the total soluble solids and antioxidant capacity of strawberry fruits was reported with the application of Fe_3O_4 NPs coated with cysteine (Azizkhani et al., 2023). Fe_2O_3 NPs in *Raphanus sativus* L. cv. Champion increased the content of chlorophylls, vitamin C, glutathione (GSH), and antioxidant capacity in leaves and roots (Garza-Alonso et al., 2023). In melon, the use of NPs of γ-Fe_2O_3 and Fe_3O_4 promoted the weight of the fruit and vitamin C content (up to 46.95%) (Wang et al., 2019).

Nano-Se, nano-TiO_2, and nano-CeO_2 strongly affect tomato fruit, especially nano-Se, which causes an increase in fruit size and weight as well as higher sugar content (fructose, sucrose, and glucose) and higher ascorbic acid and salicylic acid content, GSH, SOD, and CAT concomitant with a reduction of MDA, H_2O_2, and oxygen free radical, as well as a restriction in the production of oxaloacetate and α-ketoglutarato, affecting the synthesis of malic acid and minimizing the titratable acidity of the fruit (Liu, 2023a,b). Selenium NPs (1, 15, 30, and 45 mg L^{-1}) increased vitamin C content, $ABTS^+$ antioxidant capacity, PAL enzymatic activity, and fruit firmness in jalapeño pepper (*Capsicum annuum* L.) (Sariñana-Navarrete et al., 2023). The application of Cu NPs led to an increase in total soluble solids, vitamin C, antioxidant capacity $ABTS^+$ and DPPH in tomato fruits (Peralta-Manjarrez et al., 2023). Likewise, the use of nano-urea-controlled release fertilizers applied to tomato plants grown in clay loam soil showed positive changes in the firmness, acidity, and vitamin C content of the fruit (Helal et al., 2023).

Fruit quality is stimulated in plants under stressful conditions. In this regard, tomato plants damaged by *Clavibacter michiganensis* and treated with Cu NPs and potassium silicate showed higher PAL and GPX activity in fruit (Cumplido-Nájera et al., 2019). Tomato fruits exposed to *Alternaria solani* and treated with Se and Cu

NPs promoted an increase in the content of vitamin C, glutathione, and greater GPX activity (Quiterio-Gutiérrez et al., 2019). Tomato plants under NaCl stress presented a significant increase in soluble solids and titratable acidity, as well as greater APX, GPX, CAT, and SOD activity of the fruit with the use of Se NPs (Morales-Espinoza et al., 2019). The application of Cu NPs in tomato plants under salinity induced an increase in the content of vitamin C (80%) and GSH (81%) in the fruit (Pérez-Labrada et al., 2019). In a study carried out on bell pepper plants under saline stress conditions, it was reported that NPs of selenium, silicon, and copper (Se 10 and 50 $mg\,L^{-1}$, Si 200 and 1000 $mg\,L^{-1}$, and Cu 100 and 500 $mg\,L^{-1}$) applied to the substrate induced an increase in APX, GPX, CAT, PAL activity, and GSH in the fruits (González-García et al., 2021). Tomato plants exposed to arsenic and with SiO_2 NPs application showed an increase in vitamin C content (González-Moscoso et al., 2022). Tomato plants under different levels of electrical conductivity of the water with foliar application of selenium and copper nanofertilizer (100 $mg\,L^{-1}$) presented a high content of soluble solids, vitamin C, firmness, and improvements in PPO, CAT, and POX (Saffan et al., 2022). In the case of Fe_2O_3 nanocomposites conjugated with humic acid (Fe_2O_3 NPs-H) and boron (Fe_2O_3 NPs-B) (0.125 and 0.25 mM, respectively) applied to cucumber, they increased POX and PPO enzymatic activity (El-Batal et al., 2023). Chitosan-based Schiff base-metal complexes (CH-Fe) increased titratable acidity (30.9%) and ascorbic acid (25.8%) in pomegranate fruit under drought (Zahedi et al., 2023).

9.5.2.1 Genetic responses

The aforementioned responses may be the consequence of alterations at the genetic level. In *S. lycopersicum* L. exposed to ZnO NPs, it is reported that the genes *CAO* (Chlorophyllide a oxygenase), *CHLG* (Chlorophyll synthase), *CRD1* (Magnesium-protoporphyrin IX monomethyl ester [oxidative] cyclase), *CHLI* (Magnesium-chelatase subunit ChlI), *HEMG* (Protoporphyrinogen oxidase), *HEMB* (5-Aminolevulinate dehydratase), *HEMC* (Porphobilinogen deaminase), *SBPASE* (Sedoheptulose-1,7-bisphosphatase), *FBPASE* (Fructose- 1,6-bisphosphatase), *PSAA* (Photosystem I P700 apoprotein A1), *PSBD* (Photosystem II protein D2), *PSBF* (Photosystem II cytochrome b559 β subunit), and *PSBH* (Photosystem II phosphoprotein) showed lower relative expression in leaf upon exposure to at high concentrations of NPs (Wang et al., 2018). The same authors report a higher relative expression of the genes *PSY* (Phytoene synthase), *LYCB* (Lycopene β cyclase), *Cu/Zn2-SOD* (SOD [Cu-Zn] 2), *Fe-SOD* (SOD [Fe]), *APX2* (Cytosolic APX 2), and *CAT1* (CAT 1). The ZnO NPs applied at a rate of 3 $mg\,L^{-1}$ in tomato increased the expression of the transcription factor *R2R3MYB* (MYB transcription factors; by 2.6 folds) as well as the genes *WRKY1* (by 6.4 folds), *bHLH* (basic/helix-loop-helix; by 4.7 folds), *EREB* (ethylene-responsive element-binding protein), *HsfA1a* (heat stress transcription factor; by 2.8 folds), *MKK2*, *CAT*, and reducing *HDA3* gene expression (histone deacetylase; by 1.9 folds) (Pejam et al., 2021). The same authors point out that the reduction in the expression of the *HDA3* gene suggests a high sensitivity and epigenetic response to ZnO NPs, altering the chromatin ultrastructure.

The Ca-based nanocomposite (nanocalcium L-aspartic acid) applied in 'Huaguang' nectarine promotes the expression of genes related to sucrose synthesis and those related to sucrose transport in the stem phloem (Zhu et al., 2023). In tomato fruits, nanocomposites (nano-Se, nano-TiO_2, and nano-CeO_2) alter gene expression, particularly *Lin5* (sucrose invertase5), *SUS2* (sucrose synthase 2), *SPS1* (sucrose phosphate synthase 1), and *SPS2* (sucrose phosphate synthase 1), which showed higher expression in the fruit with the application of nano-Se, while the *ICDH* (isocitrate dehydrogenase), *mMDH* (mitochondrial malate dehydrogenase), *CS* (citrate synthase), and *PEPC2* (phosphoenolpyruvate carboxylase) genes showed a reduction in their expression in fruits with nano-Se, nano-TiO_2, and nano-CeO_2. These data suggest that nanocomposites induce genetic regulation in the sugar profile and the synthesis of organic acids in the tricarboxylic acid cycle (Liu, 2023a,b). The same authors point out that the alterations in the genes associated with the synthesis of sugars are consistent with the promotion of photosynthesis, particularly with the chlorophyll content. The use of chitosan NPs in soil regulates the expression of genes sensitive to drought (*HsfA1a*, *SlAREB1*, *LeNCED1*, and *LePIP1*) in *S. lycopersicum* plants, which could also be a mechanism that promotes fruit quality (Mohamed and Abdel-Hakeem, 2023).

In a transcriptomic analysis, enrichment was found in the pathways related to sucrose metabolism (*SlSS*, *SlTPS6*, *SlTPS11*, and *SlTKL-1*; 101.43, 131.24, 118.71, and 32.17%, respectively) and phenylalanine metabolism (*ko00360*). Likewise, *SlDHQ-SOR* (shikimic acid pathway), *SlEPSPS-1*, and *SlCM2* (phenylalanine pathway) were regulated 15.20, 15.63, and 34.99% less with $Fe_7(PO_4)_6$ nanocomposites, as well as a positive regulation of *SlMYB12* in tomato fruit (Wang et al., 2022). On the other hand, at the metabolic level, metabolites related to sucrose (UDP-glucose, trehalose, and D-glucose-6P) were identified, which were significantly promoted with nanocomposites of $Fe_7(PO_4)_6$ (Wang et al., 2022).

9.5.3 Secondary metabolites in the fruit

The use of nanocomposites has a promoting effect on secondary metabolism by stimulating the terpene or isoprenoid content, mainly carotenes and xanthophylls (Ji et al., 2023; Pejam et al., 2021) (Fig. 9.3). In this regard, tomato fruits obtained from plants exposed to *C. michiganensis* and treated with Cu NPs and potassium silicate had higher lycopene and β-carotene content (Cumplido-Nájera et al., 2019), while the content of phenols and flavonoids increased in tomato fruits exposed to *A. solani* and treated with Se and Cu NPs (Quiterio-Gutiérrez et al., 2019). The same type of NPs applied to tomatoes led to an increase in flavonoids (Hernández-Hernández et al., 2019), phenols, lycopene, and β-carotene (Morales-Espinoza et al., 2019). In habanero peppers, the spraying of ZnO NPs promoted a higher content of capsaicin, dihydrocapsaicin, and Scoville heat units, as well as total phenols and flavonoids in the fruit (García-López et al., 2019). Phenolic compounds and anthocyanins in pomegranate fruits were increased with the application of Se NPs (Zahedi et al., 2019). The same type of NPs led to an increase in the content of phenols, flavonoids, β-carotene, and yellow carotenoids in bell peppers under salt stress conditions (González-García et al., 2021). The increase in these metabolites improves fruit quality since phenolic compounds, due to their redox properties (they can donate electrons), can participate in the elimination of ROS (García-López et al., 2019).

In strawberries, the application of nanofertilizers increased the content of phenols, total flavonoids, and tannins (Rahman et al., 2021a,b). In bitter melon fruits, the application of 20 mg L^{-1} of chitosan-selenium nanoparticles resulted in an increase in total phenols, flavonoids, and anthocyanins. Additionally, there was an increase in the content of essential oils such as cucurbitacin, cucuritan, charantin, momordicin, momordin, momordicoside, karaviloside, oleic acid, stigmasterol, cis-9-hexadecenal, and gentisic acid (Sheikhalipour et al., 2021). In Flame Seedless and Red Sultana grape cultivars, the use of chitosan-phenylalanine nanocomposites and chitosan-salicylic acid nanoforms applied before veraison induced a promotion in the content of total phenols, flavonoids, and anthocyanins (oenin); the increase in these biocompounds is probably due to alterations in the metabolic pathway of their biosynthesis or gene expression (Gohari et al., 2021b; Khalili et al., 2022). In cherry tomatoes, selenium nanocomposites (75 μg kg^{-1}) improved the synthesis of tomatidine and flavonoids (luteolin, hespertin, quercetin, eriodictyol, and kaempferol) (Cheng et al., 2022a). In another study, $MnFe_2O_4$ nanocomposites (10 mg L^{-1}) induced a significant increase in the rutin content of tomato fruit (Yue et al., 2022). In contrast, foliar sprays of Fe_3O_4, $MnFe_2O_4$, $ZnFe_2O_4$, $Zn_{0.5}Mn_{0.5}Fe_2O_4$, Mn_3O_4, and ZnO nanocomposites (250 mg L^{-1}) on tomato plants showed diverse effects: lycopene was reduced with $MnFe_2O_4$, $ZnFe_2O_4$, and $Zn_{0.5}Mn_{0.5}Fe_2O_4$ at 0 days of storage, $Zn_{0.5}Mn_{0.5}Fe_2O_4$, Mn_3O_4, and ZnO reduced β-carotene, while total phenolics increased with $ZnFe_2O_4$, $Zn_{0.5}Mn_{0.5}Fe_2O_4$ and ZnO (Cantu et al., 2022). The accumulation of flavonoids naringenin (681.39%), quercetin (54.49%), and rutin (120.36%) in tomato fruit promoted by the application of 50 mg kg^{-1} of Triiron Tetrairon Phosphate ($Fe_7(PO_4)_6$) may be due to the activation of proteins that participate in the nutritional transport systems, which makes the synthesis and transport of photosynthates more efficient, as well as an improvement in their biosynthesis pathway (Wang et al., 2022). The same authors point out that the accumulation of sucrose in the fruit could signal genes related to flavonoids. In the tomato fruit exposed to $Fe_7(PO_4)_6$ nanocomposites, enrichment in the synthesis of flavonoids, phenylpropanoid (ko00940), and secondary metabolites (ko01110) was found at the transcriptomic and metabolic levels; in the case of flavonoids, there was an increase in *SI4CL1* (20.75%) and *SICHI* (123.53%) (Wang et al., 2022). On the other hand, phenols and flavonoids (quercetin glycosides) were increased in apple fruit when applying Zn NPs (Montaño-Herrera et al., 2022).

Nano-Se, nano-TiO_2, and nano-CeO_2 promote higher lycopene content in tomatoes (Liu, 2023a,b). Ca-base nanocomposite (nanocalcium L-aspartic acid) applied to 'Huaguang' nectarine promotes the expression of genes associated with the synthesis of anthocyanins in the leaf (Zhu et al., 2023). The increase in these compounds could improve the antioxidant capacity of the fruit (Liu, 2023a,b). Likewise, the use of selenium NPs *via* drench increased the content of total phenols and flavonoids in jalapeño peppers (Sariñana-Navarrete et al., 2023). In another study carried out on strawberries, Fe_3O_4 NPs coated with cysteine (0.06 and 6 mg L^{-1}) promoted the content of anthocyanins and flavonoids (Azizkhani et al., 2023). In cherry radish (*R. sativus* L. var. radculus pers), the application of engineered Se nanocomposites (10 mg L^{-1}) to be absorbed and translocated from the shoot to the root causes an increase in the metabolic pathways of the cycle of glycolysis, pyruvate, and tricarboxylic acid, which generates a high production of flavonoids, amino acids, and tricarboxylic acid (Cheng et al., 2022b). In 2023, nano-Se has been reported to increase the content of volatile compounds in tomato fruit: 3-methyl- Butanoi acid, 3-methyl butanol, (Z)-beta-Damascenone, Beta-Cyclocitral, Hexanol, Pentanol, Pentanal, and 6-methyl-5-Hepten-2-one, which suggests an improvement in the flavor of ripe tomato fruits (Liu, 2023a,b).

In pomegranate fruits under drought conditions, the application of chitosan-based Schiff metal-based complexes increases the content of total phenols (24.3%) and total anthocyanins (9.3%) (Zahedi et al., 2023). A similar response was reported in *S. lycopersicum* L. with foliarly applied CaO and ZnO NPs that promoted the total phenol content (Farooq et al., 2023). Quercetin in tomato (Feng et al., 2022) and flavonoids in cucumber (Feng et al., 2023) were promoted by the application of 10 mg kg^{-1} of CeO_2, while Cu NPs increased total carotenes, lycopene, and flavonoids in grafted tomato fruits (Peralta-Manjarrez et al., 2023).

9.5.3.1 Genetic response

Tomato fruits presented a transcriptional regulation of the *CRTISO* gene (5.9 folds) that encodes carotene isomerase with the use of Se NPs (0.3 and 10 mg L^{-1}), while in leaf tissue a transcriptional increase of *miR172* was evidenced (a miRNA that participates in the control of the development program) and of the genes *bZIP* (transcription factor) and *CRTISO* (Neysanian et al., 2020). In tomato fruit, the genes related to the synthesis of naringenin (*SIPAL*, *SIC4H*, *SI4CL*, *SICHS1*, and *SICHI*) increased their relative expression by 83.54%, 90.83%, 189.09%, 24.16%, and 105.92%, likewise, the genes *SlF3H*, *SlF3′H*, *SlFLS*, and *Sl3GT* related to enzymes encoding naringenin to quercetin and rutin increased their relative expression (54.76%, 189.07%, 94.73%, and 96.15%, respectively) when applying nanocomposites of $Fe_7(PO_4)_6$ (Wang et al., 2022).

The response of NMs and nanocomposites (oxidative stress, stimulation of the antioxidant system, and transcriptional regulation) seems to depend on the type, concentration, route of application, and variety studied since this affects their translocation and compartmentalization (Kohatsu et al., 2021; Li et al., 2020; Mosa et al., 2018; Tabatabaee et al., 2021). These compounds that penetrate, accumulate, and transport can generate biochemical and physiological changes in the plant (El-Bialy et al., 2023; García-López et al., 2019), activating and/or regulating the accumulation of antioxidant biocompounds, improving the ascorbate-glutathione cycle, which leads to the elimination and/or equilibrium with ROS, and reducing oxidative stress in the plant and fruit (Hernández-Hernández et al., 2019). In the same way, NMs can stimulate the cytoplasmic Ca^{2+} content in addition to regulating mitogen-activated protein kinase (MAPK) (Sheikhalipour et al., 2021). In this sense, it is reported that Ag NPs, when recognized by Ca^{2+} receptors, Ca^{2+} channels, and Ca^{2+}/Na^{+} ATPases, can promote ROS production and a Ca^{2+} burst in *Arabidopsis thaliana* (Sosan et al., 2016). This would imply a phosphorylation of MAPK with the consequent stress signaling and activating transcription factors, with their consequent metabolic and transcriptomic responses generating the synthesis of secondary metabolites in the leaf and fruit (Shoala et al., 2021).

9.5.4 Nutrient content of the fruit

In addition to increasing the content of biocompounds (primary and secondary), NMs improve the nutritional level of the plant (Huang et al., 2022) (Fig. 9.3), since the size gives it greater efficiency in absorption and transport systems (Tarafder et al., 2020), or through histological modifications as found with the application of ZnO NPs that improved metaxylem in tomato (Pejam et al., 2021). NMs can positively induce expression of the *LHA1*, *LHA2*, and *LHA4* genes related to the synthesis of H^{+}ATPase transmembrane proteins in roots, a protein associated with nutrient cotransport systems (Wang et al., 2022). This allows an accumulation of nutrients such as N, P, K, Mg, Ca, Fe, Cu, Zn, Mn, and Se in the fruit (Fig. 9.3).

In conjunction with the improvement in the nutrient content in the fruit, nanocomposites can reduce or prevent physiopathies. Thus, in *Prunus persica* L., fruit cracking is reduced by the use of L-aspartic acid nanocalcium, probably due to improvements in the activity of calmodulin, which increased the content of calcium pectinate in the peel, in addition to minimizing the activity of metabolic enzymes of the cell wall, accompanied by an increase in soluble sugar in the fruit (Zhu et al., 2023). The use of carbon dots encapsulated in Ca (Ca-CD) applied *via* irrigation or foliar improves the Ca content, pectin, and cellulose content in the fruit of *Malus domestica* var. "Royal Gala." This may be due to the combination of Ca-CD with pectin, which generates cross-linked Ca-pectin polymers, promoting the mechanical resistance of the cell wall and firmness (Wang, 2023a,b,c). It can also be due to alterations in cell morphology and turgor that lead to the accumulation of proteins, allowing lignification of the cell wall (Peralta-Manjarrez et al., 2023).

9.5.5 Postharvest stimulation

Another alternative studied to promote fruit quality is the use of postharvest nanocomposites (Fig. 9.4). In this regard, it is reported that the use of α-$Fe_2O_3/C_{20}H_{38}O_{11}$ nanocoating in tomato fruit (stored at $25 \pm 2°C$ and $4°C$) protects and decreases the loss of vitamin C (Guleria et al., 2022). Similarly, chitosan nanocoating (CuO NP/ ZnO NP) in guava fruits reduces weight loss (lower pectin methylesterase enzymatic activity), respiration rate, and ethylene effects as a result of increased antioxidant activity triggered by the ROS generated by the NPs (Kalia et al., 2021). The same authors point out that the lower speed of maturation may be due to a reduction in the formation of sucrose and volatile compounds, expressed by low values of soluble solids. Likewise, the increase in antioxidant activity prevents the accumulation of or eliminates free radicals and MDA, minimizing oxidative damage in addition to

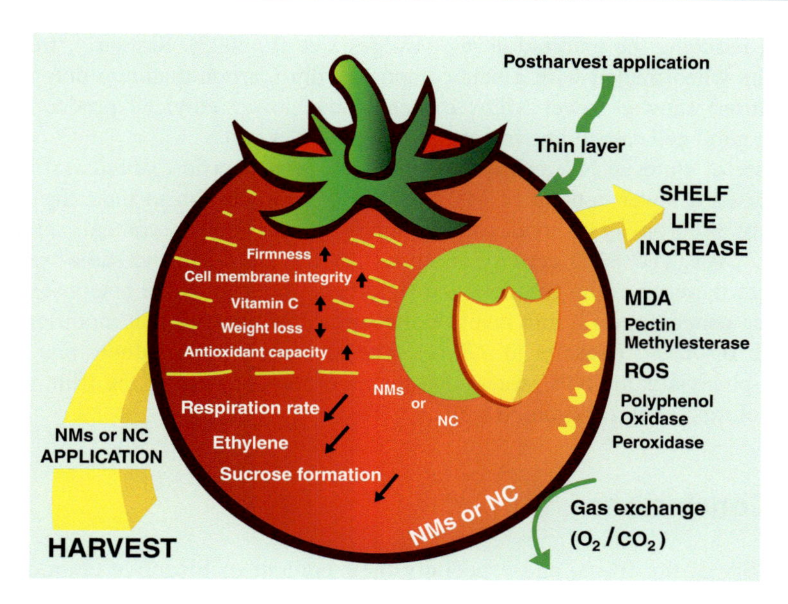

Figure 9.4 Postharvest stimulation of fruits by the application of nanocomposites or NMs. The application of NMs in the harvested fruits can induce positive responses since some physiological processes of the fruits such as respiration can be modified, or the activity of enzymes such as PPO that degrades the quality of the fruits. In addition, the content of antioxidants can be increased, and therefore, neutralize the adverse effects of reactive oxygen species.

reducing the activity of the PPO enzyme (Bahmani et al., 2022; Nasr et al., 2021). The PPO enzyme binds strongly to the chloroplast membrane, originating substrates that are stored in the vacuole when there is a high ROS content and fruit aging; however, NPs, by stabilizing the cell membrane, can reduce the activity of this enzyme. In addition, they can reduce the interaction of the fruit with oxygen, reducing the production and accumulation of ROS (Nxumalo et al., 2022).

Oranges coated with chitosan (0.5%) + nano-ZnO (0.5 g L^{-1}) (20 days at 4°C) had less loss of total soluble solids and total acidity, as well as less loss of weight, firmness, and ascorbic acid concentration, and a reduction in PPO and POX activity (Dulta et al., 2022). The application of edible coatings of bionanocomposites (based on cassava starch) reinforced with cellulose nanofibrils to strawberry fruits stored 2 ± 1°C promoted a lower respiratory rate and slow metabolism by minimizing the degradation of soluble solids and organic acids, the synthesis of anthocyanins, the activity of softening enzymes, the production of compounds with antioxidant activity, and synthesis of vanillin and resveratrol (Carvalho do Lago et al., 2023). In the case of Murcott mandarin fruits stored (5°C ± 1°C; 90%–95% relative humidity) for four months, a coating with nanosilver and a mixture of nanosilver/wax allow for maintaining the vitamin C content (Gemail et al., 2023). It has recently been reported that cinnamaldehyde nanocomposite films encapsulated in halloysite nanotube nanocarriers incorporated in polypropylene inhibit ethylene production in bananas, reducing the maturation rate (Kolgesiz et al., 2023). Similarly, freshly cut pear fruits when treated with a nanocomposite film (carbon quantum dots in chitosan solution) showed lower MDA content and a lower ethylene production rate, allowing good cell membrane integrity (Wang, 2023a,b,c).

The use of chitosan NPs coated with proline in strawberries, stored at 4°C for 12 days, caused a lower content of MDA and H_2O_2, in addition to inducing a higher content of ascorbic acid, total soluble solids, total phenolic content, antioxidant capacity, CAT, and SOD activity (Bahmani et al., 2022). In the same way, there may be an increase in the firmness and rigidity of the fruit; these responses may be due to the generation of a thin layer on the surface of the fruit that modifies the gas exchange (O_2/CO_2) (Nasr et al., 2021). The aforementioned authors cite that the chitosan-phenylalanine NPs protect and maintain the firmness of the fruit, avoiding the loss of tannins in persimmon fruits.

9.6 Conclusions

The quality of the fruits, and especially their content of bioactive compounds, is increasingly important to consumers. It is clear that consumers require not only fruits of better physicochemical quality but also fruits rich in beneficial compounds for human health. In this context, the use of technologies such as nanocomposites and NMs provides an excellent tool to meet the objective of improving the quality of the fruits as well as the content of bioactive compounds. Basically, the use of nanotechnology can be applied during the development of the crop or postharvest, since in both cases positive effects have been observed on the quality and content of bioactive compounds in the fruits. Therefore, the use of nanotechnology, through the application of NMs and nanocomposites, can be a fundamental strategy for the production of fruits that meet the quality standards and bioactive compounds required by consumers.

Although the evidence shows that the use of NMs and nanocomposites is a useful tool to improve the quality of fruits and increase the content of bioactive compounds, it is still not clear what the final destination of the NMs is. Therefore, research must be carried out to clarify what happens to the NMs that are applied to plants and to determine exactly how much of these NMs reach the fruit and can be consumed by humans. If this happens, then it is necessary to know if there may be any impact on human health from the consumption of NMs that are applied to plants. In addition, studies involving the tracking of NMs at different trophic levels should be carried out to determine the possible impact on ecosystems, since these technologies are currently being applied in different areas, not only in agriculture.

References

Abideen, Z., Hanif, M., Munir, N., & Nielsen, B. L. (2022). Impact of nanomaterials on the regulation of gene expression and metabolomics of plants under salt stress. *Plants*, *11*, 1–23.

Adefegha, S. A. (2018). Functional foods and nutraceuticals as dietary intervention in chronic diseases; novel perspectives for health promotion and disease prevention. *Journal of Dietery Supplement*, *15*, 977–1009.

Ali, A. A., & Toliba, A. O. (2018). Effect of organic calcium spraying and nano chitosan fruits coating on yield, fruit quality and storability of peach cv 'early swelling'. *Current Science International*, *7*, 737–749.

Ali, M., Ahmed, A., Shah, S. W. A., Mehmood, T., & Abbasi, K. S. (2020). Effect of silver nanoparticle coatings on physicochemical and nutraceutical properties of loquat during postharvest storage. *Journal of Food Processing and Preservation*, *44*, e14808.

Ali, M. Y., Sina, A. A. I., Khandker, S. S., Neesa, L., Tanvir, E. M., Kabir, A., Khalil, M. I., & Gan, S. H. (2021). Nutritional composition and bioactive compounds in tomatoes and their impact on human health and disease: A review. *Foods*, *10*, 45.

Amengual, J. (2019). Bioactive properties of carotenoids in human health. *Nutrients*, *11*, 1–6.

Arroyo, B. J., Bezerra, A. C., Oliveira, L. L., Arroyo, S. J., Melo, E. A. de, & Santos, A. M. P. (2020). Antimicrobial active edible coating of alginate and chitosan add ZnO nanoparticles applied in guavas (*Psidium guajava* L.). *Food Chemistry*, *309*, 125566.

Azizkhani, S., Javadi, T., Ghaderi, N., & Farzinpour, A. (2023). Replacing conventional iron with cysteine-coated Fe_3O_4 nanoparticles in soilless culture of strawberry. *Science Horticulture (Amsterdam)*, *318*, 112098.

Bahmani, R., Razavi, F., Mortazavi, S. N., Gohari, G., & Juárez-Maldonado, A. (2022). Evaluation of proline-coated chitosan nanoparticles on decay control and quality preservation of strawberry fruit (cv. Camarosa) during cold storage. *Horticulturae*, *8*, 648.

Bakhy, E. A., Zidan, N. S., & Aboul-Anean, H. E. D. (2018). The effect of nano materials on edible coating and films' improvement. *International Journal of Pharmaceutical Research and Allied Science*, *7*, 20–41.

Bose, S. K., Howlader, P., Wang, W., & Yin, H. (2021). Oligosaccharide is a promising natural preservative for improving postharvest preservation of fruit: A review. *Food Chemistry*, *341*, 128178.

Calder, P. C. (2018). Very long-chain n-3 fatty acids and human health: Fact, fiction and the future. *Proceedings of the Nutrition Society*, *77*, 52–72.

Campos Alencar Oldoni, F., Florencio, C., Brait Bertazzo, G., Aparecida Grizotto, P., Bogusz Junior, S., Lajarim Carneiro, R., Alberto Colnago, L., & David Ferreira, M. (2022). Fruit quality parameters and volatile compounds from 'Palmer' mangoes with internal breakdown. *Food Chemistry*, *388*, 132902.

Cantu, J. M., Ye, Y., Hernandez-Viezcas, J. A., Zuverza-Mena, N., White, J. C., & Gardea-Torresdey, J. L. (2022). Tomato fruit nutritional quality is altered by the foliar application of various metal oxide nanomaterials. *Nanomaterials*, *12*, 2349.

Carvalho do Lago, R., Zitha, E. Z. M., de Oliveira, A. L. M., de Abreu, D. J. M., Carvalho, E. E. N., Piccoli, R. H., Tonoli, G. H. D., & Boas, E. V. de B. V. (2023). Effect of coating with co-product-based bionanocomposites on the quality of strawberries under refrigerated storage. *Science Horticulture (Amsterdam)*, *309*, 111668.

Castelo Branco Melo, N. F., de MendonçaSoares, B. L., Marques Diniz, K., Ferreira Leal, C., Canto, D., Flores, M. A. P., Henrique da Costa Tavares-Filho, J., Galembeck, A., Montenegro Stamford, T. L., Montenegro Stamford-Arnaud, T., & Montenegro Stamford, T. C. (2018). Effects of fungal chitosan nanoparticles as eco-friendly edible coatings on the quality of postharvest table grapes. *Postharvest Biology and Technology*, *139*, 56–66.

Chen, T., Ji, D., Zhang, Z., Li, B., Qin, G., & Tian, S. (2020). Advances and strategies for controlling the quality and safety of postharvest fruit. *Engineering*, *7*, 1177–1184.

Cheng, B., Wang, C., Cao, X., Yue, L., Chen, F., Liu, X., Wang, Z., & Xing, B. (2022a). Selenium nanomaterials induce flower enlargement and improve the nutritional quality of cherry tomatoes: pot and field experiments. *Environmental Science Nano, 14*, 4190–4200.

Cheng, B., Wang, C., Chen, F., Yue, L., Cao, X., Liu, X., Yao, Y., Wang, Z., & Xing, B. (2022b). Multiomics understanding of improved quality in cherry radish (*Raphanus sativus* L. var. Radculus Pers) after foliar application of selenium nanomaterials. *Science of the Total Environment, 824*, 153712.

Cumplido-Nájera, C. F., González-Morales, S., Ortega-Ortíz, H., Cadenas-Pliego, G., Benavides-Mendoza, A., & Juárez-Maldonado, A. (2019). The application of copper nanoparticles and potassium silicate stimulate the tolerance to *Clavibacter michiganensis* in tomato plants. *Science Horticulture (Amsterdam), 245*, 82–89.

Delgado, A. M., Issaoui, M., & Chammem, N. (2019). Analysis of main and healthy phenolic compounds in foods. *Journal of AOAC International, 102*, 1356–1364.

Dulta, K., Koşarsoy Ağçeli, G., Thakur, A., Singh, S., Chauhan, P., & Chauhan, P. K. (2022). Development of alginate-chitosan based coating enriched with ZnO nanoparticles for increasing the shelf life of orange fruits (*Citrus sinensis* L.). *Journal of Polymer Environment, 30*, 3293–3306.

Dwivedi, D., Megha, K., Mishra, R., & Mandal, P. K. (2020). Glutathione in brain: Overview of its conformations, functions, biochemical characteristics, quantitation and potential therapeutic role in brain disorders. *Neurochemical Research, 45*, 1461–1480.

El-Batal, A. I., El-Sayyad, G. S., Al-shammari, B. M., Abdelaziz, A. M., Nofel, M. M., Gobara, M., Elkhatib, W. F., Eid, N. A., Salem, M. S., & Attia, M. S. (2023). Protective role of iron oxide nanocomposites on disease index, and biochemical resistance indicators against *Fusarium oxysporum* induced-cucumber wilt disease: *In vitro*, and *in vivo* studies. *Microbial Pathogenesis, 180*, 106131.

El-Bialy, S. M., El-Mahrouk, M. E., Elesawy, T., Omara, A. E. D., Elbehiry, F., El-Ramady, H., Áron, B., Prokisch, J., Brevik, E. C., & Solberg, S. (2023). Biological nanofertilizers to enhance growth potential of strawberry seedlings by boosting photosynthetic pigments, plant enzymatic antioxidants, and nutritional status. *Plants, 12*, 302.

El-Ramady, H., Hajdú, P., Törős, G., Badgar, K., Llana, X., Kiss, A., Abdalla, N., Omara, A. E. D., Elsakhawy, T., Elbasiouny, H., Elbehiry, F., Amer, M., El-Mahrouk, M. E., & Prokisch, J. (2022). Plant nutrition for human health: A pictorial review on plant bioactive compounds for sustainable agriculture. *Sustain, 14*, 8329.

Eldib, R., Khojah, E., Elhakem, A., Benajiba, N., & Helal, M. (2020). Chitosan, nisin, silicon dioxide nanoparticles coating films effects on blueberry (*Vaccinium myrtillus*) quality. *Coatings, 10*, 962.

Esyanti, R. R., Zaskia, H., Amalia, A., & Nugrahapraja, D. H. (2019). Chitosan nanoparticle-based coating as post-harvest technology in banana. *Journal of Physical Conference Series, 1204*, 012109.

Farooq, A., Javad, S., Jabeen, K., Ali Shah, A., Ahmad, A., Noor Shah, A., Nasser Alyemeni, M., F.A Mosa, W., & Abbas, A. (2023). Effect of calcium oxide, zinc oxide nanoparticles and their combined treatments on growth and yield attributes of *Solanum lycopersicum* L. *Journal of King Saudi University - Science, 35*, 102647.

Fayaz, A. M., Girilal, M., Kalaichelvan, P. T., & Venkatesan, R. (2009). Mycobased synthesis of silver nanoparticles and their incorporation into sodium alginate films for vegetable and fruit preservation. *Journal of Agriculture Food Chemistry, 57*, 6246–6252.

Fenech, M., Amaya, I., Valpuesta, V., & Botella, M. A. (2019). Vitamin C content in fruits: Biosynthesis and regulation. *Frontiers in Plant Sciene, 9*, 1–21.

Feng, Y., Wang, C., Chen, F., Cao, X., Wang, J., Yue, L., & Wang, Z. (2022). Molecular mechanisms of CeO_2 nanomaterials improving tomato yield, fruit quality, and postharvest storage performance. *Environmental Science: Nano*, *9*, 4382–4392.

Feng, Y., Wang, C., Chen, F., Cao, X., Wang, J., Yue, L., & Wang, Z. (2023). Cerium oxide nanomaterials improved cucumber flowering, fruit yield and quality: the rhizosphere effect. *Environmental Science: Nano*, *10*, 2010–2021.

Francia, V., Yang, K., Deville, S., Reker-Smit, C., Nelissen, I., & Salvati, A. (2019). Corona composition can affect the mechanisms cells use to internalize nanoparticles. *ACS Nano*, *13*, 11107–11121.

Fu, C., Khan, M. N., Yan, J., Hong, X., Zhao, F., Chen, L., Ma, H., Li, Y., Li, J., & Wu, H. (2023). Mechanisms of nanomaterials for improving plant salt tolerance. *Crop Environment*, *2*, 92–99.

Fujita, M., & Hasanuzzaman, M. (2022). Approaches to enhancing antioxidant defense in plants. *Antioxidants*, *11*, 1–5.

Gao, L., Li, Q., Zhao, Y., Wang, H., Liu, Y., Sun, Y., Wang, F., Jia, W., & Hou, X. (2017). Silver nanoparticles biologically synthesised using tea leaf extracts and their use for extension of fruit shelf life. *IET Nanobiotechnology*, *11*, 637–643.

García-Betanzos, C. I., Hernández-Sánchez, H., Bernal-Couoh, T. F., Quintanar-Guerrero, D., & Zambrano-Zaragoza, M. L. (2017). Physicochemical, total phenols and pectin methylesterase changes on quality maintenance on guava fruit (*Psidium guajava* L.) coated with candeuba wax solid lipid nanoparticles-xanthan gum. *Food Research International*, *101*, 218–227.

García-López, J. I., Niño-Medina, G., Olivares-Sáenz, E., Lira-Saldivar, R. H., Barriga-Castro, E. D., Vázquez-Alvarado, R., Rodríguez-Salinas, P. A., & Zavala-García, F. (2019). Foliar application of zinc oxide nanoparticles and zinc sulfate boosts the content of bioactive compounds in habanero peppers. *Plants*, *8*, 254.

Garza-Alonso, C. A., Cadenas-Pliego, G., Juárez-Maldonado, A., González-Fuentes, J. A., Tortella, G., & Benavides-Mendoza, A. (2023). Fe_2O_3 nanoparticles can replace Fe-EDTA fertilizer and boost the productivity and quality of *Raphanus sativus* in a soilless system. *Science Horticulture (Amsterdam)*, *321*, 112261.

Gemail, M. M., Elesawi, I. E., Jghef, M. M., Alharthi, B., Alsanei, W. A., Chen, C., El-Hefnawi, S. M., & Gad, M. M. (2023). Influence of wax and silver nanoparticles on preservation quality of murcott mandarin fruit during cold storage and after shelf-life. *Coatings*, *13*, 90.

Gohari, G., Panahirad, S., Sadeghi, M., Akbari, A., Zareei, E., Zahedi, S. M., Bahrami, M. K., & Fotopoulos, V. (2021a). Putrescine-functionalized carbon quantum dot (put-CQD) nanoparticles effectively prime grapevine (*Vitis vinifera* cv. 'Sultana') against salt stress. *BMC Plant Biology*, *21*, 1–15.

Gohari, G., Zareei, E., Kulak, M., Labib, P., Mahmoudi, R., Panahirad, S., Jafari, H., Mahdavinia, G., Juárez-Maldonado, A., & Lorenzo, J. M. (2021b). Improving the berry quality and antioxidant potential of flame seedless grapes by foliar application of chitosan–phenylalanine nanocomposites (Cs—Phe NCs). *Nanomaterials*, *11*, 2287.

Gohari, G., Zareei, E., Rostami, H., Panahirad, S., Kulak, M., Farhadi, H., Amini, M., Martinez-Ballesta, M. del C., & Fotopoulos, V. (2021c). Protective effects of cerium oxide nanoparticles in grapevine (*Vitis vinifera* L.) cv. Flame Seedless under salt stress conditions. *Ecotoxicology and Environmental Safety*, *220*, 112402.

González-García, Y., Cadenas-Pliego, G., Benavides-Mendoza, A., & Juárez-Maldonado, A. (2023). Nanomaterials as novel elicitors of plant secondary metabolites. In A. P. Ingle (Ed.), *Nanotechnology in Agriculture and Agroecosystems* (pp. 113–139). Elsevier.

González-García, Y., Cárdenas-Álvarez, C., Cadenas-Pliego, G., Benavides-Mendoza, A., Cabrera-de-la-Fuente, M., Sandoval-Rangel, A., Valdés-Reyna, J., & Juárez-Maldonado, A. (2021). Effect of three nanoparticles (Se, Si and Cu) on the bioactive compounds of bell pepper fruits under saline stress. *Plants*, *10*, 217.

González-Moscoso, M., Martínez-Villegas, N., Cadenas-Pliego, G., & Juárez-Maldonado, A. (2022). Effect of silicon nanoparticles on tomato plants exposed to two forms of inorganic arsenic. *Agronomy*, *12*, 2366.

González-Saucedo, A., Barrera-Necha, L. L., Ventura-Aguilar, R. I., Correa-Pacheco, Z. N., Bautista-Baños, S., & Hernández-López, M. (2019). Extension of the postharvest quality of bell pepper by applying nanostructured coatings of chitosan with *Byrsonima crassifolia* extract (L.) Kunth. *Postharvest Biology and Technology*, *149*, 74–82.

Granato, D., Barba, F. J., Bursać Kovačević, D., Lorenzo, J. M., Cruz, A. G., & Putnik, P. (2020). Functional foods: Product development, technological trends, efficacy testing, and safety. *Annu. Review Food Science & Technology*, *11*, 93–118.

Gudadhe, J. A., Yadav, A., Gade, A., Marcato, P. D., Durán, N., & Rai, M. (2014). Preparation of an agar-silver nanoparticles (A-AgNp) film for increasing the shelf-life of fruits. *IET Nanobiotechnology*, *8*, 190–195.

Guleria, G., Thakur, Sapna, Sharma, D. K., Thakur, Shweta, Kumari, P., & Shandilya, M. (2022). Environment-friendly and biodegradable a-$Fe_2O_3/C_{20}H_{38}O_{11}$ nanocomposite growth to lengthen the *Solanum lycopersicum* storage process. *Advanced Nature Science and Nanoscience & Nanotechnology*, *13*, 025004.

Hammam, K. A., & Shoala, T. (2020). Influence of spraying nano-curcumin and nano-rosemarinic acid on growth, fresh herb yield, chemicals composition and postharvest criteria of French basil (*Ocimum basilicum* L. var Grand Vert) plants. *Journal of Agricultural Rural Research*, *5*, 1–22.

Helal, M. I. D., El-Mogy, M. M., Khater, H. A., Fathy, M. A., Ibrahim, F. E., Li, Y. C., Tong, Z., & Abdelgawad, K. F. (2023). A controlled-release nanofertilizer improves tomato growth and minimizes nitrogen consumption. *Plants*, *12*, 1978.

Hernández-Hernández, H., Quiterio-Gutiérrez, T., Cadenas-Pliego, G., Ortega-Ortiz, H., Hernández-Fuentes, A. D., De La Fuente, M. C., Valdés-Reyna, J., & Juárez-Maldonado, A. (2019). Impact of selenium and copper nanoparticles on yield, antioxidant system, and fruit quality of tomato plants. *Plants*, *8*, 355.

Hernández-López, G., Ventura-Aguilar, R. I., Correa-Pacheco, Z. N., Bautista-Baños, S., & Barrera-Necha, L. L. (2020). Nanostructured chitosan edible coating loaded with α-pinene for the preservation of the postharvest quality of *Capsicum annuum* L. and *Alternaria alternata* control. *International Journal of Biology and Macromolecules*, *165*, 1881–1888.

Hu, X., Saravanakumar, K., Sathiyaseelan, A., & Wang, M. H. (2020). Chitosan nanoparticles as edible surface coating agent to preserve the fresh-cut bell pepper (*Capsicum annuum* L. var. Grossum (L.) Sendt). *International Journal of Biology and Macromoleules*, *165*, 948–957.

Huang, H., Lian, Q., Wang, L., Shan, Y., Li, F., Kiat, S., & Jiang, Y. (2020). Chemical composition of the cuticular membrane in guava fruit (*Psidium guajava* L.) affects barrier property to transpiration. *Plant Physiology and Biochemistry*, *155*, 589–595.

Huang, Y., Dong, Y., Ding, X., Ning, Z., Shen, J., Chen, H., & Su, Z. (2022). Effect of nano-TiO_2 composite on the fertilization and fruit-setting of litchi. *Nanomaterials*, *12*, 4287.

Hubbard, J. D., Lui, A., & Landry, M. P. (2020). Multiscale and multidisciplinary approach to understanding nanoparticle transport in plants. *Current Opinion Chemistry Engineering*, *30*, 135–143.

Hussein, Z., Fawole, O. A., & Opara, U. L. (2020). Harvest and postharvest factors affecting bruise damage of fresh fruits. *Horticulture Plant Journal*, *6*, 1–13.

Jansasithorn, R., East, A. R., Hewett, E. W., & Heyes, J. A. (2014). Skin cracking and postharvest water loss of Jalapeño chilli. *Science and Horticulture (Amsterdam)*, *175*, 201–207.

Ji, H., Wang, J., Xue, A., Chen, F., Guo, H., Xiao, Z., & Wang, Z. (2023). Chitosan–silica nanocomposites induced resistance in faba bean plants against aphids (*Acyrthosiphon pisum*). *Environmental Science: Nano*, *10*, 1966–1977.

Juárez-Maldonado, A., Ortega-Ortiz, H., González-Morales, S., Morelos-Moreno, Á., Cabrera-de la Fuente, M., Sandoval-Rangel, A., Cadenas-Pliego, G., & Benavides-Mendoza, A. (2019). Nanoparticles and nanomaterials as plant biostimulants. *International Journal of Molecular Science*, *20*, 1–19.

Juárez-Maldonado, A., Tortella, G., Rubilar, O., Fincheira, P., & Benavides-Mendoza, A. (2021). Biostimulation and toxicity: The magnitude of the impact of nanomaterials in microorganisms and plants. *Journal of Advanced Research*, *31*, 113–126.

Kalia, A., Kaur, M., Shami, A., Jawandha, S. K., Alghuthaymi, M. A., Thakur, A., & Abd-Elsalam, K. A. (2021). Nettle-leaf extract derived zno/cuo nanoparticle-biopolymer-based antioxidant and antimicrobial nanocomposite packaging films and their impact on extending the post-harvest shelf life of guava fruit. *Biomolecules*, *11*, 1–24.

Karimirad, R., Behnamian, M., Dezhsetan, S., & Sonnenberg, A. (2018). Chitosan nanoparticles loaded *Citrus aurantium* essential oil: A novel delivery system for preserving the postharvest quality of *Agaricus bisporus*. *Journal of Science and Food Agriculture*, *98*, 5112–5119.

Khalili, N., Oraei, M., Gohari, G., Panahirad, S., Nourafcan, H., & Hano, C. (2022). Chitosan-enriched salicylic acid nanoparticles enhanced anthocyanin content in grape (*Vitis vinifera* L. cv. Red Sultana) berries. *Polymers (Basel)*, *14*, 3349.

Kohatsu, M. Y., Lange, C. N., Pelegrino, M. T., Pieretti, J. C., Tortella, G., Rubilar, O., Batista, B. L., Seabra, A. B., & Jesus, T. A. de (2021). Foliar spraying of biogenic CuO nanoparticles protects the defence system and photosynthetic pigments of lettuce (*Lactuca sativa*). *Journal of Clean Production*, *324*, 129264.

Kolgesiz, S., Tas, C. E., Koken, D., Genc, M. H., Yalcin, I., Kalender, K., Unal, S., & Unal, H. (2023). Extending the shelf life of bananas with cinnamaldehyde-impregnated halloysite/polypropylene nanocomposite films. *ACS Food Science & Technology*, *3*, 340–349.

Kuang, L., Kang, Y., Wang, H., Huang, R., Lei, B., Zhong, M., & Yang, X. (2023). The roles of Salvia miltiorrhiza-derived carbon dots involving in maintaining quality by delaying senescence of postharvest flowering Chinese cabbage. *Food Chemistry*, *404*, 134704.

Kumari, K., Rana, V., Yadav, S. K., & Kumar, V. (2023). Nanotechnology as a powerful tool in plant sciences: Recent developments, challenges and perspectives. *Plant Nano Biology*, *5*, 100046.

Lakshmi, S. J., Bai, R. R. S., Sharanagouda, H., Ramachandra, C. T., Nadagouda, S., & Nidoni, U. (2018). Effect of biosynthesized zinc oxide nanoparticles coating on quality parameters of fig (*Ficus carica* L.) fruit. *Journal of Pharmacogenesis & Phytochemistry*, *7*, 10–14.

Lala, S. (2021). Nanoparticles as elicitors and harvesters of economically important secondary metabolites in higher plants: A review. *IET Nanobiotechnology*, *15*, 28–57.

Lavinia, M., Hibaturrahman, S. N., Harinata, H., & Wardana, A. A. (2020). Antimicrobial activity and application of nanocomposite coating from chitosan and ZnO nanoparticle to inhibit microbial growth on fresh-cut papaya. *Food Research*, *4*, 307–311.

Li, J., Sun, Q., Sun, Y., Chen, B., Wu, X., & Le, T. (2019). Improvement of banana postharvest quality using a novel soybean protein isolate/cinnamaldehyde/zinc oxide bionanocomposite coating strategy. *Science Horticulture (Amsterdam)*, *258*, 108786.

Li, J., Wan, F., Guo, W., Huang, J., Dai, Z., Yi, L., & Wang, Y. (2020). Influence of α- and γ-Fe_2O_3 Nanoparticles on Watermelon (*Citrullus lanatus*) Physiology and Fruit Quality. *Water, Air, & Soil Pollution*, *231*, 1–12.

Li, S., Chen, K., & Grierson, D. (2021). Molecular and Hormonal Mechanisms Regulating Fleshy Fruit Ripening. *Cells*, *10*, 1136.

Lin, M., Fang, S., Zhao, X., Liang, X., & Wu, D. (2020). Natamycin-loaded zein nanoparticles stabilized by carboxymethyl chitosan: Evaluation of colloidal/chemical performance and application in postharvest treatments. *Food Hydrocolloids*, *106*, 105871.

Liu, S., Zhang, X., Zeng, K., He, C., Huang, Y., Xin, G., & Huang, X. (2023a). Insights into eco-corona formation and its role in the biological effects of nanomaterials from a molecular mechanisms perspective. *Science of The Total Environment*, *858*, 159867.

Liu, Y., Liu, R., Cheng, L., Yu, S., Nie, Y., Zhang, H., Li, J. Q., Pan, C., Zhu, W., Diao, J., & Zhou, Z. (2023b). Improvement by application of three nanomaterials on flavor quality and physiological and antioxidant properties of tomato and their comparison. *Plant Physiology and Biochemistry*, *201*, 107834.

Lo'ay, A. A., & Ameer, N. M. (2019). Performance of calcium nanoparticles blending with ascorbic acid and alleviation internal browning of 'Hindi Be-Sennara' mango fruit at a low temperature. *Science Horticulture (Amsterdam)*, *254*, 199–207.

Lu, J., Shao, G., Gao, Y., Zhang, K., Wei, Q., & Cheng, J. (2021). Effects of water deficit combined with soil texture, soil bulk density and tomato variety on tomato fruit quality: A meta-analysis. *Agricultural Water Management*, *243*, 106427.

Lufu, R., Ambaw, A., & Opara, U. L. (2020). Water loss of fresh fruit: Influencing pre-harvest, harvest and postharvest factors. *Science Horticulture (Amsterdam)*, *272*, 109519.

Luksiene, Z., Rasiukeviciute, N., Zudyte, B., & Uselis, N. (2020). Innovative approach to sunlight activated biofungicides for strawberry crop protection: ZnO nanoparticles. *Journal of Photochemistry and Photobiology B: Biology*, *203*, 111656.

Luo, X., Wang, Z., Wang, C., Yue, L., Tao, M., Elmer, W. H., White, J. C., Cao, X., & Xing, B. (2023). Nanomaterial size and surface modification mediate disease resistance activation in cucumber (*Cucumis sativus*). *ACS Nano*, *17*, 4871–4885.

Ma, C., Han, L., Shang, H., Hao, Y., Xu, X., White, J. C., Wang, Z., & Xing, B. (2023). Nanomaterials in agricultural soils: Ecotoxicity and application. *Current Opinion Environmental Science and Health*, *31*, 100432.

Malek, N. S. A., Rosman, N., Mahmood, M. R., Khusaimi, Z., & Asli, N. A. (2020). Effects of storage temperature on shelf-life of mango coated with zinc oxide nanoparticles. *Science Letters*, *14*, 47–57.

Maringgal, B., Hashim, N., Mohamed Amin Tawakkal, I. S., & Muda Mohamed, M. T. (2020). Recent advance in edible coating and its effect on fresh/fresh-cut fruits quality. *Trends Food Science & Technology*, *96*, 253–267.

Martínez-González, M. C., Bautista-Baños, S., Correa-Pacheco, Z. N., Corona-Rangel, M. L., Ventura-Aguilar, R. I., Del Río-García, J. C., & Ramos-García, M. L. (2020). Effect of nanostructured chitosan/propolis coatings on the quality and antioxidant capacity of strawberries during storage. *Coatings*, *10*, 90.

Marzouk, N. M., Abd-Alrahman, H. A., EL-Tanahy, A. M. M., & Mahmoud, S. H. (2019). Impact of foliar spraying of nano micronutrient fertilizers on the growth, yield, physical quality, and nutritional value of two snap bean cultivars in sandy soils. *Bulletin of National Research Center*, *43*, 1–9.

Meena, M., Pilania, S., Pal, A., Mandhania, S., Bhushan, B., Kumar, S., Gohari, G., & Saharan, V. (2020). Cu-chitosan nano-net improves keeping quality of tomato by modulating physio-biochemical responses. *Science Reports*, *10*, 21914.

Meindrawan, B., Suyatma, N. E., Wardana, A. A., & Pamela, V. Y. (2018). Nanocomposite coating based on carrageenan and ZnO nanoparticles to maintain the storage quality of mango. *Food Package Shelf Life*, *18*, 140–146.

Melo, N. F. C. B., Pintado, M. M. E., Medeiros, J. A. da C., Galembeck, A., Vasconcelos, M. A. da S., Xavier, V. L., de Lima, M. A. B., Stamford, T. L. M., Stamford–Arnaud, T. M., Flores, M. A. P., & Stamford, T. C. M. (2020). Quality of postharvest strawberries: comparative effect of fungal chitosan gel, nanoparticles and gel enriched with edible nanoparticles coatings. *International Journal of Food Study*, *9*, 373–393.

Mikulic-Petkovsek, M., Veberic, R., Hudina, M., Zorenc, Z., Koron, D., & Senica, M. (2021). Fruit quality characteristics and biochemical composition of fully ripe blackberries harvested at different times. *Foods*, *10*, 1581.

Mohamed, N. G., & Abdel-Hakeem, M. A. (2023). Chitosan nanoparticles enhance drought tolerance in tomatoes (*Solanum lycopersicum* L.) via gene expression modulation. *Plant Gene*, *34*, 100406.

Montaño-Herrera, A., Santiago-Saenz, Y. O., López-Palestina, C. U., Cadenas-Pliego, G., Pinedo-Guerrero, Z. H., Hernández-Fuentes, A. D., & Pinedo-Espinoza, J. M. (2022). Effects of edaphic fertilization and foliar application of Se and Zn nanoparticles on yield and bioactive compounds in *Malus domestica* L. *Horticulturae*, *8*, 1–11.

Morales-Espinoza, M. C., Cadenas-Pliego, G., Pérez-Alvarez, M., Hernández-Fuentes, A. D., Cabrera de la Fuente, M., Benavides-Mendoza, A., Valdés-Reyna, J., & Juárez-Maldonado, A. (2019). Se nanoparticles induce changes in the growth, antioxidant responses, and fruit quality of tomato developed under NaCl stress. *Molecules*, *24*, 1–17.

Mosa, K. A., El-Naggar, M., Ramamoorthy, K., Alawadhi, H., Elnaggar, A., Wartanian, S., Ibrahim, E., & Hani, H. (2018). Copper nanoparticles induced genotoxicty, oxidative stress, and changes in superoxide dismutase (SOD) gene expression in cucumber (*Cucumis sativus*) plants. *Frontiers in Plant Science*, *9*, 1–13.

Mossa, A.-T. H., Mohafrash, S. M. M., Ziedan, E.-S. H. E., Abdelsalam, I. S., & Sahab, A. F. (2021). Development of eco-friendly nanoemulsions of some natural oils and evaluating of its efficiency against postharvest fruit rot fungi of cucumber. *Industrial Crops Production*, *159*, 113049.

Mustafa, M. A., Ali, A., Manickam, S., & Siddiqui, Y. (2014). Ultrasound-assisted chitosan-surfactant nanostructure assemblies: Towards maintaining postharvest quality of tomatoes. *Food Bioprocess Technology*, *7*, 2102–2111.

Nasr, F., Pateiro, M., Rabiei, V., Razavi, F., Formaneck, S., Gohari, G., & Lorenzo, J. M. (2021). Chitosan-phenylalanine nanoparticles (Cs-Phe Nps) extend the postharvest life of persimmon (*Diospyros kaki*) fruits under chilling stress. *Coatings*, *11*, 819.

Neysanian, M., Iranbakhsh, A., Ahmadvand, R., Ardebili, Z. O., & Ebadi, M. (2020). Comparative efficacy of selenate and selenium nanoparticles for improving growth, productivity, fruit quality, and postharvest longevity through modifying nutrition, metabolism, and gene expression in tomato; potential benefits and risk assessment. *PLoS One*, *15*, e0244207.

Nguyen, H. V. H., & Nguyen, D. H. H. (2020). Effects of nano-chitosan and chitosan coating on the postharvest quality, polyphenol oxidase activity and malondialdehyde content of strawberry (*Fragaria* × *ananassa* Duch.). *Journal of Horticulture Postharvest Research*, *3*, 11–24.

Nguyen, V. T. B., Nguyen, D. H. H., & Nguyen, H. V. H. (2020). Combination effects of calcium chloride and nano-chitosan on the postharvest quality of strawberry (*Fragaria* × *ananassa* Duch.). *Postharvest Biology Technology*, *162*, 111103.

Noori, A., Donnelly, T., Colbert, J., Cai, W., Newman, L. A., & White, J. C. (2020). Exposure of tomato (*Lycopersicon esculentum*) to silver nanoparticles and silver nitrate: physiological and molecular response. *International Journal of Phytoremediation*, *22*, 40–51.

Nxumalo, K. A., Fawole, O. A., & Oluwafemi, O. S. (2022). Evaluating the efficacy of gum arabic-zinc oxide nanoparticles composite coating on shelf-life extension of Mandarins (cv. Kinnow). *Frontiers in Plant Science*, *13*, 953861.

Oyedele, O. A., Kuzamani, K. Y., Adetunji, M. C., Osopale, B. A., Makinde, O. M., Onyebuenyi, O. E., Ogunmola, O. M., Mozea, O. C., Ayeni, K. I., Ezeokoli, O. T., Oyinloye, A. M., Ngoma, L., Mwanza, M., & Ezekiel, C. N. (2020). Bacteriological assessment of tropical retail fresh-cut, ready-to-eat fruits in south-western Nigeria. *Science African*, *9*, e00505.

Pejam, F., Ardebili, Z. O., Ladan-Moghadam, A., & Danaee, E. (2021). Zinc oxide nanoparticles mediated substantial physiological and molecular changes in tomato. *PLoS One*, *16*, 1–16.

Peralta-Manjarrez, R. M., Cabrera-De La Fuente, M., Benavides-Mendoza, A., Campos-Montiel, R. G., Ortega-Ortíz, H., & Hernández-Fuentes, A. D. (2023). Effects of copper nanoparticles on the organoleptic, physicochemical and nutraceutical properties of grafted tomato fruit. *Pakistan Journal of Botany*, *55*, 1707–1713.

Pérez-Labrada, F., López-Vargas, E. R., Ortega-Ortiz, H., Cadenas-Pliego, G., Benavides-Mendoza, A., & Juárez-Maldonado, A. (2019). Responses of tomato plants under saline stress to foliar application of copper nanoparticles. *Plants*, *8*, 1–17.

Pina-Barrera, A. M., Alvarez-Roman, R., Baez-Gonzalez, J. G., Amaya-Guerra, C. A., Rivas-Morales, C., Gallardo-Rivera, C. T., & Galindo-Rodriguez, S. A. (2019). Application of a multisystem coating based on polymeric nanocapsules containing essential oil of *Thymus vulgaris* L. to increase the shelf life of table grapes (*Vitis vinifera* L.). *IEEE Transactions Nanobioscience*, *18*, 549–557.

Posé, S., Paniagua, C., Matas, A. J., Gunning, A. P., Morris, V. J., Quesada, M. A., & Mercado, J. A. (2019). A nanostructural view of the cell wall disassembly process during fruit ripening and postharvest storage by atomic force microscopy. *Trends Food Science & Technology*, *87*, 47–58.

Prakash, S., & Deswal, R. (2020). Analysis of temporally evolved nanoparticle-protein corona highlighted the potential ability of gold nanoparticles to stably interact with proteins and influence the major biochemical pathways in *Brassica juncea*. *Plant Physiology & Biochemistry*, *146*, 143–156.

Queiroz, C., Lopes, M. L. M., Fialho, E., & Valente-Mesquita, V. L. (2008). Polyphenol oxidase: Characteristics and mechanisms of browning control. *Food Review Interntional*, *24*, 361–375.

Quiterio-Gutiérrez, T., Ortega-Ortiz, H., Cadenas-Pliego, G., Hernández-Fuentes, A. D., Sandoval-Rangel, A., Benavides-Mendoza, A., Cabrera-de la Fuente, M., & Juárez-Maldonado, A. (2019). The application of selenium and copper nanoparticles modifies the biochemical responses of tomato plants under stress by *Alternaria solani*. *International Journal of Molecular Science*, *20*, 1–16.

Rahman, M. H., Hasan, M. N., & Khan, M. Z. H. (2021a). Study on different nano fertilizers influencing the growth, proximate composition and antioxidant properties of strawberry fruits. *Journal of Agricultural Food Research*, *6*, 100246.

Rahman, M. H., Hasan, M. N., Nigar, S., Ma, F., Aly Saad Aly, M., & Khan, M. Z. H. (2021b). Synthesis and characterization of a mixed nanofertilizer influencing the nutrient use efficiency, productivity, and nutritive value of tomato fruits. *ACS Omega*, *6*, 27112–27120.

Rajpal, V. R., Prakash, S., Mehta, S., Minkina, T., Rajput, V. D., & Deswal, R. (2023). *A comprehensive review on mitigating abiotic stresses in plants by metallic nanomaterials: Prospects and concerns. Clean technologies and environmental policy*. Springer Berlin Heidelberg.

Rajput, V. D., Harish., Singh, R. K., Verma, K. K., Sharma, L., Quiroz-Figueroa, F. R., Meena, M., Gour, V. S., Minkina, T., Sushkova, S., & Mandzhieva, S. (2021). Recent developments in enzymatic antioxidant defence mechanism in plants with special reference to abiotic stress. *Biology (Basel)*, *10*, 267.

Ranjbar, S., Rahemi, M., & Ramezanian, A. (2018). Comparison of nano-calcium and calcium chloride spray on postharvest quality and cell wall enzymes activity in apple cv. Red Delicious. *Science Horticulture (Amsterdam)*, *240*, 57–64.

Rasouli, M., & Saba, M. K. (2018). Pre-harvest zinc spray impact on enzymatic browning and fruit flesh color changes in two apple cultivars. *Science Horticulture (Amsterdam)*, *240*, 318–325.

Robledo, N., Vera, P., López, L., Yazdani-Pedram, M., Tapia, C., & Abugoch, L. (2018). Thymol nanoemulsions incorporated in quinoa protein/chitosan edible films; antifungal effect in cherry tomatoes. *Food Chemistry*, *246*, 211–219.

Ruiz Rodríguez, L. G., Zamora Gasga, V. M., Pescuma, M., Van Nieuwenhove, C., Mozzi, F., & Sánchez Burgos, J. A. (2020). Fruits and fruit by-products as sources of bioactive compounds. Benefits and trends of lactic acid fermentation in the development of novel fruit-based functional beverages. *Food Research International*, 109854.

Saffan, M. M., Koriem, M. A., El-Henawy, A., El-Mahdy, S., El-Ramady, H., Elbehiry, F., Omara, A. E. D., Bayoumi, Y., Badgar, K., & Prokisch, J. (2022). Sustainable production of tomato plants (*Solanum lycopersicum* L.) under low-quality irrigation water as affected by bio-nanofertilizers of selenium and copper. *Sustainability*, *14*, 3236.

Saini, S., Kumar, P., Sharma, N. C., Sharma, N., & Balachandar, D. (2021). Nano-enabled Zn fertilization against conventional Zn analogues in strawberry (*Fragaria* × *ananassa* Duch.). *Science Horticulture (Amsterdam)*, *282*, 110016.

Salama, D. M., Abd El-Aziz, M. E., Shaaban, E. A., Osman, S. A., & Abd El-Wahed, M. S. (2022). The impact of nanofertilizer on agro-morphological criteria, yield, and genomic stability of common bean (*Phaseolus vulgaris* L.). *Science Reports*, *12*, 1–15.

Saqib, S., Zaman, W., Ayaz, A., Habib, S., Bahadur, S., Hussain, S., Muhammad, S., & Ullah, F. (2020). Postharvest disease inhibition in fruit by synthesis and characterization of chitosan iron oxide nanoparticles. *Biocatalytic Agricultural Biotechnology*, *28*, 101729.

Sariñana-Navarrete, M. de los Á., Morelos-Moreno, Á., Sánchez, E., Cadenas-Pliego, G., Benavides-Mendoza, A., & Preciado-Rangel, P. (2023). Selenium nanoparticles improve quality, bioactive compounds and enzymatic activity in jalapeño pepper fruits. *Agronomy*, *13*, 652.

Sarraf, M., Vishwakarma, K., Kumar, V., Arif, N., Das, S., Johnson, R., Janeeshma, E., Puthur, J. T., Aliniaeifard, S., Chauhan, D. K., Fujita, M., & Hasanuzzaman, M. (2022). Metal/metalloid-based nanomaterials for plant abiotic stress tolerance: An overview of the mechanisms. *Plants*, *11*, 1–31.

Selvakesavan, R. K., Kruszka, D., Shakya, P., Mondal, D., & Franklin, G. (2023). *Impact of nanomaterials on plant secondary metabolism. Nanomaterial interactions with plant cellular mechanisms and macromolecules and agricultural implications* (pp. 133–170). Cham: Springer International Publishing.

Sha, R., Zhu, S., Wu, L., Li, X., Zhang, H., Yao, D., Lv, Q., Wang, F., Zhao, F., Li, P., & Yu, K. (2022). Pre-harvest application of multi-walled carbon nanotubes improves the antioxidant capacity of 'flame seedless' grapes during storage. *Sustainability*, *14*, 9568.

Shah, S. W. A., Jahangir, M., Qaisar, M., Khan, S. A., Mahmood, T., Saeed, M., Farid, A., & Liaquat, M. (2015). Storage stability of kinnow fruit (*Citrus reticulata*) as affected by CMC and guar gum-based silver nanoparticle coatings. *Molecules*, *20*, 22645–22661.

Shahat, M., Ibrahim, M., Osheba, A., & Taha, I. (2020a). Preparation and characterization of silver nanoparticles and their use for improving the quality of apricot fruits. *Al-Azhar Journal of Agricultural Research*, *45*, 33–43.

Shahat, M., Ibrahim, M., Osheba, A., & Taha, I. (2020b). Improving the quality and shelf-life of strawberries as coated with nano-edible films during storage. *Al-Azhar Journal of Agricultural Research*, *45*, 1–13.

Shani, M. Y., Ahmed, S. R., Ashraf, M. Y., Khan, Z., Cocozza, C., De Mastro, F., Gul, N., Pervaiz, S., Abbas, S., Nawaz, H., & Brunetti, G. (2023). Nano-biochar suspension mediated alterations in yield and juice quality of kinnow (*Citrus reticulata* L.). *Horticulturae*, *9*, 521.

Sharma, R. R., Nagaraja, A., Goswami, A. K., Thakre, M., Kumar, R., & Varghese, E. (2020). Influence of on-the-tree fruit bagging on biotic stresses and postharvest quality of rainy-season crop of 'Allahabad Safeda' guava (*Psidium guajava* L.). *Crop Protection*, *135*, 105216.

Sharma, S., Katoch, V., Kumar, S., & Chatterjee, S. (2021). Functional relationship of vegetable colors and bioactive compounds: Implications in human health. *Journal of Nutritional Biochemistry*, *92*, 108615.

Sheikhalipour, M., Esmaielpour, B., Behnamian, M., Gohari, G., Giglou, M. T., Vachova, P., Rastogi, A., Brestic, M., & Skalicky, M. (2021). Chitosan–selenium nanoparticle (Cs–Se NP) foliar spray alleviates salt stress in bitter melon. *Nanomaterials*, *11*, 1–23.

Shoala, T., Al-Karmalawy, A. A., Germoush, M. O., Alshamrani, S. M., Abdein, M. A., & Awad, N. S. (2021). Nanobiotechnological approaches to enhance potato resistance against potato leafroll virus (Plrv) using glycyrrhizic acid ammonium salt and salicylic acid nanoparticles. *Horticulturae*, *7*, 402.

Song, H., Yuan, W., Jin, P., Wang, W., Wang, X., Yang, L., & Zhang, Y. (2016). Effects of chitosan/nano-silica on postharvest quality and antioxidant capacity of loquat fruit during cold storage. *Postharvest Biology Technology*, *119*, 41–48.

Sosan, A., Svistunenko, D., Straltsova, D., Tsiurkina, K., Smolich, I., Lawson, T., Subramaniam, S., Golovko, V., Anderson, D., Sokolik, A., Colbeck, I., & Demidchik, V. (2016). Engineered silver nanoparticles are sensed at the plasma membrane and dramatically modify the physiology of *Arabidopsis thaliana* plants. *Plant Journal*, *85*, 245–257.

Tabatabaee, S., Iranbakhsh, A., Shamili, M., & Oraghi Ardebili, Z. (2021). Copper nanoparticles mediated physiological changes and transcriptional variations in microRNA159 (miR159) and mevalonate kinase (MVK) in pepper; potential benefits and phytotoxicity assessment. *Journal of Environmental Chemistry Engineering*, *9*, 106151.

Taheri, A., Behnamian, M., Dezhsetan, S., & Karimirad, R. (2020). Shelf life extension of bell pepper by application of chitosan nanoparticles containing *Heracleum persicum* fruit essential oil. *Postharvest Biology Technology*, *170*, 111313.

Tang, N., An, J., Deng, W., Gao, Y., Chen, Z., & Li, Z. (2020). Metabolic and transcriptional regulatory mechanism associated with postharvest fruit ripening and senescence in cherry tomatoes. *Postharvest Biology Technology*, *168*, 111274.

Tarafder, C., Daizy, M., Alam, M. M., Ali, M. R., Islam, M. J., Islam, R., Ahommed, M. S., Aly Saad Aly, M., & Khan, M. Z. H. (2020). Formulation of a hybrid nanofertilizer for slow and sustainable release of micronutrients. *ACS Omega*, *5*, 23960–23966.

Tombuloglu, H., Slimani, Y., Tombuloglu, G., Alshammari, T., Almessiere, M., Korkmaz, A. D., Baykal, A., & Samia, A. C. S. (2020). Engineered magnetic nanoparticles enhance chlorophyll content and growth of barley through the induction of photosystem genes. *Environmental Science and Pollution Research*, *27*, 34311–34321.

Usman, M., Farooqb, M., Wakeel, A., Nawaz, A., Cheema, S. A., Rehman, H., Ashraf, I., & Sanaullah, M. (2020). Nanotechnology in agriculture: Current status, challenges and future opportunities. *Science of the Total Environment*, *721*, 137778.

Vardakas, P., Kyriazis, I. D., Kourti, M., Skaperda, Z., Tekos, F., & Kouretas, D. (2022). *Oxidative stress–mediated nanotoxicity: Mechanisms, adverse effects, and oxidative potential of engineered nanomaterials. Advanced nanomaterials and their applications in renewable energy* (pp. 179–218). Elsevier.

Venzhik, Y. V., & Deryabin, A. N. (2023). The use of nanomaterials as a plant-protection strategy from adverse temperatures. *Russian Journal of Plant Physiology*, *70*, 339–353.

Vieira, A. C. F., de Matos Fonseca, J., Menezes, N. M. C., Monteiro, A. R., & Valencia, G. A. (2020). Active coatings based on hydroxypropyl methylcellulose and silver nanoparticles to extend the papaya (*Carica papaya* L.) shelf life. *International Journal of Biology Macromolecule*, *164*, 489–498.

Vulk, M., Dragica, M., Djeki, I., Wolf, B., Zuber, J., Vogt, C., & Dragiši, J. (2023). Sugars and organic acids in 25 strawberry cultivars: Qualitative and quantitative evaluation. *Plants*, *12*, 2238.

Wang, C., Liu, X., Li, J., Yue, L., Yang, H., Zou, H., Wang, Z., & Xing, B. (2021a). Copper nanoclusters promote tomato (*Solanum lycopersicum* L.) yield and quality through improving photosynthesis and roots growth. *Environmental Pollution*, *289*, 117912.

Wang, C., Yang, H., Chen, F., Yue, L., Wang, Z., & Xing, B. (2021b). Nitrogen-doped carbon dots increased light conversion and electron supply to improve the corn photosystem and yield. *Environmental Science & Technology*, *55*, 12317–12325.

Wang, D., Lu, Q., Wang, X., Ling, H., & Huang, N. (2023a). Original article elucidating the role of SlXTH5 in tomato fruit softening. *Horticultural Plant Journal*, *9*, 777–788.

Wang, G., Liu, X., Liu, Y., Zhang, Z., You, C., Wang, X., & Zhang, S. (2023b). Efficient enhancement of calcium content in apple by calcium-encapsulated carbon dots. *Fruit Research*, *3*, 10.

Wang, Y., Qiu, W. Y., Fang, X. Q., Li, W., & Sun, Y. (2023c). Carbon dots-based reinforced hydrogen-rich water nanocomposite coating for storage quality of fresh-cut pear. *Food Bioscience*, *53*, 102837.

Wang, X. P., Li, Q. Q., Pei, Z. M., & Wang, S. C. (2018). Effects of zinc oxide nanoparticles on the growth, photosynthetic traits, and antioxidative enzymes in tomato plants. *Biology Plant*, *62*, 801–808.

Wang, Y., Wang, S., Xu, M., Xiao, L., Dai, Z., & Li, J. (2019). The impacts of γ-Fe_2O_3 and Fe_3O_4 nanoparticles on the physiology and fruit quality of muskmelon (*Cucumis melo*) plants. *Environmental Pollution*, *249*, 1011–1018.

Wang, Z., Le, X., Cao, X., Wang, C., Chen, F., Wang, J., Feng, Y., Yue, L., & Xing, B. (2022). Triiron tetrairon phosphate ($Fe_7(PO_4)_6$) nanomaterials enhanced flavonoid accumulation in tomato fruits. *Nanomaterials*, *12*, 1341.

Wei, X., Xie, D., Mao, L., Xu, C., Luo, Z., Xia, M., Zhao, X., Han, X., & Lu, W. (2019). Excess water loss induced by simulated transport vibration in postharvest kiwifruit. *Science Horticulture (Amsterdam)*, *250*, 113–120.

Xing, Y., Yang, H., Guo, X., Bi, X., Liu, X., Xu, Q., Wang, Q., Li, W., Li, X., Shui, Y., Chen, C., & Zheng, Y. (2020). Effect of chitosan/Nano-TiO_2 composite coatings on the postharvest quality and physicochemical characteristics of mango fruits. *Science Horticulture (Amsterdam)*, *263*, 109135.

Yang, M., Dong, C., & Shi, Y. (2023). Nano fertilizer synergist effects on nitrogen utilization and related gene expression in wheat. *BMC Plant Biology*, *23*, 1–13.

Yu, L., Qiao, N., Zhao, J., Zhang, H., Tian, F., Zhai, Q., & Chen, W. (2020). Postharvest control of *Penicillium expansum* in fruits: A review. *Food Bioscience*, *36*, 100633.

Yue, L., Feng, Y., Ma, C., Wang, C., Chen, F., Cao, X., Wang, J., White, J. C., Wang, Z., & Xing, B. (2022). Molecular mechanisms of early flowering in tomatoes induced by manganese ferrite ($MnFe_2O_4$) nanomaterials. *ACS Nano*, *16*, 5636–5646.

Zahedi, S. M., Hosseini, M. S., Daneshvar Hakimi Meybodi, N., & Teixeira da Silva, J. A. (2019). Foliar application of selenium and nano-selenium affects pomegranate (*Punica granatum* cv. Malase Saveh) fruit yield and quality. *South African Journal of Botany*, *124*, 350–358.

Zahedi, S. M., Hosseini, M. S., Karimi, M., Gholami, R., Amini, M., Abdelrahman, M., & Tran, L. S. P. (2023). Chitosan-based Schiff base-metal (Fe, Cu, and Zn) complexes mitigate the negative consequences of drought stress on pomegranate fruits. *Plant Physiology Biochemistry*, *196*, 952–964.

Zambrano-Zaragoza, M. L., Mercado-Silva, E., Ramirez-Zamorano, P., Cornejo-Villegas, M. A., Gutiérrez-Cortez, E., & Quintanar-Guerrero, D. (2013). Use of solid lipid nanoparticles (SLNs) in edible coatings to increase guava (*Psidium guajava* L.) shelf-life. *Food Research International*, *51*, 946–953.

Zambrano-Zaragoza, M. L., Quintanar-Guerrero, D., Del Real, A., González-Reza, R. M., Cornejo-Villegas, M. A., & Gutiérrez-Corte, E. (2020). Effect of nano-edible coating based on beeswax solid lipid nanoparticles on strawberry's preservation. *Coatings*, *10*, 253.

Zhang, Y., Fu, C., Yan, Y., Fan, X., Li, M., & Wang, Y. (2014). Foliar application of sugar alcohol zinc increases sugar content in apple fruit and promotes activity of metabolic enzymes. *HortScience*, *49*, 1067–1070.

Zhu, M., Yu, J., Wang, R., Zeng, Y., Kang, L., & Chen, Z. (2023). Nano-calcium alleviates the cracking of nectarine fruit and improves fruit quality. *Plant Physiology Biochemistry*, *196*, 370–380.

Impacts of nanocomposites on the postharvest physiology and shelf life of agricultural crops

10

Maryam Haghmadad Milani[1], *Gholamreza Gohari*[2,3], *George A. Manganaris*[3] *and Vasileios Fotopoulos*[3]
[1]Department of Biology, Faculty of Basic Sciences, University of Maragheh, Maragheh, Iran, [2]Department of Horticultural Science, Faculty of Agriculture, University of Maragheh, Maragheh, Iran, [3]Department of Agricultural Sciences, Biotechnology & Food Science, Cyprus University of Technology, Limassol, Cyprus

10.1 Introduction, global food security

Modern agriculture faces an array of challenges that threaten global food security and sustainability. In recent years, the integration of nanotechnology into agricultural practice, known as phyto-nanotechnology, has emerged as a promising solution to address these pressing issues. Phyto-nanotechnology involves the use of nanomaterials (NMs) in plants, offering unprecedented opportunities to revolutionize agriculture and ensure economic stability (Jiang et al., 2021; Shang et al., 2019). The agriculture sector, being the foundation of social and technological development, is currently facing significant challenges in ensuring sustainable production and food security. Rapid population growth, climate change, environmental pollution, and increased demands for water and energy are straining food production and distribution worldwide. To tackle these issues, nanotechnology has emerged as a promising solution in agriculture, offering practical applications to address various challenges (Lv et al., 2018; Prasad et al., 2017).

Nanotechnology is a science that deals with the manipulation and development of materials and structures at the nanoscale, ranging from 1 to 100 nm. NMs exhibit unique chemical and physical properties, including particle size, quantum effects, and surface area, making them versatile and applicable in diverse fields such as medicine, electronics, materials engineering, and agriculture (Gohari et al., 2024; Saleh, 2020). In the context of agriculture, nanotechnology has shown great potential to revolutionize the industry by providing new techniques and applications based on NMs. These NMs can be used as fertilizers, pesticides, additives, and elicitors, leading to increased yield production and the efficient generation of functional foods (Prasad et al., 2017). With continuous research and modification of NM properties, nanotechnology is poised to play a critical role in addressing pressing challenges and promoting sustainable practices in the agriculture and food sectors.

Nanocomposites for Environmental, Energy, and Agricultural Applications. DOI: https://doi.org/10.1016/B978-0-443-13935-2.00010-3

The applications of phyto-nanotechnology are diverse and extensive, with its potential reaching far beyond traditional agricultural methods. One of the key areas of focus is the enhancement of postharvest and fruit quality, which plays a pivotal role in extending shelf life, reducing postharvest losses, and ultimately meeting the growing demand for high-quality produce. This book chapter delves into the innovative use of nanocomposites to tackle postharvest challenges and improve fruit quality. Phyto-nanotechnology offers an array of potential applications and research areas that surpass the capabilities of conventional agricultural methods. By enabling targeted transport of nutrients and controlled release of agrochemicals, nanocomposites facilitate optimal nutrient uptake and bolster environmental stress tolerance, thereby significantly increasing agricultural productivity (Salama et al., 2021). The precise delivery of genes using nanocarriers has shown remarkable applicability in crop breeding, especially in the realm of plant genetic engineering (Agrawal et al., 2022). Intelligent management and resource-efficient practices are vital for the sustainable growth of agriculture. Smart phyto-nanotechnology has emerged as a pivotal contributor, providing innovative solutions to traditional agricultural methods. The implementation of nano-carriers in agriculture has paved the way for "smart" cropping systems, addressing challenges such as crop genome editing, tissue culture, regeneration, and species dependence. Additionally, NM-based biosensors and "sensing materials" have revolutionized the detection and measurement of pathogens, bacteria, and pesticides in food (Sanzari et al., 2019). A variety of nanosensors, including antibody nanosensors, carbon-based electrochemical nanosensors, nanowire nanosensors, plasmonic nanosensors, and fluorescence resonance energy transfer-based nanosensors, have proven instrumental in the food industry (Chaudhry et al., 2018).

According to the Food and Agriculture Organization, 30% of food for human consumption worldwide is lost at the postharvest stage (Elik et al., 2019). Fruit postharvest losses are estimated at 20% in developed countries and at more than 50% in nations that lack adequate storage, transportation, and handling methods (Elik et al., 2019). In addition, fruits are perishable commodities that are prone to widespread postharvest diseases, which result in tremendous economic losses (Dukare et al., 2019). Loss of agriproducts postharvest is a major concern not only for farmers but also for distributors, sellers, and consumers, as it dramatically impacts the economics of large-scale agriculture (Bhardwaj et al., 2023a,b). Fruits and vegetables postharvest are affected by a number of factors, including the temperature at which they are stored, water loss, maturity and firmness, ethylene production, mechanical damage during transportation and storage, and microorganisms and pests that cause postharvest diseases (Videira-Quintela et al., 2022). The storage of agricultural products with a short shelf life results in major postharvest losses (Sau et al., 2021). The main causes of this damage are various pathogen attacks (such as microorganisms and insects), a decline in produce quality, and exposure to a variety of atmospheric chemicals like oxygen, carbon dioxide, and water. As a result, products lose their visual appeal, nutritional content, and quality, and their shelf life is reduced (Bhardwaj et al., 2023a,b).

Nanotechnology offers a promising alternative for managing agricultural crops after harvest, particularly fruits and vegetables (Arabpoor et al., 2021). In coatings, packaging, and diagnostics, nanotechnology has provided innovative solutions with the use of nanoparticles (NPs), nanocomposites, and nanofilms (Bhardwaj et al., 2023a,b). Using NMs to increase the shelf life of food is possible through their unique properties of fine size (nano-scale), greater surface area, higher stability, and biological activity: first, by maintaining the visual characteristics along with nutritional parameters, and second, by protecting them from pathogenic microbes (Nasr et al., 2021).

The primary objective of this chapter is to provide an in-depth exploration of various types of NMs and nanocarriers. These hold significant promise as effective vehicles for delivering phytohormones and other compounds crucial for enhancing fruit quality and prolonging the shelf life of numerous horticultural crops. Additionally, we will explore the application of postharvest and preharvest treatments, with a specific focus on eco-friendly and safe bio-based materials. These interventions aim to elevate the quality of horticultural crops, consequently bolstering the economic prospects of farmers and stakeholders and expanding the scope of nanocomposite in agriculture.

10.2 Nanocomposite classification and application in agriculture

In recent years, significant progress has been made in exploring the potential of engineered NPs as innovative nanocomposites in postharvest technology (Neme et al., 2021). These NMs boast unique properties that contribute to the enhancement of plant growth, development, stress tolerance, fruit quality, shelf life, and transportation. As we delve into this realm, we classify various NMs based on their mode of action and emphasize their physiological effects on plants and fruits as preharvest and postharvest treatments. The use of nanocarriers for the intelligent delivery of different compounds, including phytohormones, amino acids, osmolytes, and other bioactive compounds, is an emerging and promising field in postharvest technology. These nanocarriers play a crucial role in improving the efficiency and targeted distribution of agrochemicals, leading to enhanced plant growth, stress resilience, fruit quality, and overall productivity (Jiang et al., 2021; Verma et al., 2017). Below are some exemplifications of diverse nanocarriers utilized for the intelligent delivery of agrochemicals:

10.2.1 *Lipid-based nanocarriers*

Liposomes: Among the lipid-based nanocarriers, liposomes stand out as spherical vesicles with lipid bilayers, making them effective tools for delivering priming agents and phytohormones. For instance, liposomes loaded with phytohormones have exhibited their capability to enhance drought tolerance in plants (Esfanjani et al., 2018). Nanoliposomes are used in the encapsulation and release of food materials. They improve the bioavailability, stability, and shelf life of ingredients.

The use of liposomes and nanoliposomes in food science also includes the delivery of flavors, and nutrients and the prevention of microbes against contamination (Reza Mozafari et al., 2008).

10.2.2 Solid lipid nanoparticles

Another noteworthy lipid-based nanocarrier is solid lipid nanoparticles (SLNs). These nanosized particles composed of solid lipids have proven to be efficient carriers for various agrochemicals. For example, SLNs loaded with amino acids, such as glycine betaine, have demonstrated their potential to improve salt stress tolerance in wheat plants. This improvement is attributed to their ability to enhance osmotic adjustment and antioxidant defense mechanisms (Nakasato et al., 2017). SLNs are also promising carriers for nonpolar substances whose mobility is restricted by lipid interactions, resulting in altered release profiles (Naseri et al., 2020; Severino et al., 2011). There are several advantages to these carrier systems, such as the reduction of active agent quantities, reduced loss due to leaching, degradation, and volatilization, and lower environmental impacts (Cea et al., 2010; Liu et al., 2016; Otto et al., 2008; Wanyika et al., 2012). A major advantage of SLNs and polymeric NPs in agriculture is their low toxicities since the matrices contain low-toxicity polymers [such as poly(ε-caprolactone)] and lipids, which are present in many organisms (Gohari et al., 2024).

10.2.3 Polymeric nanocarriers

Polymeric NPs, particularly chitosan-based NPs, have been employed to deliver priming agents effectively. Chitosan NPs loaded with biostimulants, such as melatonin, have shown promising results in enhancing salinity stress tolerance in spearmint plants. This enhancement is attributed to their ability to regulate osmotic potential and antioxidant activities, as evidenced in Gohari et al. (2023). Edible coatings offer intriguing potential among postharvest management technologies, as they serve as gas barriers on fruit surfaces, reducing permeability to O_2, CO_2, and water vapor. This leads to decreased respiration rates and extended postharvest life for horticultural products (Panahirad et al., 2021). One of the notable coating materials is chitosan (CTS), a natural polysaccharide derived from chitin after deacetylation, possessing cationic properties and a high molecular weight (Muzzarelli et al., 2012a,b). CTS stands out due to its biocompatibility, biodegradability, relative antimicrobial activity, and nontoxicity to humans and animals, making it a widely employed biopolymer for fruit packaging and coating, effectively increasing shelf life and storage time (Hernandez-Lauzardo et al., 2011).

Numerous studies have reported that chitosan coatings enhance the postharvest quality and antioxidant activity of fresh produce, such as sweet cherries, cucumbers, and plums (Hashim et al., 2018; Xin et al., 2020). In particular, coating with CTS has proven effective in preserving organoleptic properties and prolonging the shelf life of plums ("Stanley" and "Giant" cultivars) stored at 0°C–1°C (Bal, 2013). Despite its biochemical advantages, chitosan suffers from poor mechanical strength

(Gadgey & Sharma, 2020). To address this, nano-encapsulation of chitosan has garnered significant interest as an approach to improving the postharvest quality of chitosan-coated fruit. This technique enhances stability and allows for the controlled release of bioactive components. For instance, postharvest chitosan-g-salicylic acid treatment has been shown to inhibit malondialdehyde (MDA) and electrolyte leakage (EL) in cucumber fruit, thereby reducing chilling injury and maintaining fruit quality during storage (Zhang et al., 2015).

Advances in chitosan-related nanocarriers or nano-encapsulation, such as Cs loaded with thiamine, phenylalanine, Ag^{+}, and Cu, have further elevated the importance of this approach in various fields of plant science (Ortiz-Duarte et al., 2019). As researchers continue to explore and refine this technology, the utilization of nano-encapsulation holds great promise for improving the postharvest quality and preservation of fruits coated with chitosan, benefiting the agricultural industry and consumers alike.

10.2.4 Inorganic nanocarriers

Silica NPs (MSN) are highly promising inorganic nanocarriers due to their large surface area and pore volume, which makes them well-suited for delivering agrochemicals. For instance, MSNs loaded with salicylic acid have demonstrated their potential to improve disease resistance in tomato plants. This improvement is achieved through a controlled release of salicylic acid, resulting in enhanced defense responses (Derbalah et al., 2018).

Carbon nanotubes (CNTs) and graphene oxide (GO) have emerged as notable nanocarriers for agrochemicals. Functionalized CNTs, in particular, have shown promising results when loaded with amino acids. These functionalized CNTs have proven effective in improving growth and nutrient uptake in plants under stress conditions (Zahedi et al., 2023).

These inorganic nanocarriers offer several advantages, such as controlled release, protection of priming agents from degradation, and targeted delivery to specific plant tissues. These properties enable efficient and tailored delivery of agrochemicals, thereby enhancing their overall efficacy and reducing the required dosage. However, it is essential to carefully consider the biocompatibility and potential toxicity of these nanocarriers to ensure their safe and sustainable use in agricultural applications (Bueno & Ghoshal, 2022).

10.2.5 Lipid-based nanocarriers

Liposomes are self-assembled vesicles composed of phospholipid bilayers. Their unique structure allows them to encapsulate both hydrophilic and hydrophobic priming agents within their aqueous core or lipid bilayers, respectively. This characteristic provides protection to the loaded agrochemicals, enhances their stability, and enables controlled release when needed (Esfanjani et al., 2018).

SLNs are nanosized particles composed of solid lipids. They offer several advantages, including enhanced stability, prolonged release, and improved bioavailability of the loaded priming agents. SLNs have been effectively used to deliver various agrochemicals, such as amino acids and osmolytes. Notably, SLNs loaded with glycine betaine, an osmolyte, have demonstrated their ability to enhance salt stress tolerance in wheat plants by improving osmotic adjustment and antioxidant defense mechanisms (Nakasato et al., 2017).

The application of lipid-based nanocarriers in agriculture holds significant promise, as they can facilitate the efficient delivery of agrochemicals to plants, thereby enhancing their growth, quality, and stress tolerance. These carriers offer precise and targeted delivery, allowing for optimal resource utilization and reducing the overall environmental impact. However, careful research and evaluation of their safety and environmental implications are imperative to ensure responsible and sustainable agricultural practices (Kaliamurthi et al., 2019).

The examples provide compelling evidence of the versatility and promising potential of various NMs as nanocarriers for the intelligent delivery of agrochemicals. However, the selection of the most suitable nanocarrier for a specific application hinges on several critical factors. The physicochemical properties of the priming agent play a crucial role in determining the appropriate nanocarrier. Different priming agents may have distinct solubilities, charges, and molecular sizes, necessitating specific carriers that can accommodate and effectively deliver these agents. For instance, hydrophilic agrochemicals might require encapsulation within hydrophilic NPs like liposomes, while hydrophobic agents may be better suited to be housed within lipid-based carriers. Targeted delivery requirements must be considered when choosing the ideal nanocarrier.

Depending on the intended destination within the plant, the nanocarrier should possess attributes that allow it to reach the desired plant tissues or cells efficiently. Certain nanocarriers can be designed to release their cargo in response to specific environmental cues or triggers, ensuring that the agrochemicals are delivered precisely when and where they are needed most. Furthermore, the desired release kinetics of the agrochemicals are a crucial determinant in selecting the appropriate nanocarrier. Some applications may demand sustained release over an extended period, while others might require rapid and controlled release for immediate impact. Nanocarriers can be engineered to modulate the release rate of the encapsulated agrochemicals, tailoring the delivery to match the specific requirements of the plant and its growth stage (Mehrazar et al., 2015; Solanki et al., 2015).

The successful utilization of nanocarriers for the smart delivery of agrochemicals in agriculture requires a thoughtful evaluation of the physicochemical properties of the priming agent, the targeted delivery needs, and the desired release kinetics. The synergistic integration of these factors will pave the way for the development of innovative and efficient nanocarrier systems that can revolutionize crop management, enhance stress tolerance, and contribute to sustainable and improved agricultural practices. As research in this field progresses, we can expect even more sophisticated and tailored nanocarrier solutions to address the unique challenges faced by agriculture and propel us toward a greener and more productive future (Mishra & Khare, 2021; Wanyika et al., 2012; Fincheira et al., 2023).

10.3 Mechanisms of action of nanocomposites in postharvest physiology

10.3.1 Preservation of postharvest quality attributes

The constant respiration and transpiration of agricultural products, especially fruits and vegetables, result in a loss of quality and a short shelf life after harvest (Kumar et al., 2020). In the processing of agricultural products, physical damage is mainly caused by peeling, slicing, shredding, coring, bleaching, canning, etc., which contributes to cell wall-modifying enzyme activity (e.g., cellulase, polygalacturonase, and pectin methylesterase). A variety of physiological and chemical factors contribute to the ripening of fruits and vegetables, including sugar content, enzymatic cell wall degradation, chlorophyll content, respiration rate, aromatic levels, and ethylene production (Lee & Hwang, 2017). In recent years, nanotechnology has been widely used in environmental management and agricultural science to enhance the quality and quantity of agricultural products (Adeel et al., 2021; Ahmad et al., 2022). The benefits of edible coatings among the available postharvest management technologies have made them the preferred preservation choice because they act as gas barriers on fruit surfaces and reduce the permeability to oxygen, carbon dioxide, and water vapor, reducing respiration rates and extending the postharvest life of horticultural products (Hazarika et al., 2019; Panahirad et al., 2021; Salvia-Trujillo et al., 2015; Zambrano-Zaragoza et al., 2014). Among their main benefits are lower respiration and tissue-softening ratios, longer postharvest life, biodegradability, and less microbial contamination (Arnon et al., 2014). In addition, edible coatings containing active ingredients reduce tissue softening during storage. Polygalacturonate activity is delayed by these coatings, retaining membrane structural integrity and making them firm (Firdous, 2021).

Fruits and vegetables can reduce both physical and biological stress with these coatings. As well as reducing moisture and firmness, it also reduces ethylene production, maintains oxygen levels, slows metabolisms like respiration, oxidation, and chlorophyll loss, and has antimicrobial properties that prevent fruit and vegetable postharvest diseases (Eshghi et al., 2022). Depending on the coating material, CTS is a cationic polysaccharide with a high molecular weight derived from deacetylated chitin (Muzzarelli et al., 2012a,b). The unique characteristics of this biopolymer, such as its biocompatibility, biodegradability, antimicrobial activity, and nontoxicity during human and animal use, have made it widely utilized for fruit packaging and coating to increase shelf life and longevity (Kumar, 2000; Ngo et al., 2021). It has been shown that chitosan-based composites consisting of different natural antimicrobials and NMs are effective formulations for improving shelf-life and preserving postharvest product quality (Flórez et al., 2022).

A great deal of interest has been drawn to chitosan encapsulation due to its eco-friendliness and biochemical properties (Gadgey & Sharma, 2020; Oh & Hwang, 2013). In pre- and postharvest, nanochitosan is used to control plant diseases, extend shelf life, and preserve quality (Ruffo Roberto et al., 2019). In the studies conducted by Meena et al. (2020a,b), Cu-chitosan NP treatment successfully maintained qualitative and quantitative characteristics, such as retaining firmness, preventing a decrease in titrable acidity, ascorbic acid, lycopene, and phenolic compounds, and preventing increased decay in tomato fruit by increasing antioxidant activity and enhancing NP properties over a period of 21 days. A range of postharvest treatments (spraying, coating, or dipping) help maintain quality (Aaby et al., 2012). A chitosan coating has been shown to improve postharvest quality and antioxidant activity in fresh produce, such as sweet cherry (Prunus avium L.) (Xin et al., 2020), cucumber (Cucumis sativus L.) (Hashim et al., 2018), and plum (Prunus domestica L.) (Kumar et al., 2017). Additionally, CTS coating has proven to prolong the shelf life of plums ("Stanley" and "Giant" cultivars) stored at $0^{\circ}C-1^{\circ}C$ while preserving their organoleptic properties (Bal, 2013).

Studies have shown that chitosan coatings decrease weight loss, firmness, and decay and increase antioxidant activity and total soluble solids (TSS) in fresh fruits (Adiletta et al., 2019). As a composite, it can be used to stabilize postharvest crops' quality and shelf life along with NMs (Xing et al., 2019). The combination of chitosan and aloe vera has been used to delay the ripening of tomatoes for up to 42 days after harvest (Khatri et al., 2020). A chitosan coating has also been shown to maintain strawberries (Petriccione et al., 2015; Rahman et al., 2018), ber fruit (Hesami et al., 2021), apples, tomatoes, cucumbers (Duan et al., 2019), and figs (Adiletta et al., 2019) postharvest quality. Using calcium gluconate and chitosan preserved the quality of strawberries for up to 10 days while rotting occurred in more than 60% of untreated fruit (Dam et al., 2020). Researchers used quaternary chitosan films along with carboxymethylcellulose (CMC) to extend the shelf life of bananas, which showed the films contained a high level of chitosan and slowed decay (Hu et al., 2016). A chitosan treatment combined with nanosilicon reduced the red index, respiration rate, decay percentage, and weight loss of jujubes during room-temperature storage compared to a control (Yu et al., 2012). Shi et al. (2013) found that hybrid films made of chitosan and nanosilica significantly extended the shelf life of fresh longan fruit by reducing browning, weight loss, and MDA accumulation while preventing polyphenoloxidase activity and maintaining TSS, titratable acidity (TA), and ascorbic acid levels. Punjab Beauty pear fruit was coated with chitosan and SA under cold and room temperature conditions in a recent study (Sinha et al., 2022). It was demonstrated that this treatment reduced EL and MDA accumulation and prevented membrane damage. Additionally, this substance retarded fruit softening by inhibiting enzymes that degrade cell walls, including polygalacturonase, cellulase, and pectin methylesterase (Sinha et al., 2022).

A variety of fields of plant science are increasingly considering the application of chitosan-related nanocarriers or nano-encapsulation, such as CTS loaded with phenylalanine (Jirawutthiwongchai et al., 2016), thiamine (Muthukrishnan et al., 2019), Cu (Saharan et al., 2016), and Ag^{+} (Ortiz-Duarte et al., 2019). A study by Kaya et al. (2016) found that chitosan can be used as a coating on red Kiwifruit to extend its shelf life. It was reported by Deng et al. (2017) that cellulose nanocrystals enhanced the storability of postharvest pears under both ambient and cold storage conditions by strengthening chitosan coatings. Using a 0.5% chitosan coating on sweet cherries, Petriccione et al. (2015) extended their postharvest life, improved their storage stability, and increased their nutraceutical value. In a study by Kaewklin et al. (2018), nanosized titanium dioxide films coated with chitosan were found to exhibit ethylene photodegradation activity upon exposure to ultraviolet light, resulting in a delay in ripening and alterations to product quality. Hence, chitosan nanocoating films are attracting increasing attention due to their use in storing and preserving fruits and vegetables (Xing et al., 2020). Additionally, this coating preserved the fruit's antioxidant properties (Song et al., 2016). Using chitosan-clay nanocomposite coating, Taghinezhad and Ebadollahi (2017a,b) preserved lemon quality and protected it from a variety of environmental factors. This coating was shown to reduce weight loss, firmness loss, shear, and punch force loss. Laureth et al. (2018) evaluated the physiological and nutritional effects of nanocellulose-based nanocomposites applied to 'Tahiti' acid lime. Cassava starch, citrus pectin, and sodium CMC were used to prepare coatings. To synthesize these nanocomposites, cellulose nanocrystals were added to the biopolymers. All nanocomposites were used to coat fruit samples. It has been demonstrated that nanocellulose in nanocomposites improves the barrier properties of the fruit, helping to preserve its postharvest quality.

An edible coating made of chitosan clay nanocomposite also showed antifungal properties (Ruffo Roberto et al., 2019). To prevent grape quality and quantity losses after harvest, Piña-Barrera et al. (2019) used a multisystem coating containing polymeric nanocapsules and pullulan. Polymeric nanocapsules were formulated from essential oils obtained from Thymus vulgari. It was found that the coating on table grapes had antimicrobial activity because it showed no signs of microbiological damage. Using the multisystem coating reduced moisture loss and the respiratory rate due to oxidative stress. Research has shown that chitosan-based coatings inhibit polyphenol oxidase (PPO) action in pineapple, fig, lemon, grape, guava, longan, loquat, pistachio, mango, sweet apple, lemon, strawberry, sweet cherry, and cherimoya. In the study, chitosan-based nanocomposites reduced the accessibility of O_2, which is essential to initiate browning. The coatings also maintain membrane firmness by balancing reactive oxygen species and scavengers, maintaining compartments within cells, and separating PPO from phenolic substrates (Adiletta et al., 2018).

Antibrowning and antioxidant agents used in coating composition include ascorbic, sorbic, carboxylic acids, thiol-containing compounds, resorcinol, and phenolic acids (Liu et al., 2016). It is believed that these antibrowning and antioxidant components act on PPO activity in two ways: they both inactivate the enzyme by interacting with the deoxy form of PPO, and they also act as enzyme inhibitors by competing for the catalytic site (Panahirad et al., 2021). In a study by Basumatary et al. (2021), chitosan-based coatings were formulated with aloe vera gel as an antioxidant as well as zinc oxide NPs for antimicrobial activity. Comparing coated pineapples with uncoated pineapples, they found that weight loss, oxidative decay, and delayed ripening were reduced by 5%. Antioxidant activity and phenol content of chitosan films were increased by adding Zataria multiflora Boiss essential oil and grape seed extract (Moradi et al., 2012).

The nano-silicon oxides (SiO_x)-chitosan composites coated on tomato fruits prevented increases in MDA contents, total polyphenol contents, peroxidase, phenylalanine ammonia-lyase (PAL), and PPO activities (Zhu et al., 2019). The chitosan-polyvinyl alcohol-oxalic acid composition found in Lo'ay and Dawood (2017) was coated on banana fruits in a similar study. There was a notable reduction in the activity of browning enzymes, PPO, and PAL. The fruit peel suffered less browning during shelf life due to the preserved phenol content. An edible film containing glycerol, aloe vera gel, and ZnO-NPs solutions has been developed by Dubey et al. (2019). With ZnO-NPs solutions, mango fruits could be stored and maintained at normal temperatures for up to nine days with improved quality and shelf life. These nanocomposite edible films reduced mass loss, maintained lower soluble solids, increased TA and ascorbic acid concentrations in fruit, maintained pH, increased thickness, and improved transmittance. It has become popular to apply chitosan and nanochitosan coatings to crops as a means to extend their storage life (Xing et al., 2019). In recent years, few studies have explored nanostructured chitosan coatings with other compounds as a means to preserve the quality of fruit during storage (Dam et al., 2020; Hernández-López et al., 2020; Shen & Yang, 2017).

10.3.2 Inhibition of microbial growth and control of pathogens

In plant disease management, hybrid polymer nanocomposites are used to make mulch films that control weeds, as nanopesticides, and as a biostatic agent (Adisa et al., 2019). The use of chemical fungicides in orchards is widespread, but if used in large quantities, they can cause adverse health and environmental effects (Palou et al., 2015). Postharvest methods such as modified atmosphere packaging, hot water or air treatment, biocontrol agents, and wax coatings have been extensively studied in recent years to extend fruit shelf life (Duan et al., 2018; Li et al., 2013; Spadoni et al., 2014). The agricultural industry is the backbone of any nation's economy. In recent years, pesticides and insecticides have been used more frequently in agriculture (Hassaan & El Nemr, 2020; Kalia & Gosal, 2011; Meena et al., 2020a,b). Today, however, many farmers are concerned about the widespread use of chemical pesticides to combat microbes and insects. Ascomycetes (Verticillium, Alternaria, and Fusarium) and basidiomycetes (Rhizoctonia, Sclerotium) among microbes can cause significant economic losses to farmers through growth and yield deterrent effects (Shuping & Eloff, 2017). These disease manifestations can be restricted with a stringent dynamic approach encompassing smart materials with biomedical ingredients for extended efficacy and sustainability.

Phytopathogenic agents cause significant losses not only in the field but also after harvest as well. Due to their high efficacy, chemical fungicides are the main strategy to control diseases; however, their residual presence in the environment and their association with microbial resistance have been extensively studied. The use of these products has also been associated with a number of health problems. Rather than applying chemical fungicides for postharvest disease control, nanotechnology is a promising alternative. Micrometric materials are less efficient than NMs. Moreover, chitosan can be applied as an eco-friendly treatment at the nanoscale in conjunction with other control systems (González-Estrada et al., 2023). A fruit's moisture content, size, respiration rate, and texture contribute to the growth and development of pathogens at any stage between harvest and consumption, making fruits more susceptible to perish than cereals, legumes, and oil seed crops (El-Ramady et al., 2015). Several physiological processes during fruit storage increase susceptibility to pathogens, especially fungi and bacteria (Droby & Wisniewski, 2018). Hence, by avoiding mechanical damage, delaying senescence, and maintaining or promoting resistance, one can reduce postharvest decay and diseases (Zhang et al., 2020). Despite these efforts, postharvest diseases are not usually prevented by these practices (Martinez et al., 2018). The principal postharvest fruit pathogens are fungi, which enter the host tissue by three principal mechanisms: (i) wounds; (ii) natural openings (pedicel-interphase, lenticels, stem ends); and (iii) natural breaching on the cuticle (Berger et al., 2007). Plant pathogens can establish themselves almost instantly after spores enter the host tissue or remain in a quiescent state for months until favorable opportunistic conditions exist (Mendgen et al., 1996).

Silver NPs inhibit phytopathogens such as Rhizoctonia solani, Sclerotinia sclerotiorum, and Sarracenia minor effectively, as shown by Hamdi et al. (2021). Furthermore, silver NPs can break the hyphal wall membrane, causing extensive internal damage to the hyphae. The SE-MgO nanocomposite was composed of sepiolite, a magnesium silicate, and magnesium oxide, which demonstrated excellent antifungal activity against rice pathogens Fusarium verticillioides, Bipolaris oryzae, and Fusarium fujikuroi at $ED90 > 249\ \mu g\ mL^{-1}$ (Sidhu et al., 2020).

Postharvest disease is initiated by environmental factors such as temperature, moisture, and air composition (O_2 and CO_2) (Yang et al., 2020). Temperature is one of the most crucial factors in controlling postharvest diseases, which is why other methods to control the disease are sometimes classified as "supplementary alternatives" (Sommer, 1985). Postharvest disease development is therefore related to higher temperatures and humidity (Tian, 2007). The temperature rise promotes and even accelerates the growth of many serious diseases (Pan et al., 2017). A second important factor during the storage of fruits after harvest is relative humidity (Park et al., 2018). Humidity is also highly correlated with favorable temperatures for spore germination, and, unfortunately, fresh fruit needs high humidity to maintain its turgidity (Pinto et al., 2020). To promote the marketing stage, fruits are harvested before they have reached maturity (Sommer, 1985). When fruits are ripe or in senescence, they are more likely to become infected with fungi like Monilinia spp., Botrytis cinerea, Rhizopus spp., or Penicillium spp. (Durian et al., 2020; Ma et al., 2020; Wang et al., 2021). Among the many causes of injuries are insects, weather, birds, farm implements, and rodents (Kabelitz et al., 2019). It is also possible for vibration damage to occur when packaging or in long-distance transportation (Fernando et al., 2018). Because reinforcement tissue is formed on the zone, healed lesions are no longer susceptible to fungal invasion (Moyo et al., 2020). There are a number of fungi and bacteria that can cause postharvest disease in fruits, as we already mentioned. However, the most common sorts are Alternaria, Aspergillus, Botrytis, Colletotrichum, Diplodia, Fusarium, Monilinia, Penicillium, Phomopsis, Rhizopus, Mucor, and Pseudomonas. In the right conditions, airborne disseminated pathogens can cause significant decay (Mukherjee et al., 2020).

Using chitosan/nano-TiO_2 composite emulsion, Shi et al. (2008) demonstrated exceptionally high antibacterial activity against Escherichia coli, Aspergillus niger, and Candida albicans. The chitosan/nano-TiO_2 composite studied by Karthikeyan et al. (2017) proved to be an effective antimicrobial agent, while Lin et al. (2015) demonstrated that the Ag/TiO_2/chitosan nanocomposite demonstrated superior antimicrobial activity than $AgNO_3$ or nano-Ag particles at similar concentrations.

One of the most damaging postharvest diseases of tomatoes is gray mold caused by the fungi Botrytis cinerea (Wang et al., 2019). As fruits are usually ready for consumption after harvest, postharvest diseases must be carefully managed. The use of chemical fungicides can be effective in controlling gray mold in the field, but not in treating tomato fruits postharvest. In the first place, fungicide residues pose a threat to human health and the environment. In the second place, B. cinerea can be easily resistant to chemical fungicides. To control the development of postharvest gray mold on tomatoes, safe and efficient bio-approaches are needed (Seow et al., 2014). Besides microbial antagonists, plant essential oils can also be used to treat fruit crops after harvest to control postharvest diseases (Da Costa Gonçalves et al., 2021; Dukare et al., 2019). Plant phytochemicals (e.g., essential oils, phenolics, and flavonoids) are widely used to preserve food (Govindachari et al., 2000; Tao et al., 2014).

To increase the effectiveness of essential oils and extend their application, microencapsulation is an effective formulation (Perumal et al., 2021). To increase their solubility in water, essential oils can be microencapsulated in the form of oil-in-water (O/W) (Suresh Kumar et al., 2013). The use of nanotechnology can help reduce the size of microencapsulation, which will increase the bioavailability of essential oils (Donsì & Ferrari, 2016). The nanoemulsion is a nanosized microemulsion consisting of emulsion particles that are between 20 and 200 nm in size and have high stability and good biocompatibility (Heydari et al., 2020). As it does not require high shear, pressure homogenization, ultrasonics, or other processes, spontaneous emulsification can be scaled up for mass production (Singh et al., 2017). Hydrophobic components can be transferred into aqueous phases using this technique (Mansouri et al., 2021). Nanoemulsion systems have surfactants to enhance emulsion stability and reduce particle size, increasing essential oil dispersion. Microemulsions can also be controlled effectively by releasing essential oils that inhibit pathogenic fungal growth (Zhang et al., 2022). The main component of essential oil from thyme species is thymol, a monoterpene that is environmentally friendly. An antimicrobial study found that thymol disrupts and permeabilizes pathogenic microorganisms' cellular membranes to exhibit its antimicrobial effects (Tiwari et al., 2009).

The unique size and shape of engineering NPs and their optical properties make them suitable for a wide variety of agricultural applications, including improved pest and pathogen control (Camara et al., 2019; Pestovsky & Martínez-Antonio,

2017; Worrall et al., 2018). NPs such as carbon, silver, silica, and nonmetal oxides or aluminum silicates are widely used to control plant diseases. Research conducted on carbon NMs has shown that they are useful in agriculture, promoting plant growth and development as well as controlling several plant pathogens, including Xanthomonas, Aspergillus spp., Botrytis cinerea, and Fusarium spp. Research revealed that silica NPs conferred resistance to Fusarium oxysporum and Aspergillus niger in maize (Rangaraj et al., 2014). By developing nano-hybrids or composites, the action spectrum and pest/pathogen control efficacy of nano-enabled pesticides can be improved (Paek et al., 2011; Spadola et al., 2020).

There are diverse chemical origins for nanohybrids and nanocomposites, ranging from biological to inorganic and from natural/synthetic to organic–inorganic materials. In addition, these composites do not involve physical mixing of the components, so they may or may not possess additive or augmenting properties, depending on the properties of the individual components (Paek et al., 2011). Research interest in the development of potent, effective, and multifunctional antimicrobial nanohybrids has increased dramatically in the past decade. Aluminum-silicate nanoplates, for instance, have been used to develop pesticide formulations with improved biological activity and better environmental safety than engineered nanoplates (Iavicoli et al., 2017). As a result, nanoformulated particles/composites can be effective in fighting fungal pathogen outbreaks.

Due to their functional optical properties and ease of handling, nanotechnology-based pathogen diagnosis has attracted a lot of attention from the research community (Wang et al., 2016). With nanotechnology, NPs can be conjugated with nucleic acids, proteins, and other biomolecules, enabling rapid, sensitive, and reliable pathogen detection (Kashyap et al., 2017). There are various types of NMs, but quantum dots are a special type of nanocrystal that has tunable size-dependent fluorescence characteristics that are being explored in agriculture and other fields. It has been demonstrated that a quantum-dot-based nanosensor can detect Candidatus Phytoplasmaaurantifolia in lime even when there are fewer than five phytoplasma cells μL^{-1} (Rad et al., 2012). Xanthomonas axonopodis pv vesicatoria, which causes tomato and pepper spot diseases, can be detected rapidly using fluorescent silica NPs conjugated with antibodies (Yao et al., 2009).

A nanocomposite material consists of multiple phases. It may be possible for these materials to contain components with variable phase domains, with at least one continuous phase and another with nanoscale dimensions (Ashfaq et al., 2020). It is possible to generate hybrid NMs by co-synthesizing/impregnating diverse inorganic and organic elements (Winter et al., 2020). Since nanocomposites have properties of both inorganic and organic materials that interact concurrently to perform the desired activity, they have been extensively studied (Adnan et al., 2018). As a general rule, nanocomposites are synthesized from long-chain or short-chain polymers incorporating NMs. It has been observed that the derived nanocomposites exhibit improved properties compared to any of their constituent components. In most cases, polymers combined with NPs will increase their properties significantly (Fahmy et al., 2020). These nanocomposites are widely used for food processing, pest detection, food health screenings, water treatment, disease detection, drug delivery, and the improvement of sustainable agriculture (Idumah et al., 2020; Padmanaban et al., 2016; Sadeghi et al., 2017; Wei et al., 2020). Polymer composites are also beneficial as fertilizers, as they increase nutrient uptake and decrease soil toxicity (Guha et al., 2020; Kalia et al., 2020a,b). Furthermore, nanocomposites serve as antimicrobial agents as well as sensors to increase the shelf life of food products (Usman et al., 2020).

Material combinations such as polymers, metals, and ceramics can be used to produce nanocomposites (Omanović-Mikličanin et al., 2020). The polymers and inorganic/organic materials have a high aspect ratio and surface properties, making them versatile materials for diverse applications. Among the many uses of polymers in agriculture is their ability to release chemical compounds such as fungicides, insecticides, growth stimulants, and germicides (Yao et al., 2014). A polymer's main advantage is its ability to control both the release rate and biodegradability rate of the compound it encapsulates or embeds. Polymers are widely used in medicine and agrochemicals because of these characteristics (Sampathkumar et al., 2020; Vega-Vásquez et al., 2020). A variety of polymers are widely used as drugs and agrochemical delivery agents. These include polyvinyl alcohol, polyacrylamide, and polystyrene, as well as polylactic acid, polybutadiene, polylactic-glycolic acid, and polyhydroxyalkanoates (Ashfaq et al., 2020; Spadola et al., 2020).

A polymer nanocomposites application should be carefully considered before use in agriculture. Moreover, natural polymers are preferable for agriculture applications because they are degradable and have controlled release characteristics. Due to its exceptional properties, chitosan polymer is becoming a new alternative to traditional plant protection materials. At varying concentrations, nanochitosan has antimicrobial effects against bacteria and fungi. Chitosan-encapsulated or embedded metal or metal oxide NMs exhibit improved antimicrobial properties. Ag-chitosan nanocomposite, for instance, displayed a higher level of antibacterial activity. In addition, clay chitosan nanocomposites were evaluated for their fungicidal properties in vitro and in vivo against Penicillium digitatum (Youssef & Hashim, 2020). The nanocomposite formulation demonstrated superior activity compared to conventional formulations. Likewise, copper NPs (Cu NPs), zinc oxide NPs (Zn NPs), and chitosan, zinc oxide, and copper nanocomposites (CS-Zn-Cu NCs) have been synthesized and evaluated against the plant pathogenic fungi A. alternata, R. solani, and B. cinerea (Al-Dhabaan et al., 2017). Results showed that nanocomposites exhibited higher activity at a concentration of 90 $\mu g\ mL^{-1}$. Rhizoctonia solani was effectively inhibited by bimetallic blends, Zn-chitosan, and Cu-chitosan at concentrations of 30, 60, and 90 $\mu g\ mL^{-1}$. Additionally, it showed that cotton seedling damping-off could be controlled effectively under greenhouse conditions (Abd-Elsalam et al., 2018). In a study, nanocomposites of silver and chitosan exhibited incremental growth inhibitory effects against phytopathogens isolated from chickpea seeds (Kaur et al., 2012). Moreover, the individual metal and polymer components had lower inhibitory activity against Aspergillus niger than the silver/chitosan nanocomposite. Chitosan/silica nanocomposite was evaluated in vitro and in vivo (natural and artificial infections) against Botrytis cinerea (Youssef et al., 2019).

Based on the in vitro results, the nanocomposite completely inhibited fungal growth, whereas chitosan and silica NPs had inhibition factors of 72% and 76%, respectively. Moreover, the chitosan/silica nanocomposite effectively inhibited gray mold disease in Italian grapes by 59% and in Benitaka grapes by 83% under natural conditions.

It was demonstrated that chitosan-conjugated Ag NPs functionalized with 4(E)-2-(3-hydroxynaphthalene-2-yl) diazenyl-1-benzoic acid exhibited improved efficacy against A. flavus and A. niger, forming inhibition zones of 20.2 and 27.0 mm, respectively (Mathew & Kuriakose, 2013). Likewise, Mentha piperita essential oil encapsulated in chitosan hydrogel with cinnamic acid inhibits mycelia of A. flavus at an 800 mg L^{-1} concentration (Beyki et al., 2014). As a result, we can infer from the above appropriate scientific evidence that organic/inorganic-metal-polymer hybrid nanocomposites exhibit remarkable antifungal activity under both in vitro and in vivo conditions.

It is important for a promising nano-fungicide to possess equivalent or superior activity compared to the bulk metal at relatively lower concentrations. A detailed understanding of phyto- and eco-toxicity issues due to metal ions is also desirable. NMs executed antifungal activity via multiple mechanisms. NMs can have antifungal activity in the following ways: Fungal cell walls and membranes generally contain chitin, lipids, phospholipids, and polysaccharides, with a specific predominance of mannoproteins, β-1,3-D-glucan, and β-1,6-D-glucan proteins (Patil & Chandrasekaran, 2020). The NMs are internalized by three mechanisms that are as follows; (i) direct internalization in the cell wall; (ii) receptor-mediated adsorption followed by internalization; and (iii) ion transport protein-mediated internalization (Kalia et al., 2020a,b). As a result of the internalization process, NMs can inhibit β-glucan synthase, thus impacting N-acetylglucosamine [N-acetyl-D-glucose-2-amine] synthesis in fungi. A consequence of enzyme inhibition is a thickening of the cell wall, liquefaction of the membrane, dissolution or disorganization of the cytoplasmic organelles, hypervacuolization, and detachment of the cell wall from the cytoplasmic contents, which indicate incipient plasmolysis (Arciniegas-Grijalba et al., 2017).

Numerous studies have examined the antibacterial activity of various NMs against microbes that deteriorate crops, including fruits and vegetables (Ruffo Roberto et al., 2019; Bhardwaj et al., 2023a,b). In addition, NMs such as silica and silver sulfide nanocomposites demonstrate antifungal activity against Aspergillus niger (Fateixa et al., 2009), Fusarium oxysporum, Dematophora necatrix, and Colletotrichum gloeosporioides (Sharma et al., 2017). Also, researchers have found copper NPs to be effective against Fusarium solani and Penicillium digitatum (Khamis et al., 2017), copper oxide NPs, zinc oxide NPs, and their mixture to be effective against Alternaria citri strains (Sardar et al., 2022), and chitosan NPs loaded with thymol have antimicrobial effects against Xanthomonas campestris pv campestris (Sreelatha et al., 2022). A variety of deteriorating environmental factors can be addressed with effective barrier properties by using these NMs in active, eco-friendly, and sustainable packaging (Videira-Quintela et al., 2022) (Table 10.1).

Table 10.1 Effects of nanoparticles on postharvest management in different crops.

Nanocomposite	Size	Application method	Crop	Physiological effects	References
Clay/chitosan nanocomposite	–	Coating	Lemon	Reducing weight loss, TSS amount and punching force, increasing TA amount and hardness and shearing forces of the shell, as well as improving the quality characteristics of the fruit	Taghinezhad and Ebadollahi (2017a, 2017b)
Fungal chitosan nanocomposite	–	Edible coating (immersing for 3 min)	Table grapes	Antibacterial activity against *S. aureus*, *P. aeruginosa*, *L. monocytogenes*, and *E. coli*. Delaying the ripening process, decreasing weight loss, soluble solids, and sugar content	Melo et al. (2020)
Chitosan/nano-TiO_2 composite	30 nm	Coating	Mangoes	Reduction of fruit decay index, total soluble solid content (TSS), malonaldehyde (MDA), and increasing fruit firmness, peroxidase (POD) and PPO activity, total phenol and flavonoid content	Xing et al. (2020)

Chitosan and TiO_2 nanocomposite films	–	Solution casting method	Tomato	Maintained the quality of climacteric fruit by photodegradation of ethylene activity in presence of UV light, which caused delayed fruit ripening	Kaewklin et al. (2018)
Clay-chitosan nanocomposite	–	Antifungal	Valencia late orange	Complete inhibition of green mold and reduction of decay	Youssef and Hashim (2020)
Chitosan-phenylalanine NPs (Cs-Phe Nps)	< 100 nm	Coating	Persimmon	Delayed the negative effects of chilling stress, enhanced antioxidant capacity, firmness, and TSS, Lower H_2O_2 and malonaldehyde (MDA) accumulation, and total carotenoid accumulation.	Nasr et al. (2021)
Proline coated chitosan NPs (CTS-Pro NPs)	250 nm	Coating	Strawberry	Reduced *MDA* and hydrogen peroxide content and less decay and weight loss, preserved fruit quality by conserving higher levels of ascorbic acid, TSS, total phenolic content, and antioxidant capacity and enzymes.	Bahmani et al. (2022)
Glycine betaine coated chitosan NPs	150	Coating	Plum	Decreased induction of *EL*, MDA and H_2O_2 content, and chilling injury and increased PAL enzyme activity, phenols, flavonoids and anthocyanins, DPPH inhibitory capacity	Mahmoudi et al. (2022)

(Continued)

Table 10.1 (Continued)

Nanocomposite	Size	Application method	Crop	Physiological effects	References
Chitosan-*g*-salicylic acid complex	–	Coating	Cucumber	Decrease in respiration rate, *MDA* content and *EL* and cause an increase in weight, TSS, chlorophyll and ascorbic acid, endogenous salicylic acid concentration and the activity of antioxidant enzymes such as superoxide dismutase, catalase, ascorbate peroxidase, and glutathione reductase.	Zhang et al. (2015)
Chitosan–phenylalanine nanocomposites (CS–Phe NCs)	150 nm	Foliar	Flame seedless grape	Increase in pH value, ascorbic acid, total phenol content	Gohari et al. (2021)
chitosan-salicylic acid nanocomposite	70–100 nm	Foliar	grape (*Vitis vinifera* cv. " Sultana")	Improving the photosynthetic system, increasing total chlorophyll, carotenoid, GPX, SOD, APX, and reduction of H_2O_2	Aazami et al. (2023)
Chitosan-salicylic acid Nanocomposite	150	Foliar	*Mentha spicata* L.	Increasing protein content, chlorophyll index, and carotenoid content, and reduction of hydrogen peroxide and MDA content	Hassanpouraghdam et al. (2022)

Chitosan-clay nanocomposites	–	Coating	Thomson Oranges	High pH, maintaining firmness, skin color, skin moisture and texture strength	Ruffo Roberto et al. (2019)
Chitosan-silver nanoparticle (chitosan-AgNP) composite	hydrodynamic diameter range of 495–616 nm and silver NPs with the size distribution from 10 to 15 nm	Coating	Mango	Preventing the germination of conidia on coated fruits, having the potential to increase the shelf life of mango fruit during storage, improve postharvest decay and have antifungal properties against *Colletotrichum gloeosporioides*	Chowdappa et al. (2014)

10.4 Challenges and future directions

The increasing field of nanocomposites has shown huge promise across various industries, but it also faces several challenges that require careful consideration for further application and widespread adoption. One significant concern revolves around regulatory considerations and safety issues associated with the use of NMs. As nanocomposites become more prevalent in consumer products and industrial applications, it becomes imperative to establish clear guidelines and standards for their safe use. Robust risk assessment protocols and extensive toxicological studies are necessary to ensure that nanocomposites do not pose any health or environmental hazards. Overcoming these regulatory hurdles is crucial for gaining public trust and achieving widespread commercialization. Another critical aspect that demands attention is the cost-effectiveness and scalability of nanocomposite applications. While the unique properties of NMs offer enhanced performance, their high production costs can hinder large-scale implementation. Researchers and manufacturers must focus on developing efficient and cost-effective synthesis methods to enable the mass production of nanocomposites without compromising their quality and functionality. Economies of scale, coupled with advancements in NM synthesis techniques, could play a pivotal role in driving down costs and making nanocomposite technology accessible to a broader range of industries and consumers.

As nanocomposites continue to revolutionize various sectors, it becomes essential to assess their potential environmental impacts. Addressing issues such as NP release, end-of-life disposal, and the ecological consequences of extensive nanocomposite use is critical for minimizing environmental challenges. Developing eco-friendly and biodegradable nanocomposites, as well as efficient recycling processes, will be key to reducing their environmental footprint and promoting long-term sustainability. Furthermore, the future of nanocomposite applications appears promising, with several emerging trends and areas for further research. The integration of nanocomposites in the healthcare sector, for instance, shows significant potential for advanced drug delivery, tissue engineering, and medical diagnostics. Nanocomposite materials are also poised to play a pivotal role in renewable energy applications, such as high-performance energy storage devices and solar panels. Additionally, exploring nanocomposite possibilities in the aerospace and automotive industries could lead to lighter, stronger, and more fuel-efficient materials. Continued research and collaboration between academia, industry, and regulatory bodies will be crucial in unlocking the full potential of nanocomposites and shaping their future sustainably and responsibly.

Acknowledgment

This study has received funding from the European Union Horizon Europe program with acronym PRIMESOFT, entitled "Development of innovative priming technologies safeguarding yield security in soft fruit crops through a cutting-edge interdisciplinary approach" [Grant Agreement 101079119].

References

Aaby, K., Mazur, S., Nes, A., & Skrede, G. (2012). Phenolic compounds in strawberry (*Fragaria x ananassa* Duch.) fruits: Composition in 27 cultivars and changes during ripening. *Food chemistry*, *132*(1), 86–97.

Aazami, M. A., Maleki, M., Rasouli, F., & Gohari, G. (2023). Protective effects of chitosan based salicylic acid nanocomposite (CS-SA NCs) in grape (*Vitis vinifera* cv.'Sultana') under salinity stress. *Scientific Reports*, *13*(1), 883.

Abd-Elsalam, K. A., Vasil'kov, A. Y., Said-Galiev, E. E., Rubina, M. S., Khokhlov, A. R., Naumkin, A. V., Shtykova, E. V., & Alghuthaymi, M. A. (2018). Bimetallic blends and chitosan nanocomposites: Novel antifungal agents against cotton seedling damping-off. *European Journal of Plant Pathology*, *151*, 57–72.

Adeel, M., Shakoor, N., Shafiq, M., Pavlicek, A., Part, F., Zafiu, C., Raza, A., Ahmad, M. A., Jilani, G., White, J. C., & Ehmoser, E. K. (2021). A critical review of the environmental impacts of manufactured nano-objects on earthworm species. *Environmental Pollution*, *290*, 118041.

Adiletta, G., Pasquariello, M. S., Zampella, L., Mastrobuoni, F., Scortichini, M., & Petriccione, M. (2018). Chitosan coating: A postharvest treatment to delay oxidative stress in loquat fruits during cold storage. *Agronomy*, *8*(4), 54.

Adiletta, G., Zampella, L., Coletta, C., & Petriccione, M. (2019). Chitosan coating to preserve the qualitative traits and improve antioxidant system in fresh figs (*Ficus carica* L.). *Agriculture*, *9*(4), 84.

Adisa, I. O., Pullagurala, V. L. R., Peralta-Videa, J. R., Dimkpa, C. O., Elmer, W. H., Gardea-Torresdey, J. L., & White, J. C. (2019). Recent advances in nano-enabled fertilizers and pesticides: A critical review of mechanisms of action. *Environmental Science: Nano*, *6*(7), 2002–2030.

Adnan, M. M., Dalod, A. R., Balci, M. H., Glaum, J., & Einarsrud, M. A. (2018). In situ synthesis of hybrid inorganic–polymer nanocomposites. *Polymers*, *10*(10), 1129.

Agrawal, S., Kumar, V., Kumar, S., & Shahi, S. K. (2022). Plant development and crop protection using phytonanotechnology: A new window for sustainable agriculture. *Chemosphere*, *299*, 134465.

Ahmad, M. A., Deng, X., Adeel, M., Rizwan, M., Shakoor, N., Yang, Y., & Javed, R. (2022). Influence of calcium and magnesium elimination on plant biomass and secondary metabolites of *Stevia rebaudiana* Bertoni. *Biotechnology and Applied Biochemistry*, *69*(5), 2008–2016.

Al-Dhabaan, F. A., Shoala, T., Ali, A. A., Alaa, M., Abd-Elsalam, K., & Abd-Elsalam, K. (2017). Chemically-produced copper, zinc nanoparticles and chitosan-bimetallic nanocomposites and their antifungal activity against three phytopathogenic fungi. *Intenrational Journal of Agricultural Technology*, *13*(5), 753–769.

Arabpoor, B., Yousefi, S., Weisany, W., & Ghasemlou, M. (2021). Multifunctional coating composed of *Eryngium campestre* L. essential oil encapsulated in nano-chitosan to prolong the shelf-life of fresh cherry fruits. *Food Hydrocolloids*, *111*, 106394.

Arciniegas-Grijalba, P. A., Patiño-Portela, M. C., Mosquera-Sánchez, L. P., Guerrero-Vargas, J. A., & Rodríguez-Páez, J. E. (2017). ZnO nanoparticles (ZnO-NPs) and their antifungal activity against coffee fungus *Erythricium salmonicolor*. *Applied Nanoscience*, *7*, 225–241.

Arnon, H., Zaitsev, Y., Porat, R., & Poverenov, E. (2014). Effects of carboxymethyl cellulose and chitosan bilayer edible coating on postharvest quality of citrus fruit. *Postharvest Biology and Technology*, *87*, 21–26.

Ashfaq, M., Talreja, N., Chuahan, D., & Srituravanich, W. (2020). Polymeric nanocomposite-based agriculture delivery system: Emerging technology for agriculture. *Genetic Engineering Glimpse Technical Applications*, 1–16.

Bahmani, R., Razavi, F., Mortazavi, S. N., Gohari, G., & Juárez-Maldonado, A. (2022). Evaluation of proline-coated chitosan nanoparticles on decay control and quality preservation of strawberry fruit (cv. Camarosa) during cold storage. *Horticulturae*, *8*(7), 648.

Bal, E. (2013). Postharvest application of chitosan and low temperature storage affect respiration rate and quality of plum fruits. *Journal of Agricultural Science & Technology*, *15*, 1219–1230.

Basumatary, I. B., Mukherjee, A., Katiyar, V., Kumar, S., & Dutta, J. (2021). Chitosan-based antimicrobial coating for improving postharvest shelf life of pineapple. *Coatings*, *11* (11), 1366.

Berger, S., Sinha, A. K., & Roitsch, T. (2007). Plant physiology meets phytopathology: Plant primary metabolism and plant–pathogen interactions. *Journal of Experimental Botany*, *58*(15–16), 4019–4026.

Beyki, M., Zhaveh, S., Khalili, S. T., Rahmani-Cherati, T., Abollahi, A., Bayat, M., Tabatabaei, M., & Mohsenifar, A. (2014). Encapsulation of *Mentha piperita* essential oils in chitosan–cinnamic acid nanogel with enhanced antimicrobial activity against *Aspergillus flavus*. *Industrial Crops and Products*, *54*, 310–319.

Bhardwaj, K., Meneely, J. P., Haughey, S. A., Dean, M., Wall, P., Zhang, G., Baker, B., & Elliott, C. T. (2023a). Risk assessments for the dietary intake aflatoxins in food: A systematic review (2016–2022). *Food Control*, 109687.

Bhardwaj, S., Lata, S., & Garg, R. (2023b). Application of nanotechnology for preventing postharvest losses of agriproducts. *The Journal of Horticultural Science and Biotechnology*, *98*(1), 31–44.

Bueno, V., & Ghoshal, S. (2022). *Inorganic porous nanoparticles as pesticide or nutrient carriers. Inorganic nanopesticides and nanofertilizers: A view from the mechanisms of action to field applications* (pp. 363–390). Cham: Springer International Publishing.

Camara, M. C., Campos, E. V. R., Monteiro, R. A., do Espirito Santo Pereira, A., de Freitas Proença, P. L., & Fraceto, L. F. (2019). Development of stimuli-responsive nano-based pesticides: Emerging opportunities for agriculture. *Journal of Nanobiotechnology*, *17*(1), 1–19.

Cea, M., Cartes, P., Palma, G., & Mora, M. L. (2010). Atrazine efficiency in an andisol as affected by clays and nanoclays in ethylcellulose controlled release formulations. *Revista de la ciencia del suelo y nutrición vegetal*, *10*(1), 62–77.

Chaudhry, N., Dwivedi, S., Chaudhry, V., Singh, A., Saquib, Q., Azam, A., & Musarrat, J. (2018). Bio-inspired nanomaterials in agriculture and food: Current status, foreseen applications and challenges. *Microbial Pathogenesis*, *123*, 196–200.

Chowdappa, P., Gowda, S., Chethana, C. S., & Madhura, S. (2014). Antifungal activity of chitosan-silver nanoparticle composite against *Colletotrichum gloeosporioides* associated with mango anthracnose. *African Journal of Microbiology Research*, *8*(17), 1803–1812.

Da Costa Gonçalves, D., Ribeiro, W. R., Goncalves, D. C., Menini, L., & Costa, H. (2021). Recent advances and future perspective of essential oils in control Colletotrichum spp.: A sustainable alternative in postharvest treatment of fruits. *Food Research International*, *150*, 110758.

Dam, M. S., To, X. T., Le, Q. T. P., Nguyen, L. L. P., Friedrich, L., Hitka, G., Zsom, T., Nguyen, T. C. T., Huynh, C. Q., Tran, M. D. T., & Nguyen, V. D. (2020). Postharvest quality of hydroponic strawberry coated with chitosan-calcium gluconate. *Progress in Agricultural Engineering Sciences*.

Deng, Z., Jung, J., Simonsen, J., Wang, Y., & Zhao, Y. (2017). Cellulose nanocrystal reinforced chitosan coatings for improving the storability of postharvest pears under both ambient and cold storages. *Journal of Food Science*, *82*(2), 453–462.

Derbalah, A., Shenashen, M., Hamza, A., Mohamed, A., & El Safty, S. (2018). Antifungal activity of fabricated mesoporous silica nanoparticles against early blight of tomato. *Egyptian Journal of Basic and Applied Sciences*, *5*(2), 145–150.

Donsì, F., & Ferrari, G. (2016). Essential oil nanoemulsions as antimicrobial agents in food. *Journal of Biotechnology*, *233*, 106–120.

Droby, S., & Wisniewski, M. (2018). The fruit microbiome: A new frontier for postharvest biocontrol and postharvest biology. *Postharvest Biology and Technology*, *140*, 107–112.

Duan, C., Meng, X., Meng, J., Khan, M. I. H., Dai, L., Khan, A., An, X., Zhang, J., Huq, T., & Ni, Y. (2019). Chitosan as a preservative for fruits and vegetables: A review on chemistry and antimicrobial properties. *Journal of Bioresources and Bioproducts*, *4*(1), 11–21.

Duan, X., OuYang, Q., & Tao, N. (2018). Effect of applying cinnamaldehyde incorporated in wax on green mould decay in citrus fruits. *Journal of the Science of Food and Agriculture*, *98*(2), 527–533.

Dubey, P. K., Shukla, R. N., Srivastava, G., Mishra, A. A., & Pandey, A. (2019). Study on quality parameters and storage stability of mango coated with developed nanocomposite edible film. *International Journal of Current Microbiology and Applied Sciences*, *8*(4), 2899–2935.

Dukare, A. S., Paul, S., Nambi, V. E., Gupta, R. K., Singh, R., Sharma, K., & Vishwakarma, R. K. (2019). Exploitation of microbial antagonists for the control of postharvest diseases of fruits: A review. *Critical Reviews in Food Science and Nutrition*, *59*(9), 1498–1513.

Durian, G., Jeschke, V., Rahikainen, M., Vuorinen, K., Gollan, P. J., Brosche, M., Salojärvi, J., Glawischnig, E., Winter, Z., Li, S., & Noctor, G. (2020). Protein phosphatase 2A-B′ γ controls *Botrytis cinerea* resistance and developmental leaf senescence. *Plant Physiology*, *182*(2), 1161–1181.

Elik, A., Yanik, D. K., Istanbullu, Y., Guzelsoy, N. A., Yavuz, A., & Gogus, F. (2019). Strategies to reduce postharvest losses for fruits and vegetables. *Strategies*, *5*(3), 29–39.

El-Ramady, H. R., Domokos-Szabolcsy, É., Abdalla, N. A., Taha, H. S., & Fári, M. (2015). Postharvest management of fruits and vegetables storage. *Sustainable Agriculture Reviews*, *15*, 65–152.

Esfanjani, A. F., Assadpour, E., & Jafari, S. M. (2018). Improving the bioavailability of phenolic compounds by loading them within lipid-based nanocarriers. *Trends in Food Science & Technology*, *76*, 56–66.

Eshghi, S., Karimi, R., Shiri, A., Karami, M., & Moradi, M. (2022). Effects of polysaccharide-based coatings on postharvest storage life of grape: Measuring the changes in nutritional, antioxidant and phenolic compounds. *Journal of Food Measurement and Characterization*, *16*(2), 1159–1170.

Fahmy, H. M., Eldin, R. E. S., Serea, E. S. A., Gomaa, N. M., AboElmagd, G. M., Salem, S. A., Elsayed, Z. A., Edrees, A., Shams-Eldin, E., & Shalan, A. E. (2020). Advances in nanotechnology and antibacterial properties of biodegradable food packaging materials. *RSC Advances*, *10*(35), 20467–20484.

Fateixa, S., Neves, M. C., Almeida, A., Oliveira, J., & Trindade, T. (2009). Anti-fungal activity of SiO_2/Ag2S nanocomposites against Aspergillus niger. *Colloids and Surfaces B: Biointerfaces*, *74*(1), 304–308.

Fernando, I., Fei, J., Stanley, R., & Enshaei, H. (2018). Measurement and evaluation of the effect of vibration on fruits in transit. *Packaging Technology and Science*, *31*(11), 723–738.

Fincheira, P., Hoffmann, N., Tortella, G., Ruiz, A., Cornejo, P., Diez, M. C., Seabra, A. B., Benavides-Mendoza, A., & Rubilar, O. (2023). Eco-efficient systems based on nanocarriers for the controlled release of fertilizers and pesticides: Toward smart agriculture. *Nanomaterials*, *13*(13), 1978.

Firdous, N. (2021). Significance of edible coating in mitigating postharvest losses of tomatoes in Pakistan: A review. *Journal of Horticulture and Postharvest Research*, *4*(Special Issue-Fresh-cut Products), 41–54.

Flórez, M., Guerra-Rodríguez, E., Cazón, P., & Vázquez, M. (2022). Chitosan for food packaging: Recent advances in active and intelligent films. *Food Hydrocolloids*, *124*, 107328.

Gadgey, K. K., & Sharma, G. S. (2020). Investigation of mechanical properties of chitosan based films prepared from narmada riverside crab shells. *International Journal of Mechanical Engineering Techniques*, *11*(6), 21–28.

Gohari, G., Farhadi, H., Panahirad, S., Zareei, E., Labib, P., Jafari, H., Mahdavinia, G., Hassanpouraghdam, M. B., Ioannou, A., Kulak, M., & Fotopoulos, V. (2023). Mitigation of salinity impact in spearmint plants through the application of engineered chitosan-melatonin nanoparticles. *International Journal of Biological Macromolecules*, *224*, 893–907.

Gohari, G., Jiang, M., Manganaris, G. A., Zhou, J., & Fotopoulos, V. (2024). Next generation chemical priming: with a little help from our nanocarrier friends. *Trends in Plant Science*. Available from https://doi.org/10.1016/j.tplants.2023.11.024.

Gohari, G., Zareei, E., Kulak, M., Labib, P., Mahmoudi, R., Panahirad, S., Jafari, H., Mahdavinia, G., Juárez-Maldonado, A., & Lorenzo, J. M. (2021). Improving the berry quality and antioxidant potential of flame seedless grapes by foliar application of chitosan–phenylalanine nanocomposites (CS–Phe NCs). *Nanomaterials*, *11*(9), 2287.

González-Estrada, R. R., Blancas-Benitez, F. J., Hernández-Béjar, F. J., Rivas-García, T., Moreno-Hernández, C., Aguirre-Güitrón, L., Ramos-Bell, S., & Gutierrez-Martinez, P. (2023). *Chitosan: Postharvest ecofriendly nanotechnology, control of decay, and quality in tropical and subtropical fruits. Handbook of green and sustainable nanotechnology: Fundamentals, developments and applications* (pp. 73–90). *Cham: Springer International Publishing.*

Govindachari, T. R., Suresh, G., Gopalakrishnan, G., Masilamani, S., & Banumathi, B. (2000). Antifungal activity of some tetranortriterpenoids. *Fitoterapia*, *71*(3), 317–320.

Guha, T., Gopal, G., Kundu, R., & Mukherjee, A. (2020). Nanocomposites for delivering agrochemicals: A comprehensive review. *Journal of Agricultural and Food Chemistry*, *68*(12), 3691–3702.

Hamdi, R. F., Aljameel, A. I., Obaid, A. S., & Ramizy, A. (2021). Bioproduction of silver nanoparticles by *Myrtus communis* leaf extract and their effect on plant pathogenic fungi in vitro. *International Journal of Nanoscience*, *20*(06), 2150052.

Hashim, N. F. A., Ahmad, A., & Bordoh, P. K. (2018). Effect of chitosan coating on chilling injury, antioxidant status and postharvest quality of Japanese cucumber during cold storage. *Sains Malays*, *47*(2), 287–294.

Hassaan, M. A., & El Nemr, A. (2020). Pesticides pollution: Classifications, human health impact, extraction and treatment techniques. *The Egyptian Journal of Aquatic Research*, *46*(3), 207–220.

Hassanpouraghdam, M. B., Mohammadi, L., Gohari, G., & Vojodi Mehrabani, L. (2022). The Effects of foliar application of chitosan-salicylic acid nanocomposite on *Mentha spicata* L. under salinity stress in hydroponic conditions. *Journal of Vegetables Sciences*, *5*(2), 35–51.

Hazarika, T. K., Lalrinfeli, L., Lalchhanmawia, J. O. N. A. T. H. A. N., & Mandal, D. E. B. A. S. H. I. S. (2019). Alteration of quality attributes and shelf-life in strawberry (*Fragaria* × *ananassa*) fruits during storage as influenced by edible coatings. *Indian Journal of Agricultural Sciences*, *89*(1), 28–34.

Hernandez-Lauzardo, A. N., Velázquez-del Valle, M. G., & Guerra-Sanchez, M. G. (2011). Current status of action mode and effect of chitosan against phytopathogens fungi. *African Journal of Microbiology Research*, *5*(25), 4243–4247.

Hernández-López, G., Ventura-Aguilar, R. I., Correa-Pacheco, Z. N., Bautista-Baños, S., & Barrera-Necha, L. L. (2020). Nanostructured chitosan edible coating loaded with α-pinene for the preservation of the postharvest quality of *Capsicum annuum* L. and *Alternaria alternata* control. *International Journal of Biological Macromolecules*, *165*, 1881–1888.

Hesami, A., Kavoosi, S., Khademi, R., & Sarikhani, S. (2021). Effect of chitosan coating and storage temperature on shelf-life and fruit quality of *Ziziphus mauritiana*. *International Journal of Fruit Science*, *21*(1), 509–518.

Heydari, M., Amirjani, A., Bagheri, M., Sharifian, I., & Sabahi, Q. (2020). Eco-friendly pesticide based on peppermint oil nanoemulsion: Preparation, physicochemical properties, and its aphicidal activity against cotton aphid. *Environmental Science and Pollution Research*, *27*, 6667–6679.

Hu, D., Wang, H., & Wang, L. (2016). Physical properties and antibacterial activity of quaternized chitosan/carboxymethyl cellulose blend films. *LWT-Food Science and Technology*, *65*, 398–405.

Iavicoli, I., Leso, V., Beezhold, D. H., & Shvedova, A. A. (2017). Nanotechnology in agriculture: Opportunities, toxicological implications, and occupational risks. *Toxicology and applied pharmacology*, *329*, 96–111.

Idumah, C. I., Zurina, M., Ogbu, J., Ndem, J. U., & Igba, E. C. (2020). A review on innovations in polymeric nanocomposite packaging materials and electrical sensors for food and agriculture. *Composite Interfaces*, *27*(1), 1–72.

Jiang, M., Song, Y., Kanwar, M. K., Ahammed, G. J., Shao, S., & Zhou, J. (2021). Phytonanotechnology applications in modern agriculture. *Journal of Nanobiotechnology*, *19* (1), 1–20.

Jirawutthiwongchai, J., Klaharn, I. Y., Hobang, N., Mai-Ngam, K., Klaewsongkram, J., Sereemaspun, A., & Chirachanchai, S. (2016). Chitosan-phenylalanine-mPEG nanoparticles: From a single step water-based conjugation to the potential allergen delivery system. *Carbohydrate polymers*, *141*, 41–53.

Kabelitz, T., Schmidt, B., Herppich, W. B., & Hassenberg, K. (2019). Effects of hot water dipping on apple heat transfer and post-harvest fruit quality. *LWT*, *108*, 416–420.

Kaewklin, P., Siripatrawan, U., Suwanagul, A., & Lee, Y. S. (2018). Active packaging from chitosan-titanium dioxide nanocomposite film for prolonging storage life of tomato fruit. *International Journal of Biological Macromolecules*, *112*, 523–529.

Kalia, A., & Gosal, S. K. (2011). Effect of pesticide application on soil microorganisms. *Archives of Agronomy and Soil Science*, *57*(6), 569–596.

Kalia, A., Abd-Elsalam, K. A., & Kuca, K. (2020a). Zinc-based nanomaterials for diagnosis and management of plant diseases: Ecological safety and future prospects. *Journal of fungi*, *6*(4), 222.

Kalia, A., Sharma, S. P., Kaur, H., & Kaur, H. (2020b). Novel nanocomposite-based controlled-release fertilizer and pesticide formulations: Prospects and challenges. *Multifunctional hybrid nanomaterials for sustainable agri-food and ecosystems*, 99–134.

Kaliamurthi, S., Selvaraj, G., Hou, L., Li, Z., Wei, Y., Gu, K., & Wei, D. (2019). Synergism of essential oils with lipid based nanocarriers: Emerging trends in preservation of grains and related food products. *Grain & Oil Science and Technology*, *2*(1), 21–26.

Karthikeyan, K. T., Nithya, A., & Jothivenkatachalam, K. (2017). Photocatalytic and antimicrobial activities of chitosan-TiO_2 nanocomposite. *International journal of biological macromolecules*, *104*, 1762–1773.

Kashyap, P. L., Kumar, S., & Srivastava, A. K. (2017). Nanodiagnostics for plant pathogens. *Environmental Chemistry Letters*, *15*, 7–13.

Kaur, P., Thakur, R., & Choudhary, A. (2012). An in vitro study of the antifungal activity of silver/chitosan nanoformulations against important seed borne pathogens. *Int J Sci Technol Res*, *1*(6), 83–86.

Kaya, M., Česonienė, L., Daubaras, R., Leskauskaitė, D., & Zabulionė, D. (2016). Chitosan coating of red kiwifruit (*Actinidia melanandra*) for extending of the shelf life. *International Journal of Biological Macromolecules*, *85*, 355–360.

Khamis, Y., Hashim, A. F., Abd-Elsalam, K. A., Margarita, R., & Alghuthaymi, M. A. (2017). Fungicidal efficacy of chemically-produced copper nanoparticles against *Penicillium digitatum* and *Fusarium solani* on citrus fruit. *Philippine Agricultural Scientist*, *100*(1), 69–78.

Khatri, D., Panigrahi, J., Prajapati, A., & Bariya, H. (2020). Attributes of *Aloe vera* gel and chitosan treatments on the quality and biochemical traits of post-harvest tomatoes. *Scientia Horticulturae*, *259*, 108837.

Kumar, M. N. R. (2000). A review of chitin and chitosan applications. *Reactive and functional polymers*, *46*(1), 1–27.

Kumar, P., Sethi, S., Sharma, R. R., Srivastav, M., & Varghese, E. (2017). Effect of chitosan coating on postharvest life and quality of plum during storage at low temperature. *Scientia Horticulturae*, *226*, 104–109.

Kumar, S., Mukherjee, A., & Dutta, J. (2020). Chitosan based nanocomposite films and coatings: Emerging antimicrobial food packaging alternatives. *Trends in Food Science & Technology*, *97*, 196–209.

Laureth, J. C. U., Moraes, A. J. D., França, D. L. B. D., Flauzino Neto, W. P., & Braga, G. C. (2018). Physiology and quality of'Tahiti'acid lime coated with nanocellulose-based nanocomposites. *Food Science and Technology*, *38*, 327–332.

Lee, Y., & Hwang, K. T. (2017). Changes in physicochemical properties of mulberry fruits (*Morus alba* L.) during ripening. *Scientia Horticulturae*, *217*, 189–196.

Li, F., Zhang, X., Song, B., Li, J., Shang, Z., & Guan, J. (2013). Combined effects of 1-MCP and MAP on the fruit quality of pear (*Pyrus bretschneideri* Reld cv. Laiyang) during cold storage. *Scientia Horticulturae*, *164*, 544–551.

Lin, B., Luo, Y., Teng, Z., Zhang, B., Zhou, B., & Wang, Q. (2015). Development of silver/titanium dioxide/chitosan adipate nanocomposite as an antibacterial coating for fruit storage. *LWT-Food Science and Technology*, *63*(2), 1206–1213.

Liu, K., Liu, J., Li, H., Yuan, C., Zhong, J., & Chen, Y. (2016). Influence of postharvest citric acid and chitosan coating treatment on ripening attributes and expression of cell wall related genes in cherimoya (*Annona cherimola* Mill.) fruit. *Scientia Horticulturae*, *198*, 1–11.

Lo'ay, A. A., & Dawood, H. D. (2017). Minimize browning incidence of banana by postharvest active chitosan/PVA Combines with oxalic acid treatment to during shelf-life. *Scientia Horticulturae*, *226*, 208–215.

Lv, M., Liu, Y., Geng, J., Kou, X., Xin, Z., & Yang, D. (2018). Engineering nanomaterials-based biosensors for food safety detection. *Biosensors and Bioelectronics*, *106*, 122–128.

Ma, Q., Cong, Y., Wang, J., Liu, C., Feng, L., & Chen, K. (2020). Pre-harvest treatment of kiwifruit trees with mixed culture fermentation broth of *Trichoderma pseudokoningii* and *Rhizopus nigricans* prolonged the shelf life and improved the quality of fruit. *Postharvest Biology and Technology*, *162*, 111099.

Mahmoudi, R., Razavi, F., Rabiei, V., Gohari, G., & Palou, L. (2022). Application of glycine betaine coated chitosan nanoparticles alleviate chilling injury and maintain quality of plum (*Prunus domestica* L.) fruit. *International Journal of Biological Macromolecules*, *207*, 965–977.

Mansouri, S., Pajohi-Alamoti, M., Aghajani, N., Bazargani-Gilani, B., & Nourian, A. (2021). Stability and antibacterial activity of *Thymus daenensis* L. essential oil nanoemulsion in mayonnaise. *Journal of the Science of Food and Agriculture*, *101*(9), 3880–3888.

Martinez, D. A., Loening, U. E., & Graham, M. C. (2018). Impacts of glyphosate-based herbicides on disease resistance and health of crops: A review. *Environmental Sciences Europe*, *30*, 1–14.

Mathew, T. V., & Kuriakose, S. (2013). Photochemical and antimicrobial properties of silver nanoparticle-encapsulated chitosan functionalized with photoactive groups. *Materials Science and Engineering: C*, *33*(7), 4409–4415.

Meena, M., Pilania, S., Pal, A., Mandhania, S., Bhushan, B., Kumar, S., Gohari, G., & Saharan, V. (2020a). Cu-chitosan nano-net improves keeping quality of tomato by modulating physio-biochemical responses. *Scientific Reports*, *10*(1), 21914.

Meena, R. S., Kumar, S., Datta, R., Lal, R., Vijayakumar, V., Brtnicky, M., Sharma, M. P., Yadav, G. S., Jhariya, M. K., Jangir, C. K., & Pathan, S. I. (2020b). Impact of agrochemicals on soil microbiota and management: A review. *Land*, *9*(2), 34.

Mehrazar, E., Rahaie, M., & Rahaie, S. (2015). Application of nanoparticles for pesticides, herbicides, fertilisers and animals feed management. *International Journal of Nanoparticles*, *8*(1), 1–19.

Melo, N. F. C. B., de Lima, M. A. B., Stamford, T. L. M., Galembeck, A., Flores, M. A., de Campos Takaki, G. M., da Costa Medeiros, J. A., Stamford-Arnaud, T. M., & Montenegro Stamford, T. C. (2020). In vivo and in vitro antifungal effect of fungal chitosan nanocomposite edible coating against strawberry phytopathogenic fungi. *International Journal of Food Science & Technology*, *55*(11), 3381–3391.

Mendgen, K., Hahn, M., & Deising, H. (1996). Morphogenesis and mechanisms of penetration by plant pathogenic fungi. *Annual review of phytopathology*, *34*(1), 367–386.

Mishra, D., & Khare, P. (2021). Emerging nano-agrochemicals for sustainable agriculture: Benefits, challenges and risk mitigation. *Sustainable Agriculture Reviews 50: Emerging Contaminants in Agriculture*, 235–257.

Moradi, M., Tajik, H., Rohani, S. M. R., Oromiehie, A. R., Malekinejad, H., Aliakbarlu, J., & Hadian, M. (2012). Characterization of antioxidant chitosan film incorporated with zataria multiflora Boiss essential oil and grape seed extract. *LWT-Food Science and Technology*, *46*(2), 477–484.

Moyo, P., Fourie, P. H., Masikane, S. L., de Oliveira Fialho, R., Mamba, L. C., Du Plooy, W., & Hattingh, V. (2020). The effects of postharvest treatments and sunlight exposure on the reproductive capability and viability of *Phyllosticta citricarpa* in citrus black spot fruit lesions. *Plants*, *9*(12), 1813.

Mukherjee, A., Verma, J. P., Gaurav, A. K., Chouhan, G. K., Patel, J. S., & Hesham, A. E. L. (2020). Yeast a potential bio-agent: Future for plant growth and postharvest disease management for sustainable agriculture. *Applied microbiology and biotechnology*, *104*, 1497–1510.

Muthukrishnan, S., Murugan, I., & Selvaraj, M. (2019). Chitosan nanoparticles loaded with thiamine stimulate growth and enhances protection against wilt disease in Chickpea. *Carbohydrate polymers*, *212*, 169–177.

Muzzarelli, R. A., Boudrant, J., Meyer, D., Manno, N., DeMarchis, M., & Paoletti, M. G. (2012a). Current views on fungal chitin/chitosan, human chitinases, food preservation, glucans, pectins and inulin: A tribute to Henri Braconnot, precursor of the carbohydrate polymers science, on the chitin bicentennial. *Carbohydrate polymers*, *87* (2), 995–1012.

Muzzarelli, R. A., Greco, F., Busilacchi, A., Sollazzo, V., & Gigante, A. (2012b). Chitosan, hyaluronan and chondroitin sulfate in tissue engineering for cartilage regeneration: A review. *Carbohydrate polymers*, *89*(3), 723–739.

Nakasato, D. Y., Pereira, A. E., Oliveira, J. L., Oliveira, H. C., & Fraceto, L. F. (2017). Evaluation of the effects of polymeric chitosan/tripolyphosphate and solid lipid nanoparticles on germination of *Zea mays*, *Brassica rapa* and *Pisum sativum*. *Ecotoxicology and environmental safety*, *142*, 369–374.

Naseri, M., Golmohamadzadeh, S., Arouiee, H., Jaafari, M. R., & Nemati, S. H. (2020). Preparation and comparison of various formulations of solid lipid nanoparticles (SLNs) containing essential oil of *Zataria multiflora*. *Journal of Horticulture and Postharvest Research*, *3*(1), 73–84.

Nasr, F., Pateiro, M., Rabiei, V., Razavi, F., Formaneck, S., Gohari, G., & Lorenzo, J. M. (2021). Chitosan-phenylalanine nanoparticles (Cs-Phe Nps) extend the postharvest life of persimmon (*Diospyros kaki*) fruits under chilling stress. *Coatings*, *11*(7), 819.

Neme, K., Nafady, A., Uddin, S., & Tola, Y. B. (2021). Application of nanotechnology in agriculture, postharvest loss reduction and food processing: Food security implication and challenges. *Heliyon*, *7*(12).

Ngo, T. M. P., Nguyen, T. H., Dang, T. M. Q., Do, T. V. T., Reungsang, A., Chaiwong, N., & Rachtanapun, P. (2021). Effect of pectin/nanochitosan-based coatings and storage temperature on shelf-life extension of "Elephant" mango (*Mangifera indica* L.) fruit. *Polymers*, *13*(19), 3430.

Oh, D. X., & Hwang, D. S. (2013). A biomimetic chitosan composite with improved mechanical properties in wet conditions. *Biotechnology Progress*, *29*(2), 505–512.

Omanović-Mikličanin, E., Badnjević, A., Kazlagić, A., & Hajlovac, M. (2020). Nanocomposites: A brief review. *Health and Technology*, *10*, 51–59.

Ortiz-Duarte, G., Pérez-Cabrera, L. E., Artés-Hernández, F., & Martínez-Hernández, G. B. (2019). Ag-chitosan nanocomposites in edible coatings affect the quality of fresh-cut melon. *Postharvest Biology and Technology*, *147*, 174–184.

Otto, D. P., Vosloo, H. C., Liebenberg, W., & De Villiers, M. M. (2008). Development of microporous drug-releasing films cast from artificial nanosized latexes of poly(styrene-co-methyl methacrylate) or poly(styrene-co-ethyl methacrylate). *European Journal of Pharmaceutics and Biopharmaceutics*, *69*(3), 1121–1134.

Padmanaban, V. C., Giri Nandagopal, M. S., Madhangi Priyadharshini, G., Maheswari, N., Janani Sree, G., & Selvaraju, N. (2016). Advanced approach for degradation of recalcitrant by nanophotocatalysis using nanocomposites and their future perspectives. *International Journal of Environmental Science and Technology*, *13*, 1591–1606.

Paek, S. M., Oh, J. M., & Choy, J. H. (2011). A lattice-engineering route to heterostructured functional nanohybrids. *Chemistry–An Asian Journal*, *6*(2), 324–338.

Palou, L., Valencia-Chamorro, S. A., & Pérez-Gago, M. B. (2015). Antifungal edible coatings for fresh citrus fruit: A review. *Coatings*, *5*(4), 962–986.

Pan, Y. G., Yuan, M. Q., Zhang, W. M., & Zhang, Z. K. (2017). Effect of low temperatures on chilling injury in relation to energy status in papaya fruit during storage. *Postharvest biology and Technology*, *125*, 181–187.

Panahirad, S., Dadpour, M., Peighambardoust, S. H., Soltanzadeh, M., Gullón, B., Alirezalu, K., & Lorenzo, J. M. (2021). Applications of carboxymethyl cellulose-and pectin-based active edible coatings in preservation of fruits and vegetables: A review. *Trends in Food Science & Technology*, *110*, 663–673.

Park, M. H., Sangwanangkul, P., & Choi, J. W. (2018). Reduced chilling injury and delayed fruit ripening in tomatoes with modified atmosphere and humidity packaging. *Scientia Horticulturae*, *231*, 66–72.

Patil, S., & Chandrasekaran, R. (2020). Biogenic nanoparticles: A comprehensive perspective in synthesis, characterization, application and its challenges. *Journal of Genetic Engineering and Biotechnology*, *18*, 1–23.

Perumal, A. B., Li, X., Su, Z., & He, Y. (2021). Preparation and characterization of a novel green tea essential oil nanoemulsion and its antifungal mechanism of action against Magnaporthae oryzae. *Ultrasonics Sonochemistry*, *76*, 105649.

Pestovsky, Y. S., & Martínez-Antonio, A. (2017). The use of nanoparticles and nanoformulations in agriculture. *Journal of Nanoscience and Nanotechnology*, *17*(12), 8699–8730.

Petriccione, M., Mastrobuoni, F., Pasquariello, M. S., Zampella, L., Nobis, E., Capriolo, G., & Scortichini, M. (2015). Effect of chitosan coating on the postharvest quality and antioxidant enzyme system response of strawberry fruit during cold storage. *Foods*, *4*(4), 501–523.

Piña-Barrera, A. M., Álvarez-Román, R., Báez-González, J. G., Amaya-Guerra, C. A., Rivas-Morales, C., Gallardo-Rivera, C. T., & Galindo-Rodríguez, S. A. (2019). Application of a multisystem coating based on polymeric nanocapsules containing essential oil of *Thymus vulgaris* L. to increase the shelf life of table grapes (*Vitis vinifera* L.). *IEEE Transactions on NanoBioscience*, *18*(4), 549–557.

Pinto, C. A., Moreira, S. A., Fidalgo, L. G., Inácio, R. S., Barba, F. J., & Saraiva, J. A. (2020). Effects of high-pressure processing on fungi spores: Factors affecting spore germination and inactivation and impact on ultrastructure. *Comprehensive Reviews in Food Science and Food Safety*, *19*(2), 553–573.

Prasad, R., Bhattacharyya, A., & Nguyen, Q. D. (2017). Nanotechnology in sustainable agriculture: Recent developments, challenges, and perspectives. *Frontiers in Microbiology*, *8*, 1014.

Rad, F., Mohsenifar, A., Tabatabaei, M., Safarnejad, M. R., Shahryari, F., Safarpour, H., Foroutan, A., Mardi, M., Davoudi, D., & Fotokian, M. (2012). Detection of Candidatus *Phytoplasma aurantifolia* with a quantum dots fret-based biosensor. *Journal of Plant Pathology*, 525–534.

Rahman, M., Mukta, J. A., Sabir, A. A., Gupta, D. R., Mohi-Ud-Din, M., Hasanuzzaman, M., Miah, M. G., Rahman, M., & Islam, M. T. (2018). Chitosan biopolymer promotes yield and stimulates accumulation of antioxidants in strawberry fruit. *PLoS One*, *13*(9), e0203769.

Rangaraj, S., Gopalu, K., Muthusamy, P., Rathinam, Y., Venkatachalam, R., & Narayanasamy, K. (2014). Augmented biocontrol action of silica nanoparticles and *Pseudomonas fluorescens* bioformulant in maize (*Zea mays* L.). *RSC Advances*, *4*(17), 8461–8465.

Reza Mozafari, M., Johnson, C., Hatziantoniou, S., & Demetzos, C. (2008). Nanoliposomes and their applications in food nanotechnology. *Journal of Liposome Research*, *18*(4), 309–327.

Ruffo Roberto, S., Youssef, K., Hashim, A. F., & Ippolito, A. (2019). Nanomaterials as alternative control means against postharvest diseases in fruit crops. *Nanomaterials*, *9*(12), 1752.

Sadeghi, R., Rodriguez, R. J., Yao, Y., & Kokini, J. L. (2017). Advances in nanotechnology as they pertain to food and agriculture: Benefits and risks. *Annual Review of Food Science and Technology*, *8*, 467–492.

Saharan, V., Kumaraswamy, R. V., Choudhary, R. C., Kumari, S., Pal, A., Raliya, R., & Biswas, P. (2016). Cu-chitosan nanoparticle mediated sustainable approach to enhance seedling growth in maize by mobilizing reserved food. *Journal of Agricultural and Food Chemistry*, *64*(31), 6148–6155.

Salama, D. M., Abd El-Aziz, M. E., Rizk, F. A., & Abd Elwahed, M. S. A. (2021). Applications of nanotechnology on vegetable crops. *Chemosphere*, *266*, 129026.

Saleh, T. A. (2020). Nanomaterials: Classification, properties, and environmental toxicities. *Environmental Technology & Innovation*, *20*, 101067.

Salvia-Trujillo, L., Rojas-Graü, M. A., Soliva-Fortuny, R., & Martín-Belloso, O. (2015). Use of antimicrobial nanoemulsions as edible coatings: Impact on safety and quality attributes of fresh-cut Fuji apples. *Postharvest Biology and Technology*, *105*, 8–16.

Sampathkumar, K., Tan, K. X., & Loo, S. C. J. (2020). Developing nano-delivery systems for agriculture and food applications with nature-derived polymers. *Iscience*, *23*(5).

Sanzari, I., Leone, A., & Ambrosone, A. (2019). Nanotechnology in plant science: To make a long story short. *Frontiers in Bioengineering and Biotechnology*, *7*, 120.

Sardar, M., Ahmed, W., Al Ayoubi, S., Nisa, S., Bibi, Y., Sabir, M., Khan, M. M., Ahmed, W., & Qayyum, A. (2022). Fungicidal synergistic effect of biogenically synthesized zinc oxide and copper oxide nanoparticles against *Alternaria citri* causing citrus black rot disease. *Saudi journal of biological sciences*, *29*(1), 88–95.

Sau, S., Sarkar, S., Mitra, M., & Gantait, S. (2021). Recent trends in agro-technology, post-harvest management and molecular characterisation of pomegranate. *The Journal of Horticultural Science and Biotechnology*, *96*(4), 409–427.

Seow, Y. X., Yeo, C. R., Chung, H. L., & Yuk, H. G. (2014). Plant essential oils as active antimicrobial agents. *Critical Reviews in Food Science and nutrition*, *54*(5), 625–644.

Severino, P., Pinho, S. C., Souto, E. B., & Santana, M. H. (2011). Polymorphism, crystallinity and hydrophilic–lipophilic balance of stearic acid and stearic acid–capric/caprylic triglyceride matrices for production of stable nanoparticles. *Colloids and Surfaces B: Biointerfaces*, *86*(1), 125–130.

Shang, Y., Hasan, M. K., Ahammed, G. J., Li, M., Yin, H., & Zhou, J. (2019). Applications of nanotechnology in plant growth and crop protection: A review. *Molecules*, *24*(14), 2558.

Sharma, P., Sharma, A., Sharma, M., Bhalla, N., Estrela, P., Jain, A., Thakur, P., & Thakur, A. (2017). Nanomaterial fungicides: In vitro and in vivo antimycotic activity of cobalt and nickel nanoferrites on phytopathogenic fungi. *Global Challenges*, *1*(9), 1700041.

Shen, Y., & Yang, H. (2017). Effect of preharvest chitosan-g-salicylic acid treatment on postharvest table grape quality, shelf life, and resistance to *Botrytis cinerea*-induced spoilage. *Scientia Horticulturae*, *224*, 367–373.

Shi, L., Zhao, Y., Zhang, X., Su, H., & Tan, T. (2008). Antibacterial and anti-mildew behavior of chitosan/nano-TiO_2 composite emulsion. *Korean Journal of Chemical Engineering*, *25*, 1434–1438.

Shi, S., Wang, W., Liu, L., Wu, S., Wei, Y., & Li, W. (2013). Effect of chitosan/nano-silica coating on the physicochemical characteristics of longan fruit under ambient temperature. *Journal of Food Engineering*, *118*(1), 125–131.

Shuping, D. S. S., & Eloff, J. N. (2017). The use of plants to protect plants and food against fungal pathogens: A review. *African Journal of Traditional, Complementary and Alternative Medicines*, *14*(4), 120–127.

Sidhu, A., Bala, A., Singh, H., Ahuja, R., & Kumar, A. (2020). Development of MgO-sepoilite nanocomposites against phytopathogenic fungi of rice (*Oryzae sativa*): A green approach. *ACS omega*, *5*(23), 13557–13565.

Singh, Y., Meher, J. G., Raval, K., Khan, F. A., Chaurasia, M., Jain, N. K., & Chourasia, M. K. (2017). Nanoemulsion: Concepts, development and applications in drug delivery. *Journal of Controlled Release*, *252*, 28–49.

Sinha, A., Gill, P. P. S., Jawandha, S. K., & Grewal, S. K. (2022). Composite coating of chitosan with salicylic acid retards pear fruit softening under cold and supermarket storage. *Food Research International*, *160*, 111724.

Solanki, P., Bhargava, A., Chhipa, H., Jain, N., & Panwar, J. (2015). Nano-fertilizers and their smart delivery system. *Nanotechnologies in Food and Agriculture*, 81–101.

Sommer, N. F. (1985). Role of controlled environments in suppression of postharvest diseases. *Canadian Journal of Plant Pathology*, *7*(3), 331–339.

Song, H., Yuan, W., Jin, P., Wang, W., Wang, X., Yang, L., & Zhang, Y. (2016). Effects of chitosan/nano-silica on postharvest quality and antioxidant capacity of loquat fruit during cold storage. *Postharvest Biology and Technology*, *119*, 41–48.

Spadola, G., Sanna, V., Bartoli, J., Carcelli, M., Pelosi, G., Bisceglie, F., Restivo, F. M., Degola, F., & Rogolino, D. (2020). Thiosemicarbazone nano-formulation for the control of *Aspergillus flavus*. *Environmental Science and Pollution Research*, *27*, 20125–20135.

Spadoni, A., Guidarelli, M., Sanzani, S. M., Ippolito, A., & Mari, M. (2014). Influence of hot water treatment on brown rot of peach and rapid fruit response to heat stress. *Postharvest Biology and Technology*, *94*, 66–73.

Sreelatha, S., Kumar, N., Yin, T. S., & Rajani, S. (2022). Evaluating the antibacterial activity and mode of action of thymol-loaded chitosan nanoparticles against plant bacterial pathogen xanthomonas campestris pv. campestris. *Frontiers in Microbiology*, *12*, 792737.

Suresh Kumar, R. S., Shiny, P. J., Anjali, C. H., Jerobin, J., Goshen, K. M., Magdassi, S., Mukherjee, A., & Chandrasekaran, N. (2013). Distinctive effects of nano-sized permethrin in the environment. *Environmental Science and Pollution Research*, *20*, 2593–2602.

Taghinezhad, E., & Ebadollahi, A. (2017a). Potential application of chitosan-clay coating on some quality properties of lemon during storage. *Agricultural Engineering International: CIGR Journal*, *19*(3), 189–194.

Taghinezhad, E., & Ebadollahi, A. (2017b). Potential application of chitosan-clay coating on some quality properties of agricultural product during storage. *Agricultural Engineering International: CIGR Journal*, *19*(3), 189–194.

Tao, N., Jia, L., & Zhou, H. (2014). Anti-fungal activity of *Citrus reticulata* Blanco essential oil against *Penicillium italicum* and *Penicillium digitatum*. *Food Chemistry*, *153*, 265–271.

Tian, S. P. (2007). *Management of postharvest diseases in stone and pome fruit crops. In* General concepts in integrated pest and disease management (pp. 131–147). *Dordrecht*: *Springer Netherlands*.

Tiwari, B. K., Valdramidis, V. P., O'Donnell, C. P., Muthukumarappan, K., Bourke, P., & Cullen, P. J. (2009). Application of natural antimicrobials for food preservation. *Journal of Agricultural and Food Chemistry*, *57*(14), 5987–6000.

Usman, M., Farooq, M., Wakeel, A., Nawaz, A., Cheema, S. A., ur Rehman, H., Ashraf, I., & Sanaullah, M. (2020). Nanotechnology in agriculture: Current status, challenges and future opportunities. *Science of the Total Environment*, *721*, 137778.

Vega-Vásquez, P., Mosier, N. S., & Irudayaraj, J. (2020). Nanoscale drug delivery systems: From medicine to agriculture. *Frontiers in Bioengineering and Biotechnology*, *8*, 79.

Verma, D. K., Srivastava, S., Srivastav, P. P., & Asthir, B. (2017). *Nano particle based delivery system and proposed applications in agriculture. Engineering practices for agricultural production and water conservation* (pp. 363–386). Apple Academic Press.

Videira-Quintela, D., Guillén, F., Martin, O., & Montalvo, G. (2022). Antibacterial LDPE films for food packaging application filled with metal-fumed silica dual-side fillers. *Food Packaging and Shelf Life*, *31*, 100772.

Wang, C., Yuan, S., Zhang, W., Ng, T., & Ye, X. (2019). Buckwheat antifungal protein with biocontrol potential to inhibit fungal (Botrytis cinerea) infection of cherry tomato. *Journal of Agricultural and Food Chemistry*, *67*(24), 6748–6756.

Wang, P., Lombi, E., Zhao, F. J., & Kopittke, P. M. (2016). Nanotechnology: A new opportunity in plant sciences. *Trends in Plant Science*, *21*(8), 699–712.

Wang, Z., Yuan, G., Pu, H., Shan, S., Zhang, Z., Song, H., & Xu, X. (2021). 1-Methylcyclopropene suppressed the growth of *Penicillium digitatum* and inhibited the green mould in citrus fruit. *Journal of Phytopathology*, *169*(2), 83–90.

Wanyika, H., Gatebe, E., Kioni, P., Tang, Z., & Gao, Y. (2012). Mesoporous silica nanoparticles carrier for urea: Potential applications in agrochemical delivery systems. *Journal of Nanoscience and Nanotechnology*, *12*(3), 2221–2228.

Wei, X., Wang, X., Gao, B., Zou, W., & Dong, L. (2020). Facile ball-milling synthesis of CuO/biochar nanocomposites for efficient removal of reactive red 120. *ACS Omega*, *5* (11), 5748–5755.

Winter, J., Nicolas, J., & Ruan, G. (2020). Hybrid nanoparticle composites. *Journal of Materials Chemistry B*, *8*(22), 4713–4714.

Worrall, E. A., Hamid, A., Mody, K. T., Mitter, N., & Pappu, H. R. (2018). Nanotechnology for plant disease management. *Agronomy*, *8*(12), 285.

Xin, Y., Jin, Z., Chen, F., Lai, S., & Yang, H. (2020). Effect of chitosan coatings on the evolution of sodium carbonate-soluble pectin during sweet cherry softening under non-isothermal conditions. *International Journal of Biological Macromolecules*, *154*, 267–275.

Xing, Y., Li, W., Wang, Q., Li, X., Xu, Q., Guo, X., Bi, X., Liu, X., Shui, Y., Lin, H., & Yang, H. (2019). Antimicrobial nanoparticles incorporated in edible coatings and films for the preservation of fruits and vegetables. *Molecules*, *24*(9), 1695.

Xing, Y., Yang, H., Guo, X., Bi, X., Liu, X., Xu, Q., Wang, Q., Li, W., Li, X., Shui, Y., & Chen, C. (2020). Effect of chitosan/Nano-TiO_2 composite coatings on the postharvest quality and physicochemical characteristics of mango fruits. *Scientia Horticulturae*, *263*, 109135.

Yang, X., Zhang, P., Wei, Z., Liu, J., Hu, X., & Liu, F. (2020). Effects of CO_2 fertilization on tomato fruit quality under reduced irrigation. *Agricultural Water Management*, *230*, 105985.

Yao, J., Yang, M., & Duan, Y. (2014). Chemistry, biology, and medicine of fluorescent nanomaterials and related systems: New insights into biosensing, bioimaging, genomics, diagnostics, and therapy. *Chemical reviews*, *114*(12), 6130–6178.

Yao, K. S., Li, S. J., Tzeng, K. C., Cheng, T. C., Chang, C. Y., Chiu, C. Y., Liao, C. Y., Hsu, J. J., & Lin, Z. P. (2009). Fluorescence silica nanoprobe as a biomarker for rapid detection of plant pathogens. *Advanced Materials Research*, *79*, 513–516.

Youssef, K., & Hashim, A. F. (2020). Inhibitory effect of clay/chitosan nanocomposite against penicillium digitatum on citrus and its possible mode of action. *Jordan Journal of Biological Sciences*, *13*(3).

Youssef, K., de Oliveira, A. G., Tischer, C. A., Hussain, I., & Roberto, S. R. (2019). Synergistic effect of a novel chitosan/silica nanocomposites-based formulation against gray mold of table grapes and its possible mode of action. *International Journal of Biological Macromolecules*, *141*, 247–258.

Yu, Y., Zhang, S., Ren, Y., Li, H., Zhang, X., & Di, J. (2012). Jujube preservation using chitosan film with nano-silicon dioxide. *Journal of Food Engineering*, *113*(3), 408–414.

Zahedi, S. M., Abolhassani, M., Hadian-Deljou, M., Feyzi, H., Akbari, A., Rasouli, F., Koçak, M. Z., Kulak, M., & Gohari, G. (2023). Proline-functionalized graphene oxide nanoparticles (GO-pro NPs): A new engineered nanoparticle to ameliorate salinity stress on grape (*Vitis vinifera* l. cv sultana). *Plant Stress*, *7*, 100128.

Zambrano-Zaragoza, M. L., Mercado-Silva, E., Gutiérrez-Cortez, E., Cornejo-Villegas, M. A., & Quintanar-Guerrero, D. (2014). The effect of nano-coatings with α-tocopherol and xanthan gum on shelf-life and browning index of fresh-cut "Red Delicious" apples. *Innovative Food Science & Emerging Technologies*, *22*, 188–196.

Zhang, J., Hao, Y., Lu, H., Li, P., Chen, J., Shi, Z., Xie, Y., Mo, H., & Hu, L. (2022). Nano-Thymol emulsion inhibits botrytis cinerea to control postharvest gray mold on tomato fruit. *Agronomy*, *12*(12), 2973.

Zhang, X., Li, B., Zhang, Z., Chen, Y., & Tian, S. (2020). Antagonistic yeasts: A promising alternative to chemical fungicides for controlling postharvest decay of fruit. *Journal of Fungi*, *6*(3), 158.

Zhang, Z., Huber, D. J., Qu, H., Yun, Z. E., Wang, H., Huang, Z., Huang, H., & Jiang, Y. (2015). Enzymatic browning and antioxidant activities in harvested litchi fruit as influenced by apple polyphenols. *Food Chemistry*, *171*, 191–199.

Zhu, Y., Li, D., Belwal, T., Li, L., Chen, H., Xu, T., & Luo, Z. (2019). Effect of nano-SiOx/chitosan complex coating on the physicochemical characteristics and preservation performance of green tomato. *Molecules*, *24*(24), 4552.

Index

Note: Page numbers followed by "*f*" and "*t*" refer to figures and tables, respectively.

Q

R

S

Printed in the United States
by Baker & Taylor Publisher Services